T0184343

INTERNATIONAL CENTRE FOR MECHANICAL SCIENCES

COURSES AND LECTURES - No. 290

STATIC AND DYNAMIC PHOTOELASTICITY AND CAUSTICS

RECENT DEVELOPMENTS

EDITED BY

A. LAGARDE
UNIVERSITY OF POITIERS

SPRINGER-VERLAG WIEN GMBH

ISBN 978-3-211-81952-4   ISBN 978-3-7091-2630-1 (eBook)
DOI 10.1007/978-3-7091-2630-1

PREFACE

Studies on the propagation of light waves in bodies that exhibit slight optical anisotropy and on the properties of laser radiation have contributed much to the revival of two- and three-dimensional photoelasticity. This coincided with advances in recording techniques, e.g., the multiple-spark camera, and in the field of data analysis.

The new photoelastic techniques associate three optical parameters at every point of a plane slice in a three dimensional medium. These parameters enable us to obtain the two classical mechanical parameters, namely the directions and difference of secondary principal stresses.

The first technique presented in this course is that of integrated photoelasticity which seems to be particularly adapted to the study of monocrystals, of shell structures of revolution and other bodies where the axisymmetrical stress-state is predominant.

The other techniques utilize the scattered-light phenomenon in the three-dimensional medium. Scattering, itself an inner polarizer or analyzer, when combined with the new techniques it has the potential capacity for studying an important but complex class of three-dimensional problems in solid mechanics. These include the determination of the stress-state at the interior points of a turbine-blade root, or surface cracks under opening mode loading.

Two-dimensional dynamic photoelasticity studies using the high-speed multiple-spark camera will be presented with applications to stress waves propagation and dynamic fracture. The photoelastic coating technique is employed to allow the analysis of actual structural metallic elements with cracks. A new method for analyzing isochromatic fields will be discussed. It permits the determination of the characterizing parameters for a crack under mixed mode loading.

Specific methods using interferometry and holographic photoelasticity are developed for separating the principal stresses and a full field analysis is performed for projectile impact. A punctual method is presented which permits the observation of the dynamic development of principal stress at selected points. It is applied to a simulation of stress waves in a composite material.

Recent developments of the shadow optical method of caustics will be included. This method provides an opportunity for determination of the stress field around cracks in a plate under different types of fracture modes. The extension of the method so as to permit an analysis of elastoplastic material behaviour will be presented.

Alexis Lagarde

CONTENTS

The shadow optical method of caustics

INTEGRATED PHOTOELASTICITY AND ITS APPLICATIONS

Hillar Aben

Estonian Academy of Sciences

Institute of Cybernetics, Tallinn, USSR

PREFACE

Main advantage of photoelasticity lies in the possibility to determine
stresses also at internal points of a three-dimensional body. However,
classical methods used for this purpose (the frozen stress and scattered
light methods) are either labour-consuming or need complicated apparatus.

In integrated photoelasticity stresses in three-dimensional bodies
are investigated by passing polarized light through the whole body which
is usually put into an immersion bath. Change of the polarization of light
between the points of entrance and emergence on various light rays is recorded. If the state of stress has certain properties of symmetry, these
data permit to determine the stress distribution.

From the experimental point of view integrated photoelasticity is simple. However, theory of the method is rather complicated since one has to
take into account laws which govern light propagation in an anisotropic
nonhomogeneous medium. Besides, an inverse problem with many unknowns is
to be solved.

Integrated photoelasticity is mostly used for determining stresses
and checking the quality of transparent items of symmetrical form made of
glass, plastics and single crystals.

1. OPTICAL EQUATIONS OF THREE-DIMENSIONAL PHOTOELASTICITY

Optical phenomena which occur when polarized light passes through an anisotropic nonhomogeneous medium are rather complicated. In this chapter comparatively simple optical equations of three-dimensional photoelasticity are
derived. On the basis of these equations some simple particular cases are
considered.

We start with Maxwell's electromagnetic equations

$$\text{rot } \bar{H} = \frac{1}{c} \frac{\partial \bar{D}}{\partial t} \tag{1.1}$$

$$\text{rot } \bar{E} = \frac{1}{c} \frac{\partial \bar{H}}{\partial t} \tag{1.2}$$

where $\bar{E}$ is the electric vector, $\bar{H}$ is the magnetic vector, $\bar{D}$ is the electric
induction, and c is the velocity of light in vacuum.

In the case of a plane monochromatic wave of frequency ω the electromagnetic field can be expressed as

$$\bar{E} = \bar{E}'e^{i\omega t} \quad , \quad \bar{H} = \bar{H}'e^{i\omega t} \quad , \quad \bar{D} = \bar{D}'e^{i\omega t} \tag{1.3}$$

where complex vectors $\bar{E}'$, $\bar{H}'$, and $\bar{D}'$ depend only on the geometrical coordinates.

Introducing Eqs. (1.3) into Eqs. (1.1) and (1.2) and eliminating $\bar{H}'$
we obtain

$$\text{rot rot } \bar{E}' = \frac{\omega^2}{c^2} \bar{D}' \tag{1.4}$$

or for an orthogonal coordinate system, omitting primes,

$$\frac{d^2E_1}{dz^2} + \frac{\omega^2}{c^2} D_1 = 0$$

$$\frac{d^2E_2}{dz^2} + \frac{\omega^2}{c^2} D_2 = 0 \qquad (1.5)$$

Here E_i and D_i denote components of the electric vector and of the electric induction vector, respectively, along the axes x_i of an orthogonal system of coordinates x_1, x_2, x_3; $z=x_3$ is the coordinate in the direction of light propagation. Equations (1.5) describe the propagation of plane monochromatic electromagnetic waves in a nonhomogeneous dielectric medium in the general case.[1]

Using the relation

$$D_i = \sum_{j=1}^{2} \varepsilon_{ij} E_j \qquad i = 1, 2 \qquad (1.6)$$

where ε_{ij} is the dielectric tensor, Eqs. (1.5) yield

$$-\frac{d^2E_1}{dz^2} = \frac{\omega^2}{c^2} (\varepsilon_{11} E_1 + \varepsilon_{12} E_2)$$

$$-\frac{d^2E_2}{dz^2} = \frac{\omega^2}{c^2} (\varepsilon_{21} E_1 + \varepsilon_{22} E_2) \qquad (1.7)$$

The solution of Eqs. (1.7) we express as follows

$$E_j = A_j(z)e^{-ikz}, \qquad j = 1, 2, \qquad k = \frac{\omega}{c}\varepsilon \qquad (1.8)$$

where ε is the dielectric constant of the stress-free medium.

Transformation (1.8) changes neither the phase retardation nor the amplitude ratio of the components of the light vector. Therefore, the components A_j of the light vector determine the same light ellipse as the components E_j.

Introducing Eq. (1.8) into Eqs. (1.7), we have[2,3]

$$\frac{d^2A_1}{dz^2} - 2ik\frac{dA_1}{dz} + \frac{\omega^2}{c^2} (\varepsilon_{11} - \varepsilon)A_1 + \frac{\omega^2}{c^2} \varepsilon_{12} A_2 = 0$$

$$\frac{d^2A_2}{dz^2} - 2ik\frac{dA_2}{dz} + \frac{\omega^2}{c^2} \varepsilon_{21}A_1 + \frac{\omega^2}{c^2} (\varepsilon_{22} - \varepsilon)A_2 = 0 \qquad (1.9)$$

If the medium is isotropic, then A_j = constant. Optical anisotropy of the photoelastic medium is rather weak:

$$\varepsilon_{jj} - \varepsilon \cong 10^{-3} \text{ to } 10^{-4} \quad , \quad \varepsilon_{ij} \cong 10^{-3} \text{ to } 10^{-4}$$

Therefore, the functions A_j vary slowly with z.

Coefficients of the system, Eqs. (1.9), are of the following orders of magnitude:

$$k \cong 10^5 \quad , \quad \frac{\omega^2}{c^2}(\varepsilon_{jj} - \varepsilon) \cong 10^6 \text{ to } 10^7$$

$$\frac{\omega^2}{c^2}\varepsilon_{ij} \cong 10^6 \text{ to } 10^7$$

Thus, the terms with d^2A_j/dz^2 in Eqs. (1.9) can be neglected. We have[4]

$$\frac{dA_1}{dz} = -iC(\varepsilon_{11} - \varepsilon)A_1 - iC\varepsilon_{12}A_2$$

$$\frac{dA_2}{dz} = -iC\varepsilon_{21}A_1 - iC(\varepsilon_{22} - \varepsilon)A_2 \tag{1.10}$$

where

$$C = \frac{\omega}{2c\sqrt{\varepsilon}} \tag{1.11}$$

The intersection of the plane x_1, x_2 with the ellipsoid of the dielectric tensor forms an ellipse. The principal axes of this ellipse are named secondary principal values of the dielectric tensor in the plane x_1, x_2 and are denoted ε_1 and ε_2. The directions of these axes form an angle φ with the axes x_1, x_2 (Fig. 1.1) and are named secondary principal directions. If rotation of these directions takes place, then the angle φ is a function of the coordinate z.

In the plane x_1, x_2, the following relations between the components of the dielectric tensor hold

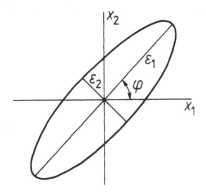

Fig. 1.1: Secondary principal directions form an angle φ with coordinate axes.

$$\varepsilon_{11} = \frac{\varepsilon_1 + \varepsilon_2}{2} + \frac{\varepsilon_1 - \varepsilon_2}{2} \cos 2\varphi$$

$$\varepsilon_{22} = \frac{\varepsilon_1 + \varepsilon_2}{2} - \frac{\varepsilon_1 - \varepsilon_2}{2} \cos 2\varphi$$

$$\varepsilon_{12} = \varepsilon_{21} = \frac{\varepsilon_1 - \varepsilon_2}{2} \sin 2\varphi \qquad (1.12)$$

Sometimes it is effective to use the secondary principal directions as coordinate axes. Denoting by A_j^s the components of the light vector in the secondary principal directions, we have

$$A_1 = A_1^s \cos \varphi - A_2^s \sin \varphi$$

$$A_2 = A_1^s \sin \varphi + A_2^s \cos \varphi \qquad\qquad (1.13)$$

Introducing Eqs. (1.12) and (1.13) into Eqs. (1.10), we obtain

$$\frac{dA_1^s}{dz} = - iC(\varepsilon_1 - \varepsilon)A^s + \frac{d\varphi}{dz} A_2^s$$

$$\frac{dA_2^s}{dz} = - \frac{d\varphi}{dz} A_1^s - iC(\varepsilon_2 - \varepsilon) A_2^s \qquad\qquad (1.14)$$

For an isotropic medium and elastic deformations the following relationship between the dielectric tensor ε_{ij} and the stress tensor σ_{ij} holds

$$\varepsilon_{ij} = \varepsilon\delta_{ij} + 2C_1\sqrt{\varepsilon}\sigma_{ij} + 2C_2\sqrt{\varepsilon} \sum_{k=1}^{3} \sigma_{kk}\sigma_{ij} \qquad\qquad (1.15)$$

where δ_{ij} is the Kronecker tensor and C_1 and C_2 are the photoelastic constants.

Taking into account Eq. (1.15), after some transformations we obtain from Eqs. (1.10) and (1.14)

$$\frac{dA_1}{dz} = - \frac{1}{2} iC_0(\sigma_{11} - \sigma_{22})A_1 - iC_0\sigma_{12}A_2$$

$$\frac{dA_2}{dz} = - iC_0\sigma_{21}A_1 + \frac{1}{2} iC_0(\sigma_{11} - \sigma_{22})A_2 \qquad\qquad (1.16)$$

and

$$\frac{dA_1^s}{dz} = -\frac{1}{2} i C_0(\sigma_1 - \sigma_2)A_1^s + \frac{d\varphi}{dz} A_2^s$$

$$\frac{dA_2^s}{dz} = -\frac{d\varphi}{dz} A_1^s + \frac{1}{2} i C_0 (\sigma_1 - \sigma_2)A_2^s \qquad (1.17)$$

Equations (1.16) and (1.17) are the simplest optical equations of photoelasticity in the case of elastic deformations.

It is possible to show[4] that Eqs. (1.17) are equivalent to the well-known Neumann equations[5]

$$\frac{d\Delta}{dz} = C(\epsilon_1 - \epsilon_2) + 2 \frac{d\varphi}{dz} \sin\Delta \cot\varkappa$$

$$\frac{d\varkappa}{dz} = -\frac{d\varphi}{dz} \cos\Delta \qquad (1.18)$$

where

$$\tan\varkappa = \frac{b}{a} \qquad (1.19)$$

and Δ is the phase retardation. Here a,b are amplitudes of the light vibration along secondary principal axes.

Equations (1.17) are also equivalent to the equations of Mindlin and Goodman.[6]

If the secondary principal directions do not vary with $z(d\varphi/dz = 0)$, while σ_1 and σ_2 vary arbitrarily, the solution of Eqs. (1.17) has the form

$$A_1^s = K_1 \exp \left[\frac{1}{2} iC_0 \mathcal{f}(\sigma_1 - \sigma_2) dz \right]$$

$$A_2^s = K_2 \exp \left[-\frac{1}{2} i \, C_0 \mathcal{f}(\sigma_1 - \sigma_2) dz \right] \qquad (1.20)$$

where K_j = constant. Plane-polarized vibrations, which coincide with the secondary principal directions, propagate independently of one another. The phase retardation is

$$\Delta = C_0 \mathcal{f}(\sigma_1 - \sigma_2) dz \qquad (1.21)$$

which is known as the integral Wertheim law. The latter law plays an important role in integrated photoelasticity.

From recent works devoted to the optical equations of photoelasticity let us mention papers by Kubo and Nagata[7] as well as by Mönch and Roengvoraphoj.[8]

Finally, let us briefly consider the analogy between photoelasticity and acoustoelasticity. In the latter case determination of stresses is based on the phenomenon of acoustical birefringence of polarized sound waves. On deriving the equations of acoustoelasticity it is useful to apply the methodology described above. That was done by Iwashimizu[9] who has derived the following equations of acoustoelasticity

$$\frac{dV_1}{dz} = \frac{1}{2} iC_a(e_{11} - e_{22})V_1 + iC_a e_{12} V_2$$

$$\frac{dV_2}{dz} = iC_a e_{21} V_1 - \frac{1}{2} iC_a(e_{11} - e_{22})V_2 \qquad (1.22)$$

where V_j are the transverse displacements, e_{ij} is the deformation tensor, and C_a is the acoustoelastic coefficient. Equations (1.22) formally coincide with Eqs. (1.16). Thus, the optical theory of photoelasticity can also be used in acoustoelasticity, even if the stresses on the wave normal are not constant.

2. OPTICAL THEORY OF NONHOMOGENEOUS PHOTOELASTIC MEDIA

The aim of this chapter is to create a conceptual framework for analyzing optical phenomena in nonhomogeneous photoelastic media.

The optical equations of photoelasticity, derived in the first chapter, can be written as

$$\frac{dE_1}{dz} = a_{11}E_1 + a_{12}E_2$$

$$\frac{dE_2}{dz} = a_{21}E_1 + a_{22}E_2 \qquad (2.1)$$

where E_j are components of the transformed electric vector whose amplitude ratio and phase retardation determine the light ellipse and a_{ij} are functions of the coordinate z.

The solution of system (2.1) can be expressed as

$$\begin{pmatrix} E_1^* \\ E_2^* \end{pmatrix} = U \begin{pmatrix} E_{10} \\ E_{20} \end{pmatrix} \qquad (2.2)$$

where U is named the matrizant of system (2.1), E_{jo} are components of the incident light vector and E_{j*} are the components of the light vector at the point where $z = z_*$ (Fig. 2.1).

It has been shown[3,4] that the matrix U is unitary, i.e. it has the

form

$$U = \begin{pmatrix} e^{i\xi} \cos\theta & e^{i\zeta} \sin\theta \\ -e^{-i\zeta}\sin\theta & e^{-i\xi} \cos\theta \end{pmatrix} \qquad (2.3)$$

where ξ, ζ, and θ are functions of the stress distribution between
the points of entrance and emergence of light. Besides, these parameters
also depend on the choice of the coordinate axes.

Due to the property of unitarity, it is possible to show that for a
light ray in a nonhomogeneous photoelastic medium there always exist two
perpendicular directions of the polarizer by which the light emerging
from the model is also linearly polarized. The corresponding directions
of the light vector at the entrance of light are named primary characteristic directions, and at the emergence of light, secondary characteristic directions. They are determined by primary (α_0) and secondary (α_*)
characteristic angles (Fig. 2.1).

Characteristic directions are determined by the formulas

$$\tan 2\alpha_0 = \frac{\sin(\zeta + \xi)\sin 2\theta}{\sin 2\xi \cos^2\theta - \sin 2\zeta \sin^2\theta} \qquad (2.4)$$

$$\tan 2\alpha_* = \frac{\sin(\zeta - \xi)\sin 2\theta}{\sin 2\xi \cos^2\theta - \sin 2\zeta \sin^2\theta} \qquad (2.5)$$

For characteristic vibrations we have

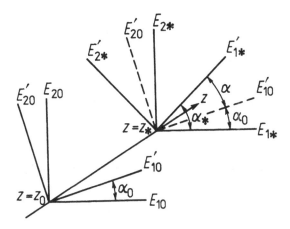

Fig. 2.1. Characteristic directions.

$$\begin{pmatrix} E'_{1*} \\ E'_{2*} \end{pmatrix} = G(\gamma) \begin{pmatrix} E'_{10} \\ E'_{20} \end{pmatrix} \qquad (2.6)$$

where

$$G(\gamma) = \begin{pmatrix} e^{i\gamma} & 0 \\ 0 & e^{-i\gamma} \end{pmatrix} \qquad (2.7)$$

The parameter 2γ is named characteristic phase retardation and it can be expressed by

$$\cos 2\gamma = \cos 2\xi \, \cos^2\theta + \cos 2\zeta \, \sin^2\theta \qquad (2.8)$$

From Eq. (2.6) it follows that

$$E'_{1*} = e^{i\gamma}E'_{10}$$

$$E'_{2*} = e^{-i\gamma}E'_{20} \qquad\qquad (2.9)$$

From the last relationship it can be concluded that if the incident light is plane-polarized in one of the two perpendicular primary characteristic directions ($E'_{10} = 0$ or $E'_{20} = 0$), then the emergent light is also plane-polarized in the corresponding secondary characteristic direction. We name the primary and the secondary characteristic directions corresponding to each other the conjugate characteristic directions. The angle α between the conjugate characteristic directions is named the characteristic angle, and it can be expressed as

$$\tan 2\alpha = \tan 2(\alpha_* - \alpha_0) = \frac{2\sin 2\theta\cos\xi\cos\zeta}{\sin^2\xi - \sin^2\zeta - \cos 2\theta(\cos^2\xi + \cos^2\zeta)} \qquad (2.10)$$

Proof of the existence of characteristic directions was based only on the unitarity of the matrix of the photoelastic medium. No assumptions concerning the principal directions of birefringence were made. Therefore, in general, the characteristic directions do not coincide with the secondary principal directions at the points of entrance and emergence of light.

Note that if the polarization direction of the incident light is parallel to one of the primary characteristic directions, inside the photoelastic medium the light is, in general, elliptically polarized (Fig. 2.2). Iwashimizu has named this phenomenon two-point polarization.[9]

The primary and the secondary characteristic directions and the characteristic phase retardation completely determine the transformation of the light ellipse in the photoelastic medium. At the point of emergence of light the vibrations along the secondary characteristic directions

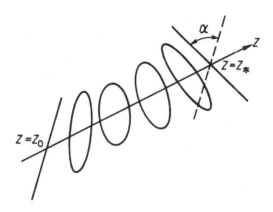

Fig. 2.2: Two-point polarization.

(the so-called characteristic vibrations) have the same amplitudes as at
the point of entrance along the primary characteristic directions. The
characteristic phase retardation does not depend on the polarization of
the incident light.

   These exceptional properties of characteristic parameters indicate
their important role in three-dimensional photoelasticity. It follows
that on every ray of light three experimental values $(\alpha_0, \alpha_*, 2\gamma)$ can be
determined.

   For a point of a two-dimensional model with the azimuth of the principal directions φ and phase retardation $2\gamma_0$ we have

$$\alpha_0 = \alpha_* = \varphi, \quad \alpha = 0, \ 2\gamma = 2\gamma_0 \qquad\qquad (2.11)$$

   From Eqs. (2.11) we may conclude that for the case of a two-dimensional photoelastic model the conjugate primary and secondary characteristic directions are collinear and coincide with the principal directions of the model. The characteristic angle is zero. The characteristic
phase retardation equals the phase retardation in the usual meaning.

Consequently, the theory of characteristic directions can be considered a natural generalization of the optical theory of two-dimensional
photoelasticity for the case of three-dimensional photoelasticity.

When light paases through a photoelastic medium in the opposite direction, the matrix U which describes the transformation of the light ellipse is transformed into its transpose $\tilde{U}$. From this it follows that the
characteristic directions preserve their properties when light passes
through the photoelastic medium in the opposite direction; only the primary characteristic directions transform into the secondary characteristic directions, and vice versa. The characteristic phase retardation does
not depend on the direction in which light passes through the medium. That
is named the principle of reversibility.[4]

From the principle of reversibility it follows that if the characteristic parameters have been determined for a direction of the light ray,
passing of light through the model in the opposite direction does not
give any new experimental information.

If the ray of light passes a model perpendicular to its plane of
symmetry, then the primary and secondary characteristic directions coin-
cide ($\alpha_0 = \alpha_*$,$\alpha = 0$). Such a media we name symmetrical photoelastic media
(Fig. 2.3). To this class of media belong also photoelastic coatings (Fig.
2.4).

Due to their exceptional properties, characteristic parameters can
be determined experimentally. That can be done for many light rays. Let
us mention that determination of characteristic parameters for different
wavelengths gives also additional independent experimental data.

On the other hand, characteristic parameters are determined by stress
distribution in the model. Let us assume that stress distribution in the
model can be described by functions which include n unknown coefficients
k_1, k_2, $\ldots$, k_n . In this case we may write

$$\alpha_0 = \alpha_0(k_1, \ldots , k_n)$$
$$\alpha_* = \alpha_*(k_1, \ldots , \kappa_n)$$
$$2\gamma = 2\gamma(k_1, \ldots , k_n)$$

(2.12)

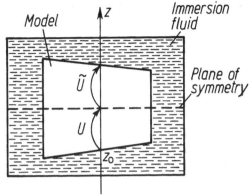
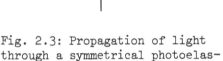

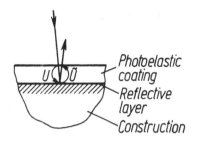

Fig. 2.3: Propagation of light through a symmetrical photoelastic model.

Fig. 2.4: Photoelastic coatings belong to symmetrical photoelastic media.

For every light ray and every wavelength we have relationships (2.12). If the number of equations is equal to or greater than n, the stress distribution can, in principle, be determined. Besides, we can use equations of compatibility of deformations, equations of equilibrium and macrostatic equilibrium conditions in addition.

However, since equations of photoelasticity have a closed form solution only in the case when rotation of the secondary principal stresses is proportional to their difference, one has to use numerical algorithms to determine the stress distribution. These algorithms are mostly based on calculating characteristic parameters for given values of the coefficients k_j.

There are two ways of calculating characteristic parameters when the stress distribution is given. First, one may numerically solve a system of differential equations for characteristic parameters[4,10]

$$\frac{d\alpha_0}{dz} = \frac{1}{2} \frac{d\Delta}{dz} \frac{1}{\sin 2\gamma} \sin 2(\varphi - \alpha_*)$$

$$\frac{d\alpha_*}{dz} = \frac{1}{2} \frac{d\Delta}{dz} \cot 2\gamma \sin 2(\varphi - \alpha_*)$$

$$\frac{d\gamma}{dz} = \frac{1}{2} \frac{d\Delta}{dz} \cos 2(\varphi - \alpha_*) \tag{2.13}$$

where $\varphi(z)$ is the azimuth of the secondary principal directions, and
$\Delta(z) = C_0(\sigma_1 - \sigma_2)$.

Secondly, a three-dimensional photoelastic model can, for a ray of
light, be represented as a pile of birefringent plates

$$U = U_n\, U_{n-1}\, \cdots\, U_j\, \cdots\, U_2\, U_1 \qquad\qquad (2.14)$$

Parameters of each matrix are determined by the coefficients k_j
which describe the stress distribution. Multiplication of the matrices is
easy to carry out with a computer. Knowing parameters of the matrix U, the
characteristic parameters can be calculated from Eqs. (2.4),(2.5) and (2.8).

Let us denote with superscript "e" characteristic parameters which are
determined experimentally, and with "c", those which are calculated. The
problem of finding the stress distribution can now be formulated as a problem of finding the coefficients k_j so that the function

$$F = \sum_{l=1}^{m} [P_1(\alpha_0^e - \alpha_0^c)^2 + P_2(\alpha_*^e - \alpha_*^c)^2 + P_3(\gamma^e - \gamma^c)^2] \qquad (2.15)$$

obtains minimum value. In Eq. (2.15) m is number of light rays on which
measurements are carried out, and P_j are weight coefficients.

The inverse problem of integrated photoelasticity is a typical ill-
determined problem of mathematical physics. It means that small experimental errors may cause great differences in the parameters to be determined.

The problem defined above may be solved by methods of nonlinear opti-
mization. However, if there is no rotation of principal axes on light rays,
it is possible to use the integral Wertheim law (1.21) and the inverse
problem is much simpler.

Optical theory of three-dimensional photoelasticity, described here,
can also be applied to the scattered light method.[11,12]

Other matrix formalisms are available for description of the polarization transformations, too.[13] Let us mention that the theory of characteristic directions is closely connected with the Poincaré equivalence
theorem.[14] The Poincaré sphere method has been widely used in optical theory of photoelasticity.[15]

3. PLATES UNDER BENDING. MAGNETOPHOTOELASTICITY

In this chapter it is shown how the Faraday effect helps to investigate stresses in plates under bending. Influence of this effect on the fringe pattern is also considered.

Let us pass polarized light normally through a bent plate (Fig. 3.1). In this case there is no rotation of the principal axes through the plate thickness, and Eqs. (1.17)

$$\frac{dA_1^s}{dz} = -\frac{1}{2} i C_0 (\sigma_1 - \sigma_2) A_1^s + \frac{d\varphi}{dz} A_2^s$$

$$\frac{dA_2^s}{dz} = -\frac{d\varphi}{dz} A_1^s + \frac{1}{2} i C_0 (\sigma_1 - \sigma_2) A_2^s \qquad (3.1)$$

reduce to

$$\frac{dA_1^s}{dz} = -\frac{1}{2} i C_0 (\sigma_1 - \sigma_2) A_1^s$$

$$\frac{dA_2^s}{dz} = \frac{1}{2} i C_0 (\sigma_1 - \sigma_2) A_2^s \qquad (3.2)$$

In Chapter 1 it was shown that from Eqs. (3.2) follows the integral Wertheim law

$$\Delta = C_0 \int (\sigma_1 - \sigma_2) dz \qquad (3.3)$$

Since stress distribution through the thickness of a bent plate is in

equilibrium, the integral in Eq. (3.3) vanishes. It means, that usual photoelastic technique does not permit to investigate stresses in bent plates.
For that purpose various special techniques have been elaborated which are
rather complicated[16-21] (Fig. 3.2).

Let us put a homogeneous plate into a magnetic field which is parallel to the wave normal (Fig. 3.1).

Due to the Faraday effect, in addition to the stress-birefringence
rotation of the planes of polarization takes place according to the equation

$$\frac{d\psi}{dz} = KH \tag{3.4}$$

where ψ is magnetic rotation of the plane of polarization, K is the Verdet

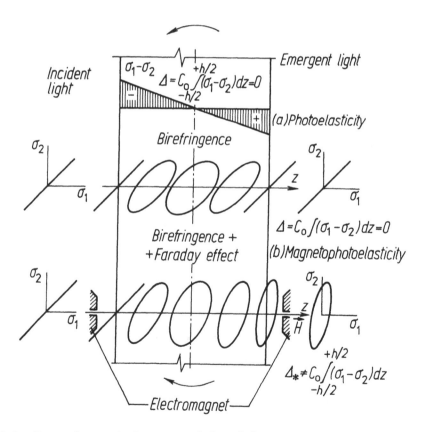

Fig. 3.1: Comparison of photoelasticity (a) and magnetophotoelasticity (b)
by investigating bent plates.

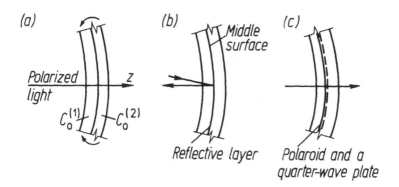

Fig. 3.2: Methods of investigating bent plates: (a) sandwich models, (b) models with a reflecting sheet in the middle plane, (c) embedded polariscope.

constant and H is the magnetic field.

Now additional terms appear in Eqs. (3.2) and the equations of magnetophotoelasticity have the form

$$\frac{dA_1^s}{dz} = -\frac{1}{2} iC_0(\sigma_1 - \sigma_2)A_1^s - \frac{d\psi}{dz} A_2^s$$

$$\frac{dA_2^s}{dz} = \frac{d\psi}{dz} A_1^s + \frac{1}{2} iC_0(_1 - _2)A_2^s \qquad (3.5)$$

Due to the additional terms in Eqs. (3.5) the integral Wertheim law (3.3) does not hold and bending stresses influence the polarization of the emergent light.

Comparison of photoelasticity and magnetophotoelasticity is shown in Fig. 3.1. In photoelasticity, due to the birefringence, the incident plane-polarized light becomes in the model elliptically polarized. Beginning with the middle surface of the plate, the light ellipse starts to narrow since the birefringence reverses its sign. The emergent light is linearly polarized again, because according to the integral Wertheim law $\Delta = 0$ (Fig.3.1a).

In magnetophotoelasticity[22,23] (Fig. 3.1,b), due to the complicated interaction of the birefringence and the magnetic rotation of the plane of

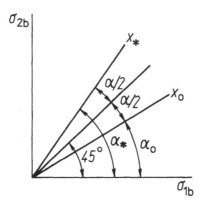

Fig. 3.3: Bisector of the characteristic angle α coincides with the bisector of the principal bending stresses σ_{1b} and σ_{2b}.

polarization, the emergent light, as a rule, is elliptically polarized, i.e. the characteristic retardation is not equal to zero. The parameters of the emergent light ellipse give us information about the bending stresses.

Calculations show that by investigating plates under bending, the following relationship holds

$$\alpha_* = \frac{\pi}{2} - \alpha_0 \qquad (3.6)$$

Thus, the bisector of the angle α forms an angle of 45° with the directions of the principal bending stresses (Fig. 3.3).

Let us denote

$$\sigma_b = \frac{360}{\lambda} C_0 h(\sigma_{1b} - \sigma_{2b}) \qquad (3.7)$$

where λ is wavelength and h is thickness of the plate.

Using discrete modelling of the photoelastic medium, the characteristic parameters α and $\Delta_* = 2\gamma$ were calculated for various values of the total angle of magnetic rotation ψ_0 and of the parameter σ_b. It is convenient to interpret the experimental data if magnetic rotation of the plane

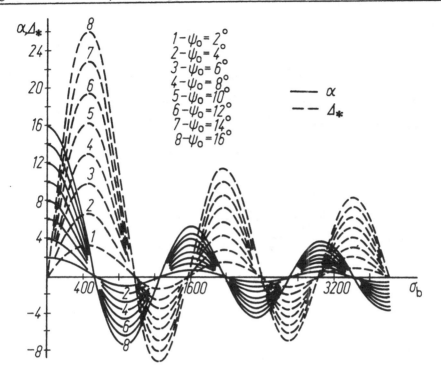

Fig. 3.4: Dependence of the characteristic parameters α and Δ_* on the bending stress σ_b for various values of the angle of rotation ψ_0 of the plane of polarization.

of polarization is uniform.

In Fig. 3.4 the dependence of the characteristic angle α and the characteristic phase retardation Δ_* on ψ_0 and σ_b is shown.

In investigating plates under bending by magnetophotoelasticity, the primary and the secondary characteristic directions are to be determined experimentally. The bisector of the angle α forms an angle of 45° with the principal bending stress directions and, knowing α, σ_b can be determined from Fig. 3.4.

In Fig. 3.5 a photoelactric magnetopolariscope is shown. Figure 3.6 shows stress distribution in a plate with a central hole by cylindrical bending (a/h = 1.00, a/b = 0.29). Results of a series of tests with plates of various geometry compare fairly well with the results published by other authors.[23,24]

Fig. 3.5: Photoelectric magnetopolariscope.

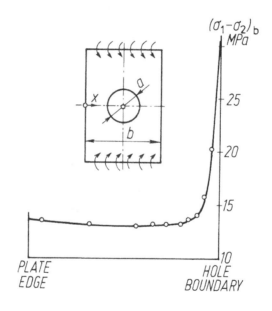

Fig. 3.6: Stress distribution in a plate with a central hole under cylindrical bending.

If there is no birefringence, Eqs. (3.5) reduce to

$$\frac{dA_1^S}{dz} = - \frac{d\psi}{dz} A_2^S$$

$$\frac{dA_2^S}{dz} = \frac{d\psi}{dz} A_1^S \tag{3.8}$$

These equations describe the Faraday effect and from them follows the integral Faraday law

$$\psi_* = \int \frac{d\psi}{dz} dz \tag{3.9}$$

It means that if $d\psi/dz$ is a function of z, the integral Faraday rotation ψ_* does not give any information about the distribution of the rotation of the plane of polarization on the light ray. If we now apply to the specimen stresses of known value, instead of Eqs. (3.8) we have Eqs. (3.5) where now the terms with $C_0(\sigma_1-\sigma_2)$ may be considered as known and the function $d\psi/dz$ is unknown. The latter function can now be determined.

The latter method may be named photoelastic magnetooptics while magnetophotoelasticity is magnetooptic photoelasticity.

Let us mention that instead of applying a magnetic field one may use as model material plastics with natural rotation of the plane of polarization, e.g. optically active polymenthylmethacrylate.[4] However, the problem of how to synthesize such polymers in blocks, sufficiently large to make photoelastic models, is not yet solved.

In the method of magnetophotoelasticity, described above, the magnetic field was introduced artificially to get rid of the integral Wertheim law. However, there may be cases when the photoelastic model is naturally situated in a magnetic field. Such a situation occurs, e.g. by investigating stresses in constructions of Tokamaks[25] by the method of photoelastic coatings.[26] In this case one has to take into account the influence of the Faraday effect on the experimental data.

If a photoelastic coating which is situated in a magnetic field is investigated in a plane polariscope, the light intensity is expressed as follows

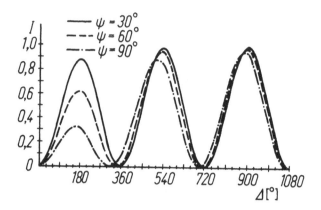

Fig. 3.7: Intensity of light in a model which is situated in a strong magnetic field; ψ denotes the total angle of Faraday rotation.

$$I = \cos^2 \frac{\Delta_*}{2} \sin^2\alpha + \sin^2 \frac{\Delta_*}{2} \sin^2(2\beta-\alpha) \qquad (3.10)$$

where β is the angle between the polarizer and the primary characteristic direction. The bigger the rotation of the plane of polarization, the more the characteristic phase retardation Δ_* differs from the phase retardation Δ, determined by stresses (Eq. (3.3)).

From Eq. (3.10) it follows that also in the presence of the magnetic field the isoclinics give directions of principal stresses. However, the light intensity on the isoclinic is not zero.

It can be shown that in a circular polariscope contrast of low order fringes gets poorer and they are shifted towards lower values of Δ (Fig. 3.7).

Finally, let us mention that if in addition to the bending stresses also membrane stresses are present (e.g., plates of large deflections, shells), then the principal stress directions are not constant through the thickness of the model. In this case stresses can be determined by the aid of the theory of characteristic directions.[27,28]

4. SHELLS OF REVOLUTION. CYLINDRICAL BODIES. FIBERS

In this chapter comparatively simple methods for determining stresses in shells of revolution and in cylindrical bodies under axisymmetric load are described. These methods are mostly based on the integral Wertheim law.

In tangential examination, the beam of light is not parallel to the normal of the middle surface. To avoid refraction, the shell is put in an immersion bath. Most informative are the light rays which pass the shell between its inner and outer surface (Fig. 4.1).

In Fig. 4.2 the stress components in cylindrical coordinates are shown. The changes in light polarization are caused by those components of the axial (σ_y), the circumferential (σ_θ), the radial (σ_r), and shear (τ_{ry}) stress which act in the plane xy perpendicular to the direction of light propagation z:

$$\sigma_x = \sigma_r \cos^2\theta + \sigma_\theta \sin^2\theta$$

$$\sigma_y = \sigma_y$$

$$\tau_{xy} = \tau_{ry} \cos\theta \qquad\qquad (4.1)$$

The secondary principal stress difference $\sigma_1-\sigma_2$ and their directions (the angle φ) are determined from

$$\sigma_1 - \sigma_2 = \sqrt{(\sigma_y-\sigma_r \cos^2\theta - \sigma_\theta \sin^2\theta)^2 + 4\tau_{ry}^2 \cos^2\theta} \qquad (4.2)$$

$$\tan 2\varphi = \frac{2\tau_{ry} \cos\theta}{\sigma_y - \sigma_r \cos^2\theta - \sigma_\theta \sin^2\theta} \qquad (4.3)$$

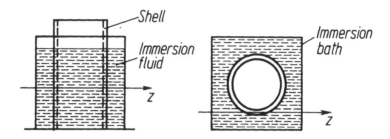

Fig. 4.1: Investigation of a cylindrical shell in an immersion bath with tangential incidence.

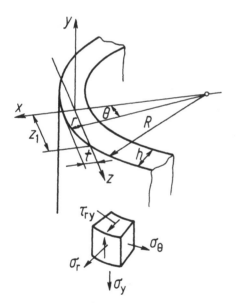

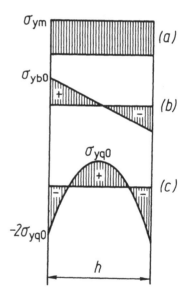

Fig. 4.2: Illustration of the tangential examination of a cylindrical shell.

Fig. 4.3: Distribution of the stress components through the shell thickness
(a) membrane stresses;
(b) bending stresses;
(c) quenching stresses.

If the shear stress τ_{ry} cannot be neglected, then, according to Eq. (4.3), rotation of the secondary principal directions takes place on the light path. In this case for determining the stress components, the characteristic parameters are to be measured on several rays and an algorithm for their interpretation is to be elaborated.

However, often τ_{ry} is negligible. Besides, $\tau_{ry} = 0$ at the point where the bending stresses reach their maximum. In this case the optical phase retardation is expressed by the integral Wertheim law (1.21) and stress analysis is considerably simpler.

Let us consider the dependence of the phase retardation on the coordinate t of the ray (Fig. 4.2) for various stress components (Fig. 4.3).

Assume that the only stress is the axial membrane stress σ_{ym} (Fig. 4.3, a) which is constant through the shell thickness. The integral Wertheim law together with Eq. (4.2) yields

$$\Delta_m = 2C_0 \int_0^{z_t} \sigma_{ym} \, dz = A\sigma_{ym} f_1(\xi) \tag{4.4}$$

where

$$z_t = \sqrt{2Rt} \ , \ A = 2C_0\sqrt{2Rt} \ , \ f_1 = \sqrt{\xi} \quad \xi = \frac{t}{h} \tag{4.5}$$

The function $f_1(\xi)$ which characterizes the dependence of the phase retardation on the dimensionless coordinate ξ is shown in Fig. 4.4.

The axial bending stresses are linearly distributed through the shell thickness (Fig. 4.4, b). In this case the integral Wertheim law yields[29,30]

$$\Delta_b = A\sigma_{ybo} f_2(\xi) \quad , \quad f_2(\xi) = \sqrt{\xi}(1 - \frac{4}{3}\xi) \tag{4.6}$$

Here σ_{ybo} is the bending stress on the outside surface of the shell. The function $f_2(\xi)$ is shown in Fig. 4.4.

During the production process of a glass specimen, there often occur quenching stresses which have a parabolic distribution through the shell thickness (Fig. 4.4, c). The distribution of the integral phase retardation in this case is[30,31]

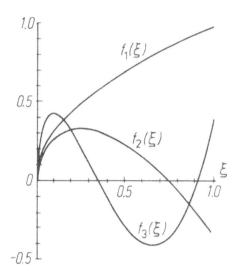

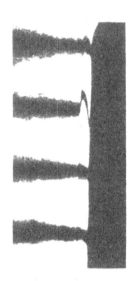

Fig. 4.4: Distribution of the integral phase retardation by tangential examination of cylindrical
shells.

Fig. 4.5: Fringe pattern by
tangential examination of a
shell photographed together
with a quartz wedge (after
Read[29]).

$$\Delta_q = - A\sigma_{yqo}f_3(\xi) \quad , \quad f_3(\xi) = \sqrt{\xi(6.4\xi^2 - 8\xi + 2)} \qquad (4.7)$$

The function $f_3(\xi)$ is shown in Fig. 4.4.

If it is known, a priori, that only one of the stress components considered above is present, then it can be determined by measuring the integral phase retardation for one value of the parameter ξ. If all the stress
components are present, the phase retardation is

$$\Delta = \Delta_m + \Delta_b + \Delta_q \qquad (4.8)$$

For determining the stresses, Δ must be measured at three values of ξ
and the following system of linear equations must be solved

$$f_1(\xi_i)\sigma_{ym} + f_2(\xi_i)\sigma_{ybo} + f_3(\xi_i)\sigma_{yqo} = \frac{\Delta(\xi_i)}{A} \qquad (4.9)$$

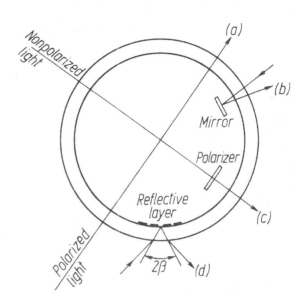

Fig. 4.6: Fringe pattern by
tangential investigation of
a quenched cylindrical shell
photographed together with a
quartz wedge (after Ritland[31]).

Fig. 4.7: Possible schemes for investigating the circumferential membrane
stresses.

Figure 4.5 presents a photograph of a cylindrical shell under tangential incidence. In the path of the light beam is placed a quartz wedge
whose horizontal fringes are seen in the left part of the photograph. The
deformed fringe in the middle of the shell matches the function $f_2(\xi)$ in
Fig. 4.4 fairly well. Fig. 4.6 shows fringes in a strongly quenched cylindrical shell.

Figure 4.7 illustrates various possibilities to determine the circumferential membrane stresses $\sigma_{\theta m}$. Knowing them, the axial bending stress
can be calculated if the shear stress vanishes, as

$$\sigma_{ybo} = \frac{\sigma_{\theta m}}{\sqrt{3/(1-\mu^2)}} \qquad (4.10)$$

There are various possibilities to determine stresses in cylindrical

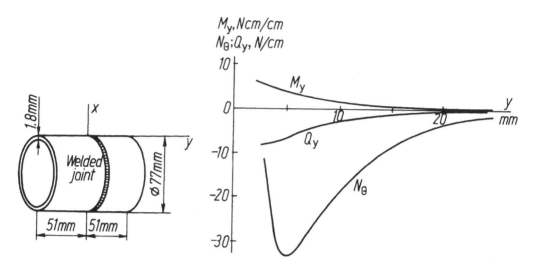

Fig. 4.8: Geometry of the   Fig. 4.9: The stress resultants in the welded
welded polycarbonate shell. shell.

shells by photoelastic measurements, using in addition equations of equilibrium.[4] In this way it is also possible to determine stresses in sandwich shells.

Figure 4.8 shows a polycarbonate shell with a welded joint, and Fig.
4.9 shows the distribution of the stress resultants in it.

Methods for investigating stresses in shells of revolution of general
shape have also been elaborated. In the latter case oblique incidence investigations may prove effective. It is also possible to take into account
rotation of the principal axes due to the shear stress.[4]

If a beam of light is passed through a cylinder, placed in an immersion bath, parallel to the z axis (Fig. 4.10), the polarization transformations are influenced by stresses which act in the xy plane

$$\sigma_x = \sigma_r \cos^2\theta + \sigma_\theta \sin^2\theta \quad , \quad \sigma_y \quad\quad\quad (4.11)$$

We have assumed that there is no stress gradient in the axial direction.

In this case one of the principal directions is parallel to the cylinder axis. Consequently, all the information about stresses consists of

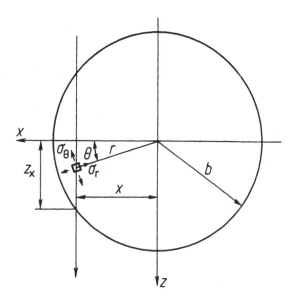

Fig. 4.10: Examination of a cylinder parallel to the z axis.

the phase retardation Δ, which may be measured at various rays

$$\Delta(x) = 2C_0 \int_0^{z_x} (\sigma_y - \sigma_r \cos^2\theta - \sigma_\theta \sin^2\theta)dz \qquad (4.12)$$

Poritsky has shown[32] that from the condition of equilibrium of a segment of a layer perpendicular to the cylinder axis follows the relationship

$$2\int_0^{z_x} (\sigma_r \cos^2\theta + \sigma_\theta \sin^2\theta)dz = 0 \qquad (4.13)$$

From Eqs. (4.12) and (4.13) it follows that the phase retardation is
due only to the axial stress σ_y

$$\Delta(x) = 2C_0 \int_0^{z_x} \sigma_y \, dz \qquad (4.14)$$

The latter relationship can be written as[33]

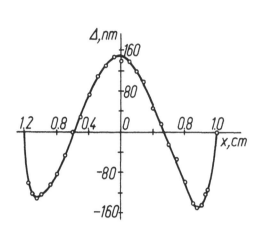

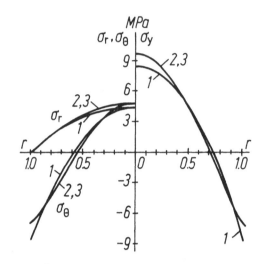

Fig. 4.11: Distribution of the integral phase retardation in a glass rod (after Brosman[34]).

Fig. 4.12: Distribution of stresses in a glass rod; the numbers indicate the number of terms in series expansion (after Brosman[34]).

$$\Delta(x) = 2C_0 \int_x^b \sigma_y \frac{r\,dr}{\sqrt{r^2 - x^2}} \qquad\qquad (4.15)$$

Using the so-called sum rule $\sigma_y = \sigma_r + \sigma_\theta$ and equation of equilibrium, equation (4.15) yields

$$\Delta(x) = 2C_0 \int_0^b \frac{d}{dr}(r^2\sigma_r) \frac{dr}{\sqrt{r^2 - x^2}} \qquad\qquad (4.16)$$

This is an Abel integral equation whose solution is known. Thermal stresses in the cylinder are

$$\sigma_r(r) = -\frac{1}{\pi C_0 r^2} \int_r^b \frac{x\Delta(x)\,dx}{\sqrt{r^2 - x^2}}$$

$$\sigma_\theta(r) = \frac{d}{dr}[r\sigma_r(r)]$$

$$\sigma_y = \sigma_r + \sigma_\theta \qquad\qquad (4.17)$$

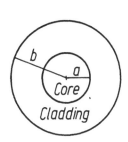

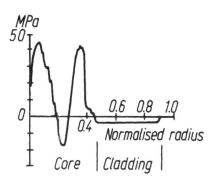

Fig. 4.13: Cross-section of a fiber.   Fig. 4.14: Axial stress profile of
                                          germanium and boron doped fiber
                                          preform (after Chu and Whitbread[36]).

Figures 4.11 and 4.12 show experimental results and stress distribution in a quenched glass rod, determined by a somewhat different method.[34]

Integrated photoelasticity may also be used to determine stresses in
optical fibers. Assume that the axial component of the residual stress is
constant over the core (σ_a) and cladding (σ_b) (Fig. 4.13). The optical retardation Δ for a diametral ray may be expressed as[35]

$$\Delta = 2C(a\sigma_a + b\sigma_b) \qquad (4.18)$$

Since axial load equals zero, we have

$$a^2\sigma_a + (b^2 - a^2)\sigma_b = 0 \qquad (4.19)$$

From Eqs. (4.18) and (4.19) follows

$$\sigma_a = \frac{\Delta \cdot (b^2 - a^2)}{2C[a(b^2 - a^2) - ba^2]}$$

$$\sigma_b = \frac{-\Delta \cdot a}{2C(b^2 - a^2 - ab)} \qquad (4.20)$$

In the general case axial stresses in core and cladding may be distributed arbitrarily. They can be determined using the general algorithm
for cylindrical bodies.[36] One example is given in Fig. 4.14.

5. AXISYMMETRIC STATES OF STRESS

Until recently it was thought that the integral fringe pattern is the only source of information while investigating a body of revolution by integrated photoelasticity. In this respect the theory of characteristic directions opens up new possibilities since additional experimental data,
the characteristic directions, are available. In this chapter it is shown
that the method of characteristic directions permits to determine an axisymmetric state of stress in the general case. To determine stresses in a
plane of symmetry a new method, integrated gradient photoelasticity, is
developed.

Determination of stresses in cylinders without axial stress gradient
was considered in the previous chapter. It is easy to determine also stress
distribution of spherical symmetry.[37] Srinath developed a method which permits to determine principal stress differences in a plane of symmetry when
the Poisson ratio μ equals 0.5.[38]

A formally general algorithm for determining stresses in an axisymmetric body was proposed by Doyle and Danyluk.[39] The only weak point of
the algorithm developed in paper[39] lies in the assumption that the characteristic directions coincide with principal stress directions at a certain
point on the light ray. This assumption can be used when rotation of secondary principal directions is small. If that is not the case, the results
may be erroneous.

Let us consider a general method of solving the axisymmetric problem.
We present the stress distribution in a transverse section, as follows:

$$\sigma_r = \sum_{k=0}^{m} a_{2k}\rho^{2k} \quad , \quad \sigma_\theta = \sum_{k=0}^{m} b_{2k}\rho^{2k}$$

$$\sigma_z = \sum_{k=0}^{m} c_{2k}\rho^{2k} \quad , \quad \tau_{rz} = \sum_{k=1}^{m} d_{2k-1}\rho^{2k-1} \quad \quad (5.1)$$

Here a_{2k}, b_{2k}, c_{2k}, d_{2k-1} are coefficients to be determined on the basis of experimental data, and ρ is dimensionless radius. Let us denote a'_{2k}, b'_{2k}, c'_{2k}, and d'_{2k-1} coefficients which determine stress distribution in a neighbouring transverse section at a distance Δz.

In a transverse section the following compatibility equation holds

$$\frac{\partial}{\partial\rho}[\sigma_\theta - \mu(\sigma_r + \sigma_z)] - (1 + \mu)\frac{\sigma_r - \sigma_\theta}{\rho} = 0 \quad \quad (5.2)$$

Introducing expressions (5.1) into Eq. (5.2) reveals

$$\sum_{k=1}^{m} 2kb_{2k}\rho^{2k-1} - \mu\sum_{k=1}^{m} 2k(a_{2k}+c_{2k})\rho^{2k-1} - (1+\mu)\sum_{k=0}^{m}(a_{2k}-b_{2k})\rho^{2k-1} = 0 \quad (5.3)$$

From the latter equation follows

$$b_{2k} = \frac{2k\mu}{2k + 1 + \mu}c_{2k} + \frac{1 + (2k + 1)\mu}{2k + 1 + \mu}a_{2k} \quad , \quad k=0,1,2,\ldots \quad (5.4)$$

Thus, Eqs. (5.4) permit to eliminate coefficients b_{2k} and b'_{2k}. Equation of equilibrium for the radial direction

$$\frac{\partial\sigma_r}{\partial\rho} + \frac{\sigma_r - \sigma_\theta}{\rho} + \frac{\partial\tau_{rz}}{\partial} = 0 \quad \quad (5.5)$$

permits to eliminate the coefficients a_{2k} and a'_{2k}

$$a_{2k} = \frac{2k\mu}{4k(k + 1)}c_{2k} - \frac{2k + 1 + \mu}{4k(k + 1)} \cdot \frac{d'_{2k-1} - d_{2k-1}}{\Delta z} \quad \quad (5.6)$$

From the equation of equilibrium for the z direction

$$\frac{\partial \sigma_z}{\partial z} + \frac{\partial \tau_{rz}}{\partial \rho} + \frac{\tau_{rz}}{\rho} = 0 \tag{5.7}$$

follows

$$c'_{2k} = c_{2k} + (k + 1)(d_{2k-1} + d'_{2k-1})\Delta z \tag{5.8}$$

Thus, all the unknown coefficients are expressed through three sets of coefficients: c_{2k}, d_{2k-1}, and d'_{2k-1}.

By experimental invextigation we may determine on every ray of light two characteristic parameters, α_0 and Δ_*, since due to symmetry $\alpha_* = \alpha_0$. Therefore, making measurements in the two neighbouring transverse sections, we have four sets of experimental data: $\alpha_0(x)$, $\Delta_*(x)$, $\alpha'_0(x)$, and $\Delta'_*(x)$. Four sets of experimental data permit to determine three sets of unknown coefficients c_{2k}, d_{2k-1}, and d'_{2k-1} in Eq. (5.8).

Unfortunately, the relationship between the stress parameters and the characteristic parameters cannot be expressed analytically. Therefore, one has to use a numerical technique for solving the problem. For example, one may try to determine the coefficients c_{2k}, d_{2k-1}, and d'_{2k-1} from the condition of minimum of the function

$$\sum_x [(\Delta_*^e \cos 2\alpha_0^e - \Delta_*^c \cos 2\alpha_0^c)^2 + (\Delta_*^e \sin 2\alpha_0^e - \Delta_*^c \sin 2\alpha_0^c)^2 +$$

$$+ (\Delta_*^{'e} \cos 2\alpha_0^{'e} - \Delta_*^{'c} \cos 2\alpha_0^{'c})^2 + (\Delta_*^{'e} \sin 2\alpha_0^{'e} - \Delta_*^{'c} \sin 2\alpha_0^{'c})^2] \tag{5.9}$$

where index "e" indicates experimental, and index "c", calculated values.

That is a problem of nonlinear minimization. It can be solved by well-known methods and program packages.[40] Further investigations have to show how effectively this method can be applied in practice.

Stresses in a plane of symmetry can also be determined by another method which is named integrated gradient photoelasticity. In gradient photoelasticity[41] developed by Pindera and his coauthors it is taken into account that in nonhomogeneous photoelastic models a bending of light rays

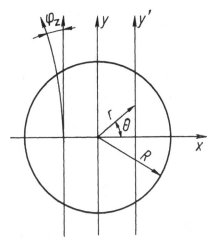

Fig. 5.1: Passing of light through a plane of symmetry of an axisymmetric body.

takes place. Up to now this phenomenon has been considered mostly as a source of errors. However, bending of light rays is also a source of new information about the stress distribution.

If we pass polarized light through a plane of symmetry (Fig. 5.1), there is no rotation of the secondary principal directions and the integral Wertheim law yields

$$\Delta(\xi) = 2C_0 R \int_0^{\sqrt{1-\xi^2}} (\sigma_z - \sigma_r \cos^2\theta - \sigma_\theta \sin^2\theta)d\eta \qquad (5.10)$$

where $C_0 = C_1 - C_2$, $\xi = x/R$, $\eta = y/R$, and C_i are the stress-optic coefficients.
Introducing polynomials (5.1) into (5.10) yields

$$\frac{\Delta(\xi)}{2C_0 R} = \sum_{k=0}^{m} (c_{2k} G_{2k} - a_{2k} F_{2k} - b_{2k} H_{2k}) \qquad (5.11)$$

where

$$G_0 = \sqrt{1 - \xi^2} \quad , \quad F_0 = \xi \text{ arc cos } \xi \quad , \quad H_0 = G_0 - F_0$$

$$G_{2k} = \frac{\sqrt{1-\xi^2}}{2k+1} + \frac{2k}{2k+1} \xi^2 G_{2k-2}$$

$$F_{2k} = \xi^2 G_{2k-2} \quad , \quad H_{2k} = G_{2k} - \xi^2 G_{2k-2} \qquad (5.12)$$

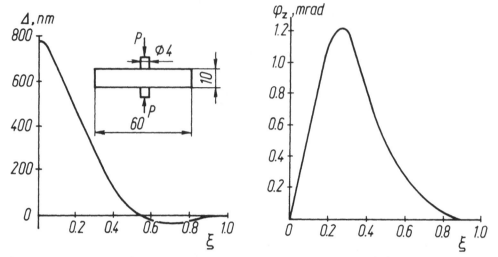

Fig. 5.2: Distribution of integral phase retardation $\Delta(\xi)$ and of angle of deflection $\varphi_z(\xi)$.

We may also use Eqs. (5.4) which follow from the compatibility condition (5.2).

The refraction index n_z can be expressed as

$$n_z = n_0 + C_1\sigma_z + C_2(\sigma_r + \sigma_\theta) \tag{5.13}$$

where n_0 is the refraction index of the stress free material.

The angle of deflection of light φ_z with vibrations along z-axis (Fig. 5.1) is

$$\varphi_z = \frac{1}{n_0} \int \frac{\partial n_z}{\partial x} \, dy \tag{5.14}$$

Introducing Eq. (5.13) into Eq. (5.14) and taking into account polynomials (5.1) yields

$$\frac{n_0\varphi_z(\xi)}{4C_1} = \sum_{k=1}^{m} (qa_{2k} + qb_{2k} + c_{2k})K_{2k} \tag{5.15}$$

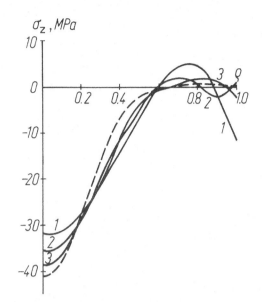

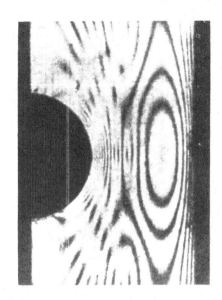

Fig. 5.3: Stress distribution
in the case shown in Fig. 5.2.

Fig. 5.4: Integral fringe pattern of
a thick tube with internal groove (after Khayyat and Stanley[43]).

where $q = C_2/C_1$, $K_2 = k\xi G_{2k-2}$.

Measuring integral phase retardation $\Delta(\xi)$ and the angle of deflection
$\varphi_z(\xi)$ for many rays of light we can determine two sets of unknown coefficients from the system of equations (5.11) and (5.15) (Eqs. (5.4) permitted
to eliminate one set of coefficients). Thus, the coefficients a_{2k}, b_{2k},
and c_{2k}, which determine the stress distribution, can be calculated.

This method was used to determine stresses in a plate loaded as shown
in Fig. 5.2. Experimental data are shown in Fig. 5.2. Distribution of the
stress σ_z for various values of k in the polynomial for σ_z (Eqs. (5.1)) is
shown in Fig. 5.3. Analytical solution[42] is given by the dotted curve.

Sometimes it is effective to use integrated photoelasticity together
with the scattered light method. Such a combination permits to reduce time-
consuming scattered light measurements. Khayyat and Stanley[43] used this
method to determine thermal stresses in a thick tube with an internal
groove (Fig. 5.4).

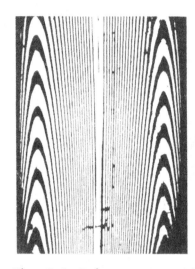

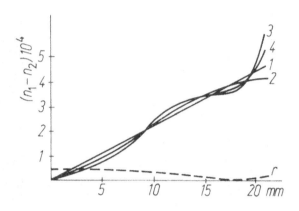

Fig. 5.5: Fringe pattern of
a cylindrical plastic scintillator.

Fig. 5.6: Radial distribution of the optical
anisotropy in a cylindrical plastic scintillator. The numbers indicate the number of
terms in expansion (5.16).

Often, especially when investigating specimen made of plastics, birefringence cannot be directly related to stresses. In this case one has
to determine the distribution of birefringence. That can also be done by
integrated photoelasticity.

In Fig. 5.5 the integral fringe pattern of the middle part of a cylindrical plastic scintillator is shown. Distribution of the integral phase
retardation Δ_* and that of the primary characteristic angle $\alpha_0 = \alpha_*$ in the
radial direction were measured experimentally. It is of interest to note
that except for a narrow region near the cylinder axis $\alpha_0 \cong 45^\circ$. From this
fact it follows that one axis of the index ellipsoid forms an angle of 45°
with the cylinder axis and the index ellipsoid is an ellipsoid of revolution.

Thus, in this case we have only one unknown function of the radial coordinate which can be expressed as

$$n_1 - n_2 = a_1 r + a_3 r^3 + a_5 r^5 + \ldots \tag{5.16}$$

where a_i are undetermined coefficients. The latter coefficients were determined from the experimental results by the method of least squares. The
radial distribution of the optical anisotropy according to various numbers
of terms in expansion (5.16) is shown in Fig. 5.6.

6. CUBIC SINGLE CRYSTALS OF CYLINDRICAL AND PRISMATIC FORM

Cubic single crystals are widely used as elements of infrared optics, windows of high-power lasers, scintillators, acoustic crystals, light modulators of large aperture, etc. During the growing process considerable residual stresses may arise in the crystal (Fig. 6.1). The stresses influence infavourably the optical properties of the crystals. In this chapter it is shown how integrated photoelasticity permits to determine stresses in cubic single crystals of simple form.

Methods of calculating residual stresses in single crystals seldom permit to obtain trustworthy results since one has no exact data about the change in the elastic and physical properties of the single crystals during the growing process.

Up to now by experimental investigation of single crystals of cylindrical form, a plate is cut out of the crystal perpendicular or parallel to the cylinder axis and its stresses are determined by the aid of two-dimensional photoelasticity.[45] With certain assumptions the stresses in the cylinder can now be calculated. This method is destructive and may give inaccurate results.

Fig. 6.1: Fringes in a garnet crystal, photographed together with a quartz wedge (after Indenbom[44]).

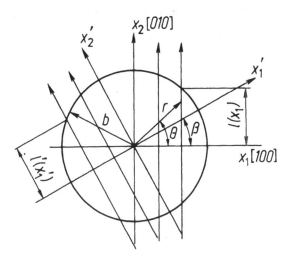

Fig. 6.2: Passing of light through a transverse section of a cylindrical crystal.

.

Let us consider passing of light through a transverse section of a cylindrical crystal with no axial stress gradient (Fig. 6.2). Since there is no rotation of the secondary principal directions, we may apply the integral Wertheim law which yields[46]

$$\Delta'(x_1') = \frac{1}{2} n_0^3 \int [(\pi_{11} - \pi_{12})(\sigma_{11}\cos^2\beta +$$

$$+ \sigma_{22}\sin^2\beta - \sigma_{33}) + \pi_{44}\sigma_{12}\sin 2\beta]dx_2' \qquad (6.1)$$

Here n_0 is the refraction index of the crystal and π_{ij} are piezooptic coefficients in matrix notation. Let us point out that in the case of a cubic single crystal there are three piezooptic coefficients instead of two in the case of an isotropic body.

From the equilibrium condition of a layer of the segment of the cylinder in the x_1' direction follows

$$\int_{-l'(x_1')}^{l'(x_1')} (\sigma_{11}\cos^2\beta + \sigma_{22}\sin^2\beta + \sigma_{12}\sin 2\beta)dx_2' = 0 \qquad (6.2)$$

where $l'(x_1')$ denotes half of the light path length (Fig. 6.2).

Putting Eq. (6.2) into Eq. (6.1) and taking $\beta = 0$ yields

$$\Delta(x_1) = -n_0^3(\pi_{11} - \pi_{12})\int_0^{l(x_1)} \sigma_z \, dx_2 \tag{6.3}$$

Equation (6.3) shows that when light is passed through the transverse section of the crystal parallel to a crystallographic axis [010], then the integral phase retardation depends only on the axial stress σ_z. Distribution of σ_z can be determined when $\Delta(x_1)$ is measured at many values of x_1, similarly to isotropic cylinders.

To determine also radial and circumferential stresses σ_r and σ_θ we need additional experimental data. Let us pass light through the transverse section of the crystal parallel to the x_2' direction (Fig. 6.2). Taking into account Eq. (6.2), Eq. (6.1) yields (for $\beta = 45°$)

$$\Delta'(x_1') = \frac{1}{2} n_0^3 \int_{-l'(x_1')}^{l'(x_1')} \{[\pi_{44} - (\pi_{11} - \pi_{12})](\sigma_r - \sigma_\theta)\sin\theta \cos\theta - (\pi_{11} - \pi_{12})\sigma_z\} dx_2' \tag{6.4}$$

The stress components to be determined can be expressed as

$$\sigma_r - \sigma_\theta = \sum_{k=0}^{m} b_{2k}\, \rho^{2k} \quad , \quad \sigma_z = \sum_{k=0}^{m} c_{2k}\, \rho^{2k} \tag{6.5}$$

We have assumed that the stress distribution is axisymmetric. That is permissible since the mechanical anisotropy of the crystals is assumed to be weak, and during the growing process the temperature field is axisymmetric.

If Eqs. (6.5) are introduced into Eqs. (6.3) and (6.4), a system of linear equations is obtained for determining the coefficients b_{2k} and c_{2k}. The stresses can be separated by numerical integration of the equilibrium equation.

As an example, Fig. 6.3 shows experimental data (a) and stress distribution (b) in a NaCl single crystal.

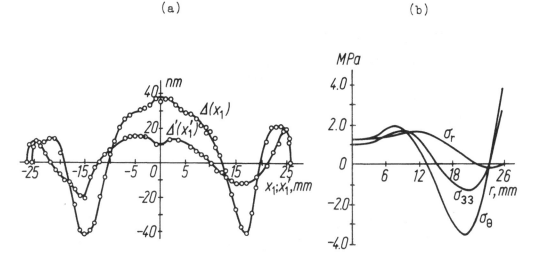

Fig. 6.3: Experimental data (a) and stress distribution (b) in a cylindrical NaCl crystal.

The integral retardation distributions were measured point-by-point with a compensator. As immersion fluid a mixture of medical vaseline oil (n = 1.48) and α-monobromnaphtaline (n = 1.66) was used (n_0 = 1.55).

It is to be seen in Fig. 6.3 that the distribution of the axial stress σ_z deviates considerably from a simple parabola. Such an "anomalous" stress distribution has been theoretically predicted by Reznikov.[47]

A method for investigating stresses in cylinders with axial stress gradient has also been elaborated.

Let us consider now investigation of stresses in a prismatic crystal of quadratic transverse section. It is possible to show that if light passes the transverse section (Fig. 6.4) parallel to the crystallographic axis [010] (β=0), then the integral phase retardation $\Delta(x_1)$ is expressed as

$$\Delta(x_1) = -\frac{1}{2} n_0^3 (\pi_{11} - \pi_{12}) \int_{-a/2}^{a/2} \sigma_{33} \, dx_2 \tag{6.6}$$

In the general case (β ≠ 0) we have

$$\Delta'(x_1') = \frac{1}{2} n_0^3 \int_{-1_1'(x_1')}^{1_2'(x_1')} \{[\pi_{44} - (\pi_{11} - \pi_{12})]\sigma_{12} \sin 2\beta -$$

$$- (\pi_{11} - \pi_{12})\sigma_{33}\} dx_2' \qquad (6.7)$$

The stress components may be expressed as

$$\sigma_{33} = \sum_{k=0}^{m} \sum_{l=0}^{m} a_{kl} \cos k\pi x_1 \cos l\pi x_2 \qquad (6.8)$$

$$\sigma_{12} = \sum_{k=1}^{n} \sum_{l=1}^{n} b_{kl} \sin k\pi x_1 \sin l\pi x_2 \qquad (6.9)$$

If Eqs. (6.8) and (6.9) are introduced in Eqs. (6.6) and (6.7), a system of linear equations is obtained for determining the coefficients a_{kl} and b_{kl} on the basis of integral retardation measurements. The state of stress can be completely determined by numerical integration of the equilibrium equations.

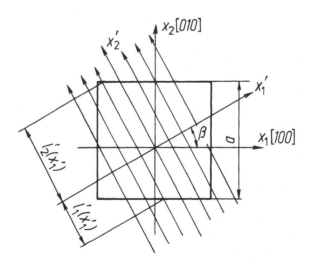

Fig. 6.4: Passing of light through a quadratic transverse section of a prismatic crystal.

If the axial stress gradient is present, but rotation of the secondary principal stress directions is small, we may use the following approximate expressions

$$\Delta(x_1)\cos 2\alpha_0(x_1) = \frac{1}{2} n_0^3(\pi_{11} - \pi_{12}) \int_{-a/2}^{a/2} (\sigma_{11} - \sigma_{33})dx_2 \tag{6.10}$$

$$\Delta(x_1)\sin 2\alpha_0(x_1) = n_0^3 \pi_{44} \int_{-a/2}^{a/2} \sigma_{31} \, dx_2 \tag{6.11}$$

and for $\beta=45^\circ$

$$\Delta'(x_1')\cos 2\alpha_0'(x_1') = \frac{1}{2} n^3 \int_{-\frac{\sqrt{2}}{2} a+x_1'}^{\frac{\sqrt{2}}{2} a-x_1'} \{ \frac{1}{2}(\pi_{11} - \pi_{12})x$$

$$x[(\sigma_{11} - \sigma_{33}) + (\sigma_{22} - \sigma_{33})] + \pi_{44}\sigma_{12}\} \, dx_2' \tag{6.12}$$

$$\Delta'(x_1')\sin 2\alpha_0'(x_1') = \frac{\sqrt{2}}{2} n_0^3 \pi_{44} \int_{-\frac{\sqrt{2}}{2} a+x_1'}^{\frac{\sqrt{2}}{2} a-x_1'} (\sigma_{23} + \sigma_{31})dx_2' \tag{6.13}$$

All the stress components are to be expressed in the form of Fourier expansions which are to be put into Eqs. (6.10) to (6.13). If the experimental measurements are carried out in two adjacent transverse sections, all the stress components can be determined.

Figures (6.5), (6.6), (6.7), and (6.8) show experimental data for a prismatic KCl crystal of quadratic transverse section of the size 60 x 60 x 480 mm. The stress distribution is given in Fig. 6.9.

A more rigorous treatment of the rotation of the secondary principal axes leads, similarly to Chap. 5, to a nonlinear minimization problem.[40]

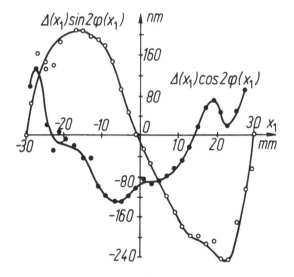

Fig. 6.5: Experimental data for a prismatic KCl crystal (section 1, wave normal in the x_2 direction).

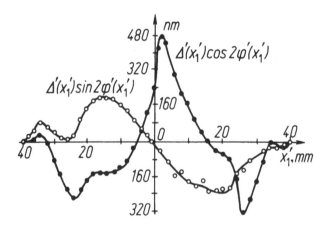

Fig. 6.6: Experimental data for a prismatic KCl crystal (section 1, wave normal in the x_2' direction with $\beta = 45°$).

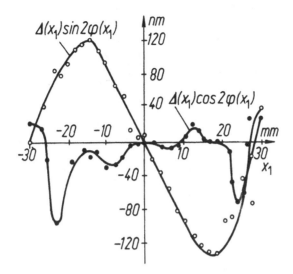

Fig. 6.7: Experimental data for a prismatic KCl crystal (section 2, wave normal in the x_2 direction).

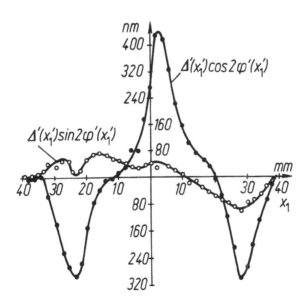

Fig. 6.8: Experimental data for a prismatic KCl crystal (section 2, wave normal in the x_2' direction with $\beta = 45^\circ$).

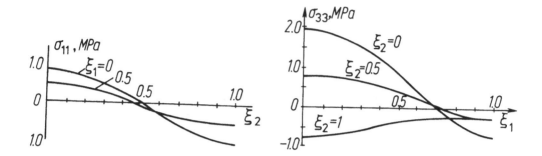

Fig. 6.9: Stress distribution in the prismatic KCl crystal ($\xi_i = 2x_i/a$).

7. INTEGRATED PHOTOELASTICITY AS OPTICAL TOMOGRAPHY OF THE STRESS FIELD

The question may arise whether integrated photoelasticity is something particular, or we can find some analogy in other branches of science. In this chapter it is shown that integrated photoelasticity can be considered as a special kind of tomography. However, an essential feature of integrated photoelasticity is that it investigates tensor fields while in traditional tomography only scalar fields are considered.

Tomography (in Greek: τυμε – section, γραφη – picture) is a technique to restore images of sections on the basis of integrated measurements.[48]

Some radiation is passed through the body in the plane xy to be investigated (Fig. 7.1) and integral data are recorded on the screen -E -E. The body may be a human body, a supernova, a part of the atmosphere, a fiber, etc. The radiation may be x-rays, ions, protons, ultrasound, visible light, etc. The data that are recorded are mostly distribution of intensity on the screen for various directions of the radiation. The image to be restored may be distribution of the coefficient of attenuation, coefficient of acoustical refraction, coefficient of absorption, etc.

Image reconstruction from projections is the process of producing an image of a two-dimensional distribution from estimates of its line integral along a finite number of lines of known locations. Since application of computers by image reconstruction is inevitable, the method is also named computer tomography. As an example, a x-ray tomographic image of human abdomen is shown in Fig. 7.2.

Image reconstruction technique is based on the Radon transform. Radon transform of a function f is defined for real number pairs $(1, \theta)$ as follows

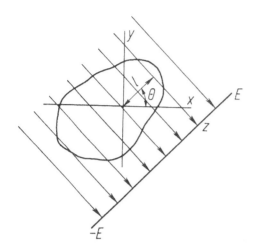
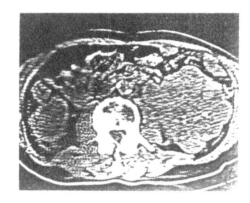

Fig. 7.1: Illustration of tomography.   Fig. 7.2: Image of an abdomen.

$$[Rf](1,\theta) = \int\limits_{-\infty}^{\infty} f[\sqrt{1^2 + z^2} \quad , \quad \theta + \text{arc tan}(z/1)]dz \qquad (7.1)$$

From Fig. 7.1 it follows that $[Rf](1,\theta)$ is the line integral of f
along a certain line z.

The Radon transform associates with a function f of two polar variables another function Rf of two variables. We are looking for an operator R^{-1}, which is such that $R^{-1}Rf$ is f. Such an operator is the following

$$[R^{-1}p](r,\phi) = \frac{1}{2\pi^2} \int\limits_{0}^{\pi} \int\limits_{-E}^{E} \frac{\partial p}{\partial l} \frac{dl d\theta}{r \cos(\theta - \phi) - 1} \qquad (7.2)$$

A lot of numerical procedures for implementation of the inverse Radon
transform (7.2) on a computer, have been elaborated.[48] Let us mention that
the Abel transform which was used to solve some problems in Chap. 4, is a
special type of the Radon transform.

In traditional tomography every point of the two-dimensional field
under investigation is characterized by a single scalar (attenuation co-

efficient, refractive index, etc.). In integrated photoelasticity every point of the field is characterized by a second rank tensor which is in general case determined through six scalars. Graphically the difference is shown in Fig. 7.3. Therefore, integrated photoelasticity is much more complicated than traditional tomography.

From Fig. 7.3 follows an important difference of integrated photo-elasticity from traditional tomography. In the latter case the influence on the passing radiation of a point does not depend on the direction of the radiation, since the point is characterized by a scalar which is a spherical tensor. In integrated photoelasticity a point is characterized by an ellipsoid (index ellipsoid, stress ellipsoid) and its influence on the passing radiation depends on the direction of the latter according to the laws of transformation of the tensor components.

In photoelasticity one is usually interested in the stress tensor. However, that is not always so. For example, in a specimen made of plastics only part of the birefringence may be caused by elastic stresses. In addition there may be present birefringence due to anisotropic atom groups in molecules – the so-called orientation birefringence. For estimation of the quality of such specimen it is important to determine the distribution of the dielectric tensor or refractive index tensor. The latter can also be considered as a tensor due to the weakness of the birefringence.

Another difference of integrated photoelasticity from traditional to-mography is that polarized radiation is used. Therefore, experimental da-ta in integrated photoelasticity contain more information.

Let us assume first that on the light ray there is no rotation of the secondary principal axes. In this case it is possible to measure absolute phase retardations Δ_x and Δ_y for two vibration directions x,y. Analog-ously to the integral Wertheim law (1.21) we have

$$\Delta_x = \int n_x \, dz \quad , \qquad \Delta_y = \int n_y \, dz \tag{7.3}$$

We may also measure the relative phase retardation Δ

$$\Delta = \Delta_x - \Delta_y = \int (n_x - n_y) \, dz \tag{7.4}$$

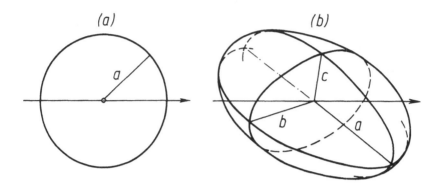

Fig. 7.3: A scalar may be represented by a sphere (a), a tensor – by an ellipsoid (b).

One has to bear in mind that only two of the phase retardations Δ_x, Δ_y, and Δ give independent information about the distribution of the refraction index.

The question may arise, whether measurement of the deflection of the light rays may give additional information about the distribution of bi-refringence.

From Eq. (5.14) follows

$$\varphi_x = \frac{1}{n_0^2} \int \frac{\partial n_x}{\partial x} \, dz \tag{7.5}$$

Differentiating the first Eq. (7.3), we have

$$\frac{\partial \Delta_x}{\partial x} = \frac{\partial}{\partial x} \int n_x \, dz = \int \frac{\partial n_x}{\partial x} \, dz \tag{7.6}$$

Comparison of Eqs. (7.5) and (7.6) shows that if we have measured absolute phase retardations, measurement of the deflection of light rays does not give additional information about the distribution of the index tensor.

Thus, if there is no rotation of the secondary principal directions,

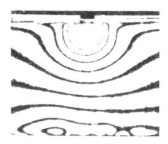

Fig. 7.4: Interferogram of a block of perspex with central concentrated
load on the upper surface (after Vest and Ural[49]).

on every ray of light we have two samples of experimental data.

Assume that we investigate a plane of symmetry of an axisymmetric body. Then, analogously to Eq. (5.10), the integral phase retardation can
be expressed as

$$\Delta(\xi) = 2R \int_{0}^{\sqrt{1-\xi^2}} (n_z - n_r \cos^2\theta - n_\theta \sin^2\theta)d\eta \qquad (7.7)$$

In addition we may measure the deflection of light rays which have
vibrations along z axis. In this case Eq. (5.14) permits to determine the
radial distribution of n_z. However, Eq. (7.7) contains two more unknown
functions, n_r and n_θ, which cannot be determined separately on the basis
of a sole experimental function $\Delta(\xi)$.

Figure 7.4 shows an interferogram of a block of perspex with a concentrated load. Due to low birefringence the scalar refraction index is
proportional to the sum of the principal stresses. The latter is the only
unknown function in this case which can be determined by the aid of one
experimental function (distribution of the absolute phase retardation).[49]

In the case of spherical symmetry we have $n_\theta = n_z$, and the distribution of refractive indices can be fully determined.

In the general case one may pass polarized light through the body in

many directions. However, in this case rotation of the secondary principal
axes of birefringence takes place. One may measure the characteristic angles α_0, α_* and the characteristic phase retardation Δ_*. Unfortunately,
these values cannot be expressed through integrals like Eq. (7.7). One
has to use complicated numerical algorithms as mentioned in Chap. 5.

Two additional differences of integrated photoelasticity from traditional tomography follow. Firstly, in the general case the tensor field
cannot be separately determined for a single section. One has to consider
the field in the whole three-dimensional body. Secondly, in the general
case experimental data cannot be expressed through the field parameters
via integrals.

In traditional tomography the scalar function to be determined may
be arbitrary, i.e., this function may have discontinuities, etc. For example, in x-ray tomography at the boundary of a bone there is a stepwise
change of the attenuation coefficient.

The situation may be the same while investigating a dielectric tensor field. For example, in crystals discontinuities may occur in the planes
of shift.

The situation is different while investigating a stress field. If we
may assume that the birefringence is caused by stresses, by interpreting
the experimental data we may use relationships of the theory of elasticity: equilibrium and compatibility equations and macrostatic equilibrium
conditions. That makes determination of a stress field easier. For example, in Chap. 5, by elaborating algorithm for stress determination in an
axisymmetric body we reduced the number of unknowns by the aid of compatibility and equilibrium equations until the experimental data were sufficient to determine the remaining unknown coefficients.

Another specific feature of the stress field tomography is that sometimes it is reasonable to consider two adjacent parallel sections together.
In Chap. 5, by elaborating a general algorithm for the axisymmetric state
of stress application of the equilibrium equation in the radial direction
made it inevitable to consider two adjacent transverse sections at the
same time.

Finally, let us mention that traditional tomography as well as integrated photoelasticity in particular cases lead to the inverse problem for a system of Fredholm integral equations like Eq. (7.1). This problem is an ill-determined problem in the sense of Adamard. It means that small changes in experimental data may lead to large changes in the parameters to be determined. To achieve a stable solution special methods, as e.g. methods of regularization,[50] are to be used.

The experience we have for today in the field of traditional tomography and integrated photoelasticity encourage us to express the hope that the class of problems, which can be effectively solved by integrated photoelasticity, can be widened.

REFERENCES

1. Гинзбург, В.Л., Об исследовании напряжений оптическим методом, *Журнал техн. физ.*, 14, 181, 1944.

2. Абен, Х., О применении метода фотоупругости для исследования выпученных пластинок, *Изв. АН ЭССР, сер. техн. и физ.-мат. наук*, 6, 28, 1957.

3. Aben, H.K., Optical phenomena in photoelastic models by the rotation of principal axes, *Exp. Mech.*6, 13, 1966.

4. Aben, H., *Integrated Photoelasticity*, McGraw-Hill, New York, London, 1979.

5. Neumann, F.E., Die Gesetze der Doppelbrechung des Lichts in komprimierten oder ungleichförmig erwärmten unkrystallinischen Körpern, *Abh. Kön. Akad. Wiss. Berlin*, 2, 3, 1843.

6. Mindlin, R.D. and Goodman, L.E., The optical equations of three-dimensional photoelasticity, *J. Appl. Phys.*, 20, 89, 1949.

7. Kubo, H. and Nagata, R., Considerations of propagation of light waves in photoelastic materials and crystals, *Optik*, 52, 37, 1978/79.

8. Mönch, E., Roengvoraphoj, S., Beitrag zur Theorie der Spannungsoptik im elastischen, statischen Bereich bei beliebig veränderlichem Spannungszustand, *Ing.-Archiv*, 50, 1, 1981.

9. Iwashimizu, Yu., Ultrasonic wave propagation in deformed isotropic elastic materials, *Intern. J. Solids and Struct.*, 7, 419, 1971.

10. Kuske, A., *Einführung in die Spannungsoptik*, Wissenschaftliche Verlagsgesellschaft, Stuttgart, 1959.

11. Srinath, L.S., *Scattered Light Photoelasticity*, McGraw-Hill, New Delhi, 1983.

12. Brillaud, J., Lagarde, A., Methode ponctuelle de la photoélasticité

tridimensionnelle, *Proc. Seventh Internat. Conf. on Exp. Stress Anal.*, *Haifa*, 329, 1982.

13. Theocaris, P.S., Gdoutos, E.E., *Matrix Theory of Photoelasticity*, Springer-Verlag, Berlin, Heidelberg, New York, 1979.

14. Poincaré, H., *Théorie mathématique de la lumière*, *II*, Paris, 1892.

15. Robert, A., The application of Poincaré's sphere to photoelasticity, *Int. J. Solids and Struct.*, 6, 423, 1970.

16. Mindlin, R.D., An extension of the photoelastic method of stress measurement to plates in transverse bending, *J. Appl. Mech.*, 8, A187, 1941.

17. Favre, H., Gilg, B., Sur une méthode purement optique pour la mesure directe des moments dans les plaques minces fléchies, *Schweiz. Bauzeitung*, 68, 253, 265, 1950.

18. Kuske, A., Die Auswertung von ebenen spannungsoptischen Versuchen. Scheiben und Platten nach dem Zweischichtverfahren, *Forsch. Ingenieurwes.*, 18, 113, 1952.

19. Schwieger, H., Ein Auswerteverfahren bei der spannungsoptischen Untersuchung elastischer Platten, *Bauplan.-Bautechn.*, 8, 174, 1954.

20. Goodier, J.N., Lee, G.H., An extension of the photoelastic method of stress measurement to plates in transverse bending, *J. Appl. Mech.*, 8, 27, 1941.

21. Lerchenthal, Ch.H., Photoelastic analysis of plates by bonding polaroid foils between plastics, *Bull. Res. Council of Israel*, GC, 202, 1958.

22. Aben, H.K., Magnetophotoelasticity – photoelasticity in a magnetic field, *Exp. Mech.*, 10, 97, 1970.

23. Aben, H.K., Idnurm, S., Stress concentration in bent plates by magnetophotoelasticity, *Proc. Fifth Intern. Conf. on Exp. Stress Anal.*, *Udine*, 4.5, 1974.

24. Peterson, R.E., *Stress Concentration Factors*, John Wiley and Sons, New York, London, Sydney, Toronto, 1974.

25. Spalding, I.J., High power lasers – their industrial and fusion applications, *Optics and Laser Technology*, 12, 187, 1980.

26. Aben, H., Idnurm, S., Josepson, J., Tatarinov, A., Spannungsoptische Effekte im starken Magnetfeld, *Wiss. Beiträge, Ingenieurhochschule Zwickau, Sonderheft zur Thema: III Kolloquium Eigenspannungen und Ober-*

flächenverfestigung, 185, 1982.

27. Kuske, A., *Spannungsoptische Untersuchung von Schalen*, Komissarverlag Hubert Hövelborn, Niederkassel-Mondorf, 1973.

28. Laermann, K.-H., Das Prinzip der integrierten Photoelastizität, angewandt auf die experimentelle Analyse von Platten mit nichtlinearen Formänderungen, *Proc. Seventh Internat. Conf. on Exp. Stress Anal., Haifa*, 301, 1982.

29. Read, W.T., An optical method for measuring the stress in glass bulbs, *J. Appl. Phys.*, 21, 250, 1950.

30. Preston, F.W., Synthesizing polariscopic strain patterns, *J. Soc. Glass Technol.*, 35, 496, 1951.

31. Ritland, H.N., Stress measurement in cylindrical vessels, *J. Amer. Ceram. Soc.*, 40, 153, 1957.

32. Poritsky, H., Analysis of thermal stresses in sealed cylinders and the effect of viscous flow during anneal, *Phys.*, 5, 406, 1934.

33. O'Rourke, R.C., Saenz, A.W., Quenching stresses in transparent isotropic media and the photoelastic method, *Quart. Appl. Math.*, 8, 303, 1950.

34. Бросман Э., К определению температурных напряжений в цилиндрах методом интегральной фотоупругости, *Изв. АН ЭССР, физика, математика*, 25, 418, 1976.

35. Saunders, M.J., Determination of the stress in optical fibers by means of a polariscope, *Rev. Sci. Instrum.*, 47, 496, 1976.

36. Chu, P.L., Whitbread, T., Measurement of stresses in optical fiber and preform, *Appl. Optics*, 21, 4241, 1982.

37. O'Rourke, R.C., Three-dimensional photoelasticity, *J. Appl. Phys.*, 22, 872, 1951.

38. Srinath, L.S., Principal-stress differences in transverse planes of symmetry, *Exp. Mech.*, 11, 130, 1971.

39. Doyle, J.F., Danyluk, H.T., Integrated photoelasticity for axisymmetric problems, *Exp. Mech.*, 18, 215, 1978.

40. Бросман, Э., Полль, В., Тяхт, К., Уточненный учет вращения квазиглавных направлений в интегральной фотоупругости кубических монокристаллов, *Изв. АН ЭССР, физика, математика*, 32, 179, 1983.

41. Hecker, F.W., Pindera, J.T., Influence of stress gradient on direction

of light propagation in photoelastic specimens, *VDI-Berichte*, N° 313, 745, 1978.

42. Sneddon, I.N., *Fourier Transforms*, McGraw-Hill, New York, 1951.

43. Khayyat, F.A., Stanley, P., The photoelastic determination of thermal stress concentrations in grooved cylinders with a radial temperature gradient, *J. Strain Anal.*, 14, 95, 1979.

44. Инденбом, В.Л., К теории образования напряжений и дислокаций при росте кристаллов, *Кристаллография*, 9, 74, 1964.

45. Hornstra, J., Penning, P., Birefringence due to residual stress in silicon, *Philips Res. Repts.*, 14, 237, 1959.

46. Aben, H., Brosman, E., Integrated photoelasticity of cubic single crystals, *VDI-Berichte*, N° 313, 45, 1978.

47. Резников, Б.А., О механизме образования "аномального" распределения остаточных напряжений при тепловой обработке монокристаллов, *Физика твердого тела*, 5, 2526, 1963.

48. Herman, G.T., *Image Reconstruction from Projections*, Academic Press, New York, 1980.

49. Vest, C.M., Ural, E.A., The role of interferometry and tomography in stress analysis of transparent media, *Proc. SESA Spring Meeting, Dearborn*, 242, 1981.

50. Tichonov, A., Arsenine, V., *Méthodes de Résolution de problèmes Mal Posés*, Mir, Moscou, 1976.

MODERN NONDESTRUCTIVE METHODS OF COHERENT LIGHT PHOTOELASTICITY WITH APPLICATIONS IN TWO AND THREE DIMENSIONAL PROBLEMS IN STATICS, CONTACT STRESSES, FRACTURE MECHANICS AND DYNAMIC IMPULSE

Alexis Lagarde
Laboratoire de Mécanique des Solides
Poitiers CEDEX, France

PREFACE

In this section we present several important ideas for new optical methods in mechanics of solids.

We introduce the basic knowledges of Fourier optics, scattered light and the speckle phenomenum that is necessary to understand the new methods that are presented.

The theoretical work on the propagation of light, developed in the previous section, is taken into account in the development of punctual and full field methods for nondestructive optical slicing in three dimensional media.

The important concepts that are necessary to measure the optical characteristics at each point of the slice are developed. They permit us to specify the elastic stress tensor at internal points. The reader's attention is drawn to the fact that the same approach constitutes a tool for non linear photomechanics.

We add examples that show the efficiency of these methods, either static or dynamic in practical applications.

CHAPTER 1

PRELIMINARIES

It is important to include, in this chapter, some fundamental
concepts that must be known by every mechanical engineer interested in
the application of optical methods. Moreover these concepts will aid in
the understanding of some developments. We shall specify the representation of light beams in usual conditions with the concepts of temporal or
spatial coherence and the light scattering phenomenon. These concepts are
the foundations of new and promising methods.

1   Representation of the beams of light within the physical optics.
    Approximations

      Consider a transparent, homogeneous, isotropic dielectric medium
without electric charges or currents. Then, the Maxwell equations are
uncoupled and each component of the electric field $\vec{E}$ and of the magnetic
field $\vec{H}$ is the solution of a scalar equation of propagation.

      Even under these conditions, the rigorous study of the diffracting
phenomenon within the electromagnetic theory is difficult. In practice,
most of the diffracting phenomena may be adequately represented by only
one complex component of the electric field (the optical scalar approximation). Then it is possible to find an integral solution for sinusoidal
vibrations. Note also that in describing polarization we must, of course,
take into account the vectorial nature of the electric field, and use
both of its complex components.

      Within the usual conditions of small apertures and small angles of in-
cidence of the beams, the complex component of the electric field is given
at a long distance R from the diaphragm plane by Fourier's integral Fig. 1.

      When the light is monochromatic and propagates along the direction
$\vec{n}$ defined by the direction cosines (u, v) on the $\vec{x}$ and $\vec{y}$ axes, the

electric field component has the following expression |1| :

$$U\left(\frac{u}{\lambda}, \frac{v}{\lambda}\right) = -\frac{j}{\lambda R} \text{ F.T. } \{a(x, y)\}$$

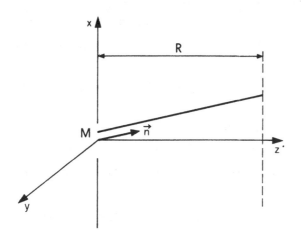

Fig. 1   Diffracting in a diaphragm

F.T. is Fourier's transform λ is the wavelength $j^2 = -1$

a(x,y) represents the complex component of the electric field suppo-
sedly known in the diaphragm plane (and is non-existent outside the aperture). Anticipating a definition which will be given later, assume that
there is spatial coherence in the area limited by the diaphragm.

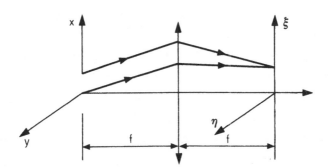

Fig. 2   Convergence in the Fourier plane

The reciprocal variables of x and y are respectively $\frac{u}{\lambda}$ and $\frac{v}{\lambda}$.

When a convergent lens is used, the infinite point considered as the
center of curvature of the wave plane is brought to a finite distance at
a point on the image focal plane Fig.2. This observation plane is called

the plane of infinite or Fourier's plane or spectral plane. The light amplitude on this plane is proportional to the Fourier transform of the distribution of the light amplitude on the diaphragm. In this case, we shall speak of diffraction at the infinite point.

In the case where the object plane and the focal image plane are symmetrically located with respect to the diaphragm, the reciprocal variables x and y respectively are $\xi/\lambda F$ and $\eta/\lambda F$.

When the boundary conditions are not precisely defined, as is the case with most commonly used light sources, it is necessary to use a statistical description of the data at the level of the diaphragm |2|. For sufficiently simple situations it is possible to develop a theory of the optical coherence based on methods of random analysis of the images. Within these conditions, it is possible to define the concepts of complete or partial spatial coherence and temporal coherence. We shall introduce them starting from one of the two complex components of the vibration.

The concept of temporal coherence can be defined only for plane or spherical waves, emitted from a point light source. Monochromatic light is not required.

There is complete temporal coherence when the phase $\psi(t)$ of the complex component of the vibration can be represented at any stage by a fixed law

$$A(t) \; e^{j\left[\bar{\omega}t + \psi(t)\right]}$$

where $\bar{\omega}$ means on average frequency (there is a limited spectral domain and it is possible to define an average value of the frequency).

As an example, refer to the vibration due to a pulse produced by a laser with synchronized modes. In this case the real part $\mathcal{R}(t)$ of the vibration can be represented as indicated Fig. 3.

To improve the legibility of the figure, the curves of the pseudo-sinusoïd representing the optical vibration of carrier frequency, have been darkened.

The temporal coherence can result in temporal interferences which causes beats between spectral components of different frequencies. For instance, there is temporal coherence between the vibration reflected or scattered by a moving particle and the incident vibration in spite of the fact

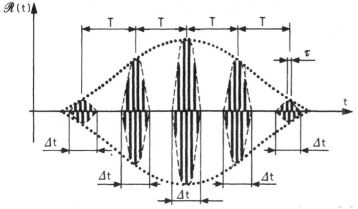

$T = \dfrac{2\pi}{\omega}$ is nano second

order (10^{-8} to 10^{-9} s).

Δt is about ten pico

seconds (10×10^{-12})

τ is hundredth of pi-

co seconds (10^{-14} s)

Fig. 3   Impulse produced by a laser
         with synchronized modes

that their frequencies are different owing to the Doppler's effect.

Some velocity meters are based on this phenomenon.

There is partial temporal coherence when the term $\psi(t)$ is the sum of a fixed part and of a random part. There is incoherence when the random part can introduce fluctuations which are, on average, larger than 2π.

In the classical theory of partial temporal coherence, the light has stationary statistic properties of the second order, i.e. that can be represented with the random superpositon of identical temporally coherent structures (set of waves). This description is well fitted to light issued from thermic sources or a non-synchronous mode laser. For each set of waves the phase relation is well defined. There is partial temporal coherence between the given vibrations (resulting from the random superposition of sets of waves) at two moments, the interval of which is shorter than the emission duration of a set of waves which is called "coherence time". This duration is the same as the "correlation time" of the given random stationary vibration, i.e., the time interval on which is situated the central maximum of the temporal auto-correlation function of the vibration. This justifies the notion of emission duration of a set of waves. It is important to emphasize that the concept of coherence time fails when the vibration has no stationary statistic properties with time.

For example, consider a vibration linearly modulated in frequency and of constant amplitude along its emission duration T. Of course, it is tem-

porally coherent during the whole time T. Its correlation time is equal to
the one on which the frequency changes of 1/T. It is not necessary to define a coherence time in this example.

It is known that the auto correlation function of a quantity is the
Fourier transform of its spectral density of energy (called here temporal spectral density of energy) which is the square of the modulus of the
spectrum. This result is due to the theorem of wiener Khintchine |3| or of
Van Cittert and Zernike |4| in the case of spatial coherence.

From this theorem the relation of geometric uncertainty is deduced :
the product of the length of coherence L, by the width of the spectrum
line $\Delta\lambda$ is equal to the square of the average wavelength

$$\Delta\lambda.L = \lambda^2$$

L being the length of the set of waves sent out during the coherence time
$\Delta\tau$, $\Delta\tau.c = L$, c being the light speed. The finer the spectrum line the larger
the coherence length. This coherence length may reach several meters for a laser, it amounts to about several millimeters for a sodium low-pressure lamp.

Now consider two complex components of the radiation. The concept of
coherence time allows to precisely define the polarization. The fixed parts
of the two phase terms are defined during the coherence time. The radiation
is called unpolarized when the amplitudes of the two components are equal
and the fixed parts of the phase terms are independent from one coherence
interval to another. For example, take thermical sources, impulse lasers,
and gas or dye lasers whose output windows are normal to the beam. When
the two amplitudes of the components are not equal, the beam is called partially polarized.

Return to the fact that only one complex component of the electric
field is required to specify the spatial coherence. It can be defined in
the particular case of partial temporal coherence when the radiation is
quasi-monochromatic.

There is spatial coherence in a defined domain of a plane, x, y, called coherence area, a, when the phase ϕ (x, y) is defined in a fixed manner
at every point of this domain.

The complex component of the radiation is then expressed by

$$A_o(x, y) \, e^{j[\omega t + \phi(x,y)]}$$

There is partial spatial coherence when the term ϕ (x, y) is the sum
of a fixed part and of a random part in a domain of the plane. There is
incoherence in this domain when the random part can introduce fluctuations
larger, on the average, than 2π.

The theory of partial spatial coherence is analogous, in the spatial
domain, to the theory of partial temporal coherence in the temporal domain.
We now consider radiation whose statistical properties are stationary at
the second order with respect to the geometrical variables x, y i.e. which
can be schematized by the random superposition of identical spatially
coherent structures (coherence cells). This type of description is well
fitted only to the radiations issued from thermical sources or unstable
irregular modal structure laser. For each coherence cell, the phase relation is well defined. There is partial spatial coherence for the given
vibrations in two points inside the coherence cell call "coherence area".
This area is the correlation area of the given stationary random vibration,
i.e. the area on which is situated the central maximum of the auto-correlation function of the vibration given by the Van Cittert and Zernike
theorem quoted above.

From this theorem there results the following uncertainty relation :
the product of the coherence area, a, by the solid angle, Ω, under which
the source is seen is equal to the square of the wavelength.

$$a.\Omega = \lambda^2$$

The spatial coherence permits the formation of light interferences,
independent of time.

The classical concepts developed here are well fitted to the representation of the radiation for which the spectral density of energy is
given by a Gaussian curve. It is the case for the thermic sources and for
the continuous lasers. In general, the laser radiations cannot be well
characterized with the concepts of spatial and temporal coherence alone.
The statistical representation should involve higher order correlations,
and this is currently a very active research topic.

We have presented the classical concepts because, in spite of their
insufficiency, they permit to describe some properties of the

radiation important in applications to the optical methods of measurement. Moreover, these concepts can immediately be applied in the analysis of the speckle phenomena which occurs when the coherent light is used. Though they are spatially coherent, the speckle structures are too elaborate to be defined in an analytical manner. Therefore a statistical approach similar to that described previously is required.

2   The phenomenon of scattering of light

The phenomenon of scattering of light in a transparent medium is characterized by the fact that part of the radiation of a principal beam is redistributed in all directions. Moreover the scattered radiation is partially or totally polarized.

A spectral study of this radiation shows that the emitted radiation has, in general the same frequency as the "principal" or incident wave ; however, in some cases "spectral lines" of several wavelengths or even continuous spectral bands can be observed.

The scattering phenomena, without wavelength change is called "elastic" scattering ; it is the most interesting in photoelasticity. Thus, we shall give only a brief account of the causes of scattering phenomenon with wavelength change :

- The fluorescence due to the electronic properties of atoms which give a continuous spectral emission of wavelengths generally smaller than those of the principal beam, but whose intensity can be important ;

- The Brillouin scattering assigned to the density fluctuations caused by the thermical perturbation which results in two close spectrum lines on both sides of the principal frequency

- The Roman scattering assigned to the proper frequencies of molecu-les which can be explained by quantical energy balances also gives spectrum lines.

For a mechanical engineer, it is important to recognize that, the two previous phenomena have a negligible intensity with respect to that of "elastic" scattering. This "elastic" scattering is due to the scattering from impurities and from molecules. The intensity of the first is much greater than that of the second.

The scattering due to the "impurities" |5| take place when the transparent medium has many small particles which introduce local variations of transparency or of index when they are not optically identical to the medium in which they are bedded. These particles permit one to identify the path of the principal beam. Other examples of this phenomenon are common. The scattering of light permits one to see tobacco-smoke in the sun light or the small particles of silica in a transparent "plastic" material. In the case of a colloidal suspension in a fluid, large variations of index may take place. This scattering phenomenon has been discovered in 1869 |6|, it is called "Tyndall effect" ; it is also called "muddy medium" scattering |7|. When the size of particles is small with respect to the wavelength we deal with "Rayleigh scattering" |8|.

The molecular scattering |9| is due to the molecules of the medium in which the light is propagating which act as scattering particles. The optical properties of the medium must present small fluctuations so that scattered light could be observed. A transparent crystal, without either defects (dislocations) or inclusions, whose atoms are perfectly arranged scatters but the scattered light beams cancel due to interference |10|. Molecular scattering is only observable when there are inhomogeneities where the dielectric properties change from one point to another. The scattered intensity on a given axis can be calculated with the fluctuation theory |11|.

The photoelastic methods using scattered light are based upon the scattering by impurities and for this reasons we shall develop this topic.

In a liquid or solid medium we assume that the scattering is produced by particles distributed in a random manner, one being sufficiently far from the others, so we can neglect the secondary scattering phenomenon. In other words, the light scattered by a particle reaches the observer without appreciable change produced by other scattering particles.

The theory used by the mechanical engineer has been elaborated by Rayleigh in 1871 |8|.

Let us consider one particle whose material is dielectric homogeneous, isotropic, and with an index different from the index of the medium. This particle is placed into a sine electromagnetic field whose frequency is ω and whose uniform amplitude is E_0. Since the particle is assumed to be small with respect to the wavelength λ of the light, it is possible to assume, that the field is uniform at any time. The particle can be represented by an electric dipole in which the incident electric field induces, by dielectric polarization, an electric moment

$$\vec{P} = \vec{P}_o \; e^{-j\omega t}$$

with $\vec{P}_o = \alpha \; V \; \vec{E}_o$

where V is the volume of the particle and α depends on the shape of the particle and on the relative index with respect to the surrounding medium.

The field radiated by such a dipole at a distance, r, on the $\vec{u}$ direction is

$$\vec{E}' = \frac{\alpha V}{r} \cdot \frac{4\pi^2}{\lambda^2} \cdot \vec{u} \wedge \vec{E}_o \wedge \vec{u}$$

This expression permits the determination of intensity and polarization properties of the light scattered in different directions when those of the incident light are known.

Consider the reference axes Oxyz, the associated spherical coordinates (r, θ, ϕ) and the local basis $\vec{u}, \vec{v}, \vec{w}$, at P in Fig. 4. The incident light propagating along the $\vec{z}$ direction is the superposition of two orthogonal vibrations $E_x \; \vec{x}$ and $E_y \; \vec{y}$.

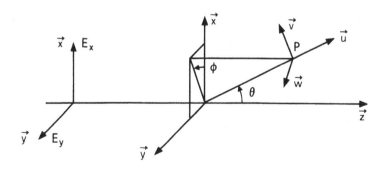

Fig. 4   Representation of a local basis

The light scattered at P consists in two vibrations, the first along $\vec{v}$, the other along $\vec{w}$. For a complete study of the phenomenon with respect to the symmetry around z we have only to observe, for instance, in the yOz plane, i.e. to put $\phi = \frac{\pi}{2}$

- The component along $\vec{v}$ of the scattered light is then :

$$\frac{\alpha V}{r} \cdot \frac{4\pi^2}{\lambda^2} \cdot E_y \cos \theta$$

- The component along $\vec{w}$ is $\frac{\alpha V}{r} \cdot \frac{4\pi^2}{\lambda^2} \cdot E_x$

The following properties are deduced :

- The scattered energy varies as $1/\lambda^4$ (this explains the blue colour of the sky) ;

- In the general case, whatever the observation direction will be, the scattered light is always polarized if the incident light is so. Particularly, it remains linearly polarized if the incident light is linearly polarized.

- In the particular case where $\theta = \pi/2$, the scattered light is then linearly polarized along $\vec{w}$ and its amplitude is proportional to E_x.

For the three-dimensional photoelasticity, the last result from the Rayleigh theory, is important. It can be interpreted as follows : when the observation is made at 90° to the principal beam, the scattered light is linearly polarized along a direction orthogonal to the plane formed with the principal beam and the observation axes. Moreover, its amplitude is proportional to those of the projection of the incident light on the polarization direction of the scattered light.

Note that, when the size of the particles become large with respect to the wavelength, the characteristics of the scattered radiation change considerably. A theory has been developed by Mie |12| for spherical particles which gives the exact solution of the boundary problem of the scattering with the Maxwell equations. It yields the Rayleigh theory as a limiting case and gives more complex results, especially about the polarization properties of the light. Even in the direction orthogonal to the principal beam, the light is not linearly polarized.

Similarly if the particles, are small, and optically anisotropic, the

polarization of the scattered light is no longer complete for an observation at 90° to the principal beam. It only reaches a maximum at this angle.

3 Use of the scattering phenomenon as Polarizer or Analyzer

Weller |13| in 1939 was the first to use the light scattering
phenomenon, inside a medium, as polarizer on the scattered beam or as
analyzer on the principal beam Fig. 5.

So as not to be disturbed by the refraction at the imput or at the
output of the model, the model is immersed in a tank, generally with
parallel sides, filled with a liquid having the same index as that of the
photoelastic material.

If we want to use the scattering phenomenon as a polarizer, the medium
study is performed, from a point P of the incident beam propagating in the
direction $\vec{z}$, upon the scattered beam along a $\vec{u}$-direction orthogonal to $\vec{z}$.

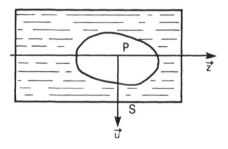

Fig. 5 Scheme of the immersion tank

According to Rayleigh's law, the vibration scattered at P along the $\vec{u}$-
direction is linearly polarized along $\vec{u} \wedge \vec{z}$. The incident beam is unpolarized which permits the amplitude of the scattered light to be constant (neglecting absorption) along its path. Elliptically polarized light depending
on the path PS is received at S.

Now if we want to use the scattering phenomenon as an analyzer, we study the medium with the principal beam completely polarized (linearly or circularly) at the input point Q Fig. 6.

At a point P of the beam, we obtain an elliptically polarized light
which depends on the QP path. If we observe at P along the $\vec{u}$-direction or-

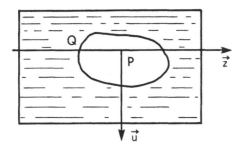

Fig. 6 Scheme of immersion tank

thogonal to z, Rayleigh's law gives the amplitude of the radiation proportional to the component along $\vec{u} \wedge \vec{z}$.

Experimentally Rayleigh's law is not exactly verified, because it is
difficult to predict the size of the scattering particles of the photoelastic materials. Even along the direction orthogonal to the direction of
the principal beam, the scattered light is not completely polarized (we do
not observe extinction with an analyzer put at $\pi/2$). We can render an
account of the observed phenomenon by adding a linear orthogonal component
with amplitude (a) and located in the wave plane of the scattered radiation
to the linear vibration with amplitude A given by Rayleigh's law. The two
vibrations having random phases. This representation results in superposing
a natural light (i.e. unpolarized) of intensity $2a^2$ to a linearly polarized
vibration obeying Rayleigh's law of intensity $A^2 - a^2$. The depolarization
ratio a^2/A^2 can be measured with accuracy $|14|$ without rotating the medium
around the considered point. It is small enough for many photoelastic
materials but it cannot be neglected in some measurement methods. Notice
also that the depolarized component only depends on intensity and not on
the nature of the polarization of the principal beam $|15|$. Moreover, the
law in $1/\lambda^4$ must be completed with a study of absorption.

References

1. Duffieux P.M., "L'intégrale de Fourier et ses applications à l'optique"
    Masson Paris 1970.

2. Enançon M., Slausky S., "Cohérence en optique" Edition C.N.R.S. 1965.

3. Born N., Wolf E., "Principles of optics" Pergamon Oxford 1970.

4. Blanc Lapierre A., Picinbonnot B., "Propriétés statistiques du bruit
    de Fond" Masson Paris 1960.

5. Van De Hulst H.C., "Light scattering by small particles" Wiley New-
    York 1957.

6. Tyndall, Phil. Mag. Vol. 37 p. 384, Vol. 38, p. 156, 1869.

7. Kastler A., "La diffusion de la lumière par les milieux troubles"
    Hermann Paris 1952.

8. Rayleigh, Phil. Mag. Vol. 41, p. 107, 274, and 447 ; 1871.

9. Fabelinskii, "Molecular scattering of light" Plenum. New-York, 1968.

10. Burhat G., Optique p. 313, Masson Paris 1965.

11. Karker M., "The scattering of ligth" Academic Press, New-York, 1969.

12. Mie G., "Ann Physik 25" p. 377, 1908.

13. Weller A., "New method for photoelasticity in three dimensions"
    Journal of Appl. Phys. 1939, 10, n° 4, p. 266.

14. Brillaud J., Lagarde A. "Mesure des formes de lumière en diffusion et
    applications C.R.Ac.Sc. Paris t. 287, Sept. 1978.

15. Brillaud J., "Contribution au développement des méthodes ponctuelles
    en photoelasticimétrie bidimensionnelle". Thèse de 3e Cycle, Poitiers
    1978.

CHAPTER 2

OPTICAL PARAMETERS FOR THREE-DIMENSIONAL PHOTOELASTICITY
BY MECHANICAL OR OPTICAL SLICING

In the last fifty years experimental techniques have been developed
for three dimensional problems where sufficient data is provided for
evaluation from isochromatic and isoclinic patterns measured within a thin
sheet or slice in the body.

After recalling the basic concepts on the optical parameters of
importance for mechanical studies, the fundamentals of the current three-
dimensional techniques are developed. Significance of the novel
non-destructive methods of optical slicing providing three optical
parameters is emphasized. The point-wise method of linear detection and
the whole-field method of optical slicing are both in this category. The
further aim is to describe the approach by these two methods in the view
of some relatively recent works on the propagation of light waves. We
deduce from these the significance of our measurements.

1   The current scheme

First recall a basic hypothesis that light is propagating in a
photoelastic medium and assume the medium to be isotropic (indeed current
photoelastic materials are slightly anisotropic). It follows that for a
ray of light propagating along the $\vec{z}$ direction, the direction planes for
the component waves (x, y) are orthogonal to $\vec{z}$.

For a photoelastic medium it can be shown that the principal directions
of the index of light refraction tensor and the stress tensor coincide and
that the change of index is related to stress in the following form
(referred to a principal coordinate system).

$$n' - n_o = c_1 \sigma' + c_2 (\sigma'' + \sigma_z)$$

$$n'' - n_o = c_1 \sigma'' + c_2 (\sigma' + \sigma_z)$$

where n' and n'' denote the principal secondary indices in the wave-plane (x, y) and σ', σ'' are corresponding secondary principal stresses ; c_1, c_2 are independant constants for a photoelastic material.

In three-dimensional photoelasticity it is usually assumed that the directions of secondary principal stresses and their values are constant through the thickness dz of a slice having its parallel faces normal to $\vec{z}$. This assumption allows to consider this slice as a birefringence plate characterized with the two following parameters.

- secondary principal angle $\alpha = (\alpha, \sigma')$
- angular birefringence $\phi = \dfrac{2\pi\delta}{\lambda}$,

    $\delta = dz (n' - n'') = C (\sigma' - \sigma'') dz \qquad C = c_1 - c_2$

    (c being a photoelastic constant)

The two quantities namely α and $\sigma' - \sigma''$ are of interest to an engineer as $\sigma_x - \sigma_y$ and τ_{xy} may be determined from the following relation ships :

$$\sigma' - \sigma'' = \sqrt{(\sigma_x - \sigma_y)^2 + 4 \tau_{xy}^2}$$

$$tg\ 2\ \alpha = \frac{2\ \tau_{xy}}{\sigma_x - \sigma_y}$$

using the well known notations.

When α and δ are measured for three series of three mutually orthogonal planes it is possible to integrate the equilibrium equations using the finite difference calculus.

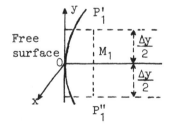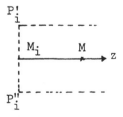

Fig. 1   Integration path of σ_z

Thus σ_z is determined at the point M from the equilibrium equation

$$\frac{\partial \sigma_z}{\partial z} + \frac{\partial \tau_{yz}}{\partial y} + \frac{\partial \tau_{xz}}{\partial x} = 0$$

Where the integration is performed along Oz, with point O on the boundary where the normal stress component σ_z (O) is assumed to be known. (it is zero for a free boundary).

$$\sigma_z (M) = \sigma_z (O) - \int_0^M \frac{\partial \tau_{yz}}{\partial y} \, dz - \int_0^M \frac{\partial \tau_{xz}}{\partial x} \, dz$$

The partial derivative $\dfrac{\partial \tau_{yz}}{\partial y}\Big|_{M_i}$ is obtained from the function τ_{yz} determined as indicated above in the plane Oyz, at the points P_i' and P_i'' in Fig. 1.

$$\frac{\partial \tau_{yz}}{\partial y}\Big|_{M_i} \simeq \frac{\Delta \tau_{yz}}{\Delta y}\Big|_{M_i} = \frac{\tau_{yz} (P_i') - \tau_{yz} (P_i'')}{\Delta y}$$

The same procedure gives $\dfrac{\partial \tau_{xz}}{\partial x}\Big|_{M_i}$ from the values of τ_{xz} determined at Q_i' and Q_i'' :

$$\frac{\partial \tau_{xz}}{\partial x}\Big|_{M_i} \simeq \frac{\Delta \tau_{xz}}{\Delta x}\Big|_{M_i} = \frac{\tau_{xz} (Q_i') - \tau_{xz} (Q_i'')}{\Delta x}$$

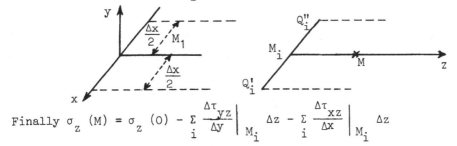

Finally $\sigma_z (M) = \sigma_z (O) - \sum_i \dfrac{\Delta \tau_{yz}}{\Delta y}\Big|_{M_i} \Delta z - \sum_i \dfrac{\Delta \tau_{xz}}{\Delta x}\Big|_{M_i} \Delta z$

The preceding measurements also give $\sigma_y - \sigma_z$ and $\sigma_x - \sigma_z$; then the σ_y and σ_x can be determined since σ_z is known.

The third shear stress component remains to be determined . As is indicated above, it is obtained from the inspection of a slice parallel to the x-y plane. The latter serves as a useful check in allowing direct measurement of $\sigma_x - \sigma_y$.

For axisymetric problems the information on the meridian slice is
sufficient to determine the unknown stresses, $|1|$. Indeed, the equilibrium
equation in cylindrical, (z, r, θ) - coordinates, become (the problem is
axixymetric with respect to z - axis)

$$\frac{\partial \sigma_z}{\partial z} + \frac{\partial \tau_{rz}}{\partial r} + \frac{\tau_{rz}}{r} = 0 \qquad ; \qquad \frac{\partial \sigma_r}{\partial r} + \frac{\partial \tau_{rz}}{\partial z} + \frac{\sigma_r - \sigma_\theta}{r} = 0$$

There are four unknowns stress components σ_r, σ_θ, σ_z and τ_{rz} and the
photoelastic measurements taken at the meridian slice provide sufficient
data for their determination. The measurements yield at each point the
difference of principal stresses $\sigma_1 - \sigma_2$ and principal stress directions.
Mohr's circle representation yields

$$\tau_{rz} = \frac{1}{2} (\sigma_1 - \sigma_2) \sin 2 \alpha$$

Then σ_z can be calculated from the first equilibrium equation provided
that its boundary value is known. Next one obtains σ_r from :

$$\sigma_r - \sigma_z = (\sigma_1 - \sigma_2) \cos 2 \alpha$$

and by substituting σ_r and τ_{rz} into the second equilibrium equation, one
calculates σ_θ.

Note in sect. 3 that the representation of a slice by a simple
homogeneous birefringent plate scheme is not satisfactory if rotation of
secondary principal directions occurs through the thickness of the slice.

2   General state in three-dimensional photoelasticity

This section reviews briefly some essential concepts on the evolution
of the methods where a slice is represented by a homogeneous birefringent
plate.

We recall that the photoelastic method of stress analysis was
established at the beginning of the twentieth century. It was in 1901 that
Mesnager $|2|$ applied it for the first time to design the bridge on the
Balme river. Of course, the analysis at that early date was two-dimensional.

By 1930, Favre $|3|$ proposed a solution to a three-dimensional problem
by manufacturing a composite model containing a thin photoelastic slice
enclosed on both sides by optically insensitive material. Two certain

different glasses were used which had the same mechanical properties. Today, the use of two types of plexiglas is more practical. The transmission technique was applied with an immersion tank containing a liquid of the same refractive index and a classical two-dimensional photoelastic analysis was used. Despite the advantage of simplicity in observing the slice, several disadvantages are associated with this technique : difficulties in obtaining two materials of the same mechanical properties, and the need to manufacture a model for each slice to be analysed. Furthermore the inactive material does not have a photoelastic constant precisely equal to zero, and a distance traversed by a ray of light in it is greater than the one through the active slice. This can produce errors in the measurements.

It is worth to note that in 1961 H. TRAMPOSH and G. GERARD |4| introduced a variant of the method discussed above with a single model material. A thin slice to be observed is inserted in the model but it is optically isolated by two polaroïd sheets placed in the glue line at its interfaces. The sheets introduced may be plane-polarizing or may constitue a circular polariscope. This procedure eliminates the source of error of birefringence measurements but for a given model only one isoclinic direction can be measured.

Those methods based on the insertion of a slice have been employed extensively in the stress analysis of axisymetric problems.

For the last fifty years there has been further extensive research on optical parameters determination in a three-dimensional medium.

Two basic concepts were developed.

The first one can be credited to Opel |5| who discovered some interesting phenomena characterizing a class of polymeric thermosetting materials. A model made from an optically sensitive polymer is heated to a temperature of about 120°C (in any case this temperature level is such that the material may behave like an elastic solid and not as a viscoelastic one).

The model is then slowly cooled to room temperature before the loads are removed. The optical response (birefringence) and the deformations the model was subjected to, under loading, are locked in the model on a molecular scale. This phenomenon which Opel improperly called "stress-freezing effect" can be explained by the diphase internal structure of polymers.

It can be explained by the locking of molecular chains by secondary bonds which reappear upon cooling, and by the fact that the modulus of elasticity at room temperature is higher (approximately 100 times for Epoxy) than that at the stress-freezing temperature. The unloading of the model is accompanied by some stress relief and a slight variation in the state of deformation and birefringence. The model is then machined in slices which can be studied by well known techniques. Thus, in 1936 Oppel was able to give for the first time a complete shear-stress map for a three-dimensional model.

This method was recognized to be a useful tool of experimental stress analysis and is still applied in many laboratories in spite of some disadvantages :

- The birefringence obtained are three to four times lower than those for the same deformations at ambient temperature. This requires more involved techniques for measuring low birefringence, or the introduction of higher deformations with the risk of perturbing boundary conditions.

- The elements used for mechanic loading which are in contact with a photoelastic material rarely exhibit identical thermal expansion coefficient. Considerable disturbances may result from that incompatibility.

- A strong limitation of the stress-freezing method is that the model materials have a Poisson's ratio of 0.5 which is equivalent to elastic incompressibility (Most of structural metals exhibit the values between 0.30 and 0.35). This mismatch does not satisfy the conditions of similitude.

The other basic idea is due to Weller |6| who in 1939 proposed the use of the phenomenon of light scattering, which plays the role of polarizer for the scattered beam and that of analyser for the principal beam. We have discussed these points in detail in chapter 1.

Weller and later other researchers found this way attractive for its simplicity ; we may cite Cheng | 7 |, Robert and Guillement |8| Gross and Paterson |9|, Srinath and Sarna |10|,Brillaud and Lagarde |11|, Robert Royer |12|, Desailly and Lagarde |13|.

Another basic concept using integrated photoelasticity procedure is discussed elsewhere in this book.

We can state that until recently no effective solution was obtained. Only the stress-freezing method has become classic, despite its liberties with similitude, a fact easily overlooked. We do not recommence there our recent analysis |14| which describes the stages of development.

3 Some theoretical results

The problem of propagation of light waves was studied by many authors |15, 16, 17, 18|. Here, the essential concepts from Aben's works will be introduced.

Aben in 1966 considered the basic equations of three-dimensional photoelasticity and analyzed the matrix representation of the solutions of those equations. He showed that when rotation of principal axes was present, there were always two pairs of perpendicular conjugate "characteristic directions" (fig. 2). He distinguished the primary characteristic directions at the entrance of light (Δ'_e, Δ''_e), and the secondary direction (Δ'_s, Δ''_s) for the light emerging from the medium. The light linearly polarized at the entrance along one of the primary directions emerges as linearly polarized along the conjugate secondary direction. We will denote by R the angle determined by two such directions, $R = (\Delta'_e, \Delta'_s)$ and by α^* the angle (x, Δ'_e). Aben, has thus generalized the concept of "isoclinic" of the plane photoelasticity to the general case when a rotation of the secondary principal directions takes places |19|. The characteristic directions are generally different from the secondary principal directions of the stress tensor (or those of the principal indices). At the entrance we have (σ'_e, σ''_e) and at emergence we have (σ'_s, σ''_s) from the medium. We denote $\alpha_o = (\sigma'_e, \sigma'_s)$.

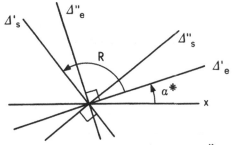

Fig. 2 : Orientation of the "characteristic directions" at the entrance, and at the emergence from the medium.

For the case where $d\alpha/dz$ and $\sigma' - \sigma''$ are constant through a thickness, important conclusions follow (Aben) :

The bisecting lines for the angles formed by two associated "characteristic directions" coincide with those for the angles formed by the associated secondary principal directions at the entrance and at the exit, (fig. 3).

Remark :

The bisectors mentioned correspond to the secondary principal directions (mechanical or optical) at mid thickness ; so their directions are defined by the angles $\pm \bar{R}/2$ from the characteristic directions :

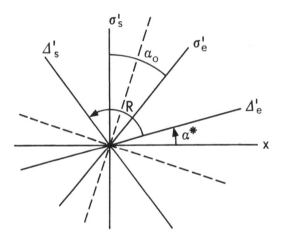

Fig. 3 : Orientation of the characteristic Δ' directions and the σ' directions at the entrance and the emergence.

The phase difference ϕ^{*} characteristic to the medium traversed by the light wave along two characteristic orthogonal directions is generally different from the angular birefringence ϕ

$\phi = \frac{2\pi \, dz}{\lambda}(n' - n'')$, which would result in the absence of rotation R.

- The quantities R, ϕ^{*}, ϕ, and α_o obey the following relationships

$$tg \, R = \frac{tg\alpha_o - \frac{\alpha_o}{X} tgX}{1 + \frac{\alpha_o}{X} tg\alpha_o tgX}$$

$$\cos \phi^* = 1 - \frac{\phi^2}{2X^2} \sin^2 X$$

$$\text{with } X = \frac{\sqrt{\phi^2 + 4\alpha_o^2}}{2}$$

Note that for $R = 0$, $\alpha = \alpha_o$ and $\phi = \phi^*$, one finds again the classical scheme of homogeneous birefringence representing a slice dissected from a model.

Using the expressions above, nomograms have been drawn which give ϕ and α_o versus physical quantities ϕ^* and R. One can thus determine the difference of secondary principal stresses $\sigma' - \sigma''$ and secondary principal directions (through the angle α) if the angle α^* describing the orientation of the characteristic directions at the entrance is known.

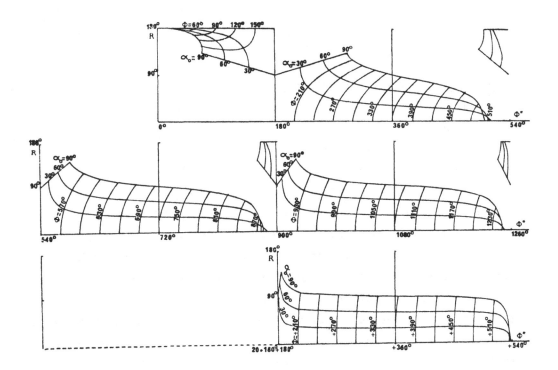

Nomograms relating ϕ^*, R, α_o, ϕ

It should be noted that a computer programme which evaluates the two
mechanical parameters $\sigma' - \sigma''$ and α as a function of three optical parameters ϕ^*, R, α^* requires approximatly 2 seconds on a programmable hand computer. The optical parameters under consideration can be accurately deter-
mined by the point method with linear detection |21, 22|.

The nomogram illustrate the following particular situations.

- ϕ^* exceeds ϕ if $\phi > \pi$. This is consistent with the result of
Drucker and Midlin |15| who indicated that "the rotation of secondary prin-
cipal directions increased the number of isochromatic fringes.

- ϕ^* is less than ϕ if $\phi < \pi$ and in the neighbourhood of this value
the error becomes important (for $\alpha_o = 30°$, $\phi^* = 140°$ for $\phi = 180°$).

-If the ratio α_o/ϕ is small, $\phi = \phi^*$ and there is coïncidence of the
secondary principal directions and the characteristic ones.

Thus we conclude that the medium under consideration can be represented in general as a birefringent plate (having the axes (Δ'_e, Δ''_e)
located by $\alpha^* = $ (x, Δ'_e) and characterized by an angular birefringence ϕ^*)
followed or preceded by a rotatory power R.

4   Representation of the photoelastic medium

Recall the work of Aben, which allowed the determination of two parameters of interest from the three physical parameters and note that this
result is based on the hypothesis that the rotation rate of principal
secondary axes and the shear-stresses are constant through the thickness.
This hypothesis is perfectly admissible for a thin slice.

To our knowledge our group was the first to represent a sheet schematically by an birefringent plate followed by a rotatory power (or inversely
with a birefringent whose orientation is shifted of the value of the rotatory power). Thus we were able to take Aben's results into account in the
new methods that were developed |13, 21|.

The hypothesis above is, however, very restrictive for representing the
behaviour of a slightly anisotropic medium with a large thickness traversed
by light .

In this case, one can resort to some discretization of the body into the
series of thin slices, each one being represented by a birefringent plate

followed by a rotatory power. In this context, the matrix formalism by Aben insures that the entire body can be represented as the combination of a bire-fringent plate body followed by a rotatory power. In this way be generalized Poincare's theorem established for series of birefringent plates. This theorem has been employed by Robert and Guillemet |12| and Robert and Royer |11| to give the same representation based on a set of discrete birefringent plates.

5   Analysis of a thin sheet

5.1 Point-wise analysis

The analysis of a slice removed from the stress-frozen photoelastic model by a point-wise plane photoelastic method where the slice is regarded as a birefringent plate can lead to considerable errors if rotation of secondary principal directions takes place. To detect rotation one can use an ellipsometry technique which employs circular incident light and permits one to determine ϕ^*. It is known now (see sect. 3) that errors made can reach $40°$ for the rotation $\alpha_o = 30°$ when ϕ is close to π. The same thing happens for the orientation of secondary principal stresses which do not correspond with the reference directions. Consequently, large errors may result for the stress components when integrating the equilibrium equations. This is a well-known situation in plane photo-elasticity.

We will not dwell up on the extent of errors which are inherent to application of the freezing and slicing method or to certain non-destructive techniques of slice analysis within the model. It seems much more important to point out that for both situations, we have developed well-adapted linear detection methods |11, 25, 27| which permits one to determine with precision the three optical parameters ϕ^*, α^*, and R.

5.2 Whole-field analysis with a plane polariscope |23, 24, 26|

Here, the analysis with a light-field polariscope is presented as it coresponds to the whole-field method of optical slicing. On can conduct an analogous study for a dark polariscope.

Let us examine a slice (which should be obtained by freezing and slicing) in a plane light – field (rectlinear) polariscope. The slice is represented

represented by a birefringent plate, and a rotatory power.

Let I_o designate uniform light-field illumination and x the polarizing axis of the polarizer. Then the light intensity is

$$I = I_o \; (\cos^2 R - \sin 2\alpha^* \sin^2 (\alpha^* + R) \sin^2 \frac{\phi^*}{2})$$

The extremum values for intensity distribution correspond to :

$$\alpha^* = -\frac{R}{2} + K \frac{\pi}{4} \qquad ; \qquad K = 0, 1, 2...$$

Two cases can be distinguished. The first one correspond to :

$$\alpha^* = -\frac{R}{2} + K \frac{\pi}{2}$$

and the maximum intensity is expressed by :

$$I_{max.} = I_o \; (1 - \sin^2 R \cos^2 \frac{\phi^*}{2})$$

The second case corresponds to $\alpha^* = -\frac{R}{2} + \frac{\pi}{4} \; (\frac{\pi}{2})$ and the minimum intensity is given by :

$$I_{min.} = I_o \; (\cos^2 R \cos^2 \frac{\phi^*}{2})$$

In order to specify the conditions of analysis of fringe patterns we plotted the variations of I_{max} and I_{min} versus ϕ for different values of α_o obtained following the relation ships given in sect. 3. As an example, curves were plotted for $\alpha_o = \frac{\pi}{9}$ in fig. 4.

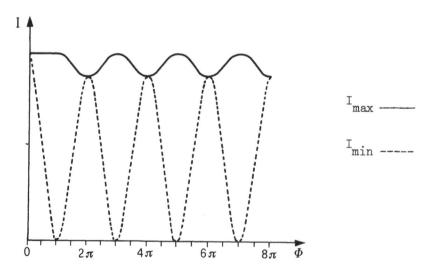

Fig. 4 : Variation of I_{max} and I_{min} as a function of ϕ ($\alpha_o = \frac{\pi}{9}$)

The foregoing analysis indicates that for small values of α_o ($\alpha_o < \frac{\pi}{9}$) the maximum intensity I_{max} shows a reduced modulation. Thus, it characterizes an isoclinic zone which permits one to locate the secondary principal stress directions (or those of the indices) in the median plane of a slice. This zone corresponds to $\alpha^* = -\frac{R}{2}$ Mod $\frac{\pi}{2}$ (see remark sect. 3). The orientation of the polarizer then coincides with one of the secondary principal directions in the median plane. This interesting result is analogous to the one established in 1957 by Hickson |28| for a dark-field polariscope.

It should be emphasized, that in order to avoid errors during the numerical integration procedure, the discretization points should lie on the median planes of the slices.

As α_o increases, the I_{max}-modulation increases and it becomes very pronounced for $\alpha_o = \frac{\pi}{3}$. In this case the isoclinic zone disappears although it should be noticed that the isoclinics are discernible up to the α_o value of $\frac{\pi}{6}$.

The term I_{min} which is strongly modulated for α_o close to $\pi/6$, characterizes the isochromatic pattern. The extremum values occur for $\phi = K\pi$ ($K = 1, 2...$) and it follows that localization of fringes is practically independent of the rotation of secondary principal axes.

We can now conclude by noting the following result : investigation of a slice within the plane (rectilinear) polariscope allows one to determinate the secondary principal stress directions in the median plane (without resorting to rotatory power measurements) and the angular birefringence ϕ for the multiple π-values when the rotation of the secondary principal axes is less than $\frac{\pi}{6}$.

We should point out that the condition on the rotation of the secondary principal axes is not very limiting since one is able to choose the slice-thickness for the non-destructive optical slicing method.

5.3 Analysis within a circular polariscope (whole-field analysis) |23, 24, 26|

In the circular polariscope there is no rotatory power effect so in the case of the light-field pattern the light intensity is given by

$$I = I_o \cos^2 \frac{\phi^*}{2}$$

The dark fringes correspond to $\phi^* = (2K + 1)\pi$. To specify the condition of analysis of fringe patterns, we plotted I versus ϕ for different values of α_0 using the relations of sect. 3. We can be see (Fig. 5) that for the values of α_0 less than $\frac{\pi}{6}$ the minimum values of I correspond practically to $\phi = (2K + 1)\pi$ therefore, the dark fringes represent isochromatics.

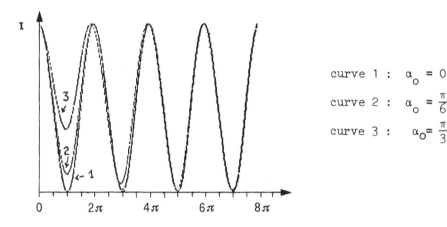

curve 1 : $\alpha_0 = 0$

curve 2 : $\alpha_0 = \frac{\pi}{6}$

curve 3 : $\alpha_0 = \frac{\pi}{3}$

Fig. 5 : Variation of I versus ϕ (for different values of α_0).

Conclusion

The basic concepts for representation of a thin slice are discussed.

The starting point is the uniformity assumption for the difference of secondary principal stresses and for the rotation rate of their directions through the thickness. This hypothesis is admissible for a thin slice. Using this hypothesis theorical studies on light wave propagation lead to the conclusion that a slice can be represented by a birefringent plate and rotatory power, characterized by three physical parameters. The latter can be determined with the point-wise method of linear detection, and allow the determination of two parameters of interest to the mechanician.

An examination of a slice cut from a stress-frozen model within a plane and/or circular polariscope gives correct data for obtaining the isoclinics and isochromatic fringes over the whole-field if the rotation of the secondary principal axes is limited to about $\frac{\pi}{6}$. This is a satisfactory result and explains why these classical techniques are reliable and were always favoured by engineers. On the contrary, the point-wise methods of two-dimensional elasticity are not, in general, suitable for the foregoing purpose.

The more exact representation of a photoelastic slice discussed here is being taken into account by the recent non-destructive techniques by optical slicing of a model, developed with the aid of laser properties. We can name here the point-wise linear detection method and the whole-field method, both permitting the measurement of parameters of interest to the mechanician. These scattered light slicing procedures can be applied to a model having a Poisson's ratio close to that of the prototype material of an actual structural part.

References

1. Kuske A. "Separation of principal stresses by means of an improved shear-difference method" Experimental mechanics 6-1 1966.

2. Mesnager A. "Mesure des efforts intérieurs dans les solides et applications" int. Ass. Test. Mat. Budapest (1901).

3. Favre M. "Sur une méthode optique de détermination des tensions intérieures dans les solides à 3 dimensions". C.R. Acad. Sc., t. 190, 1930 p. 1182-1184.

4. Tranposh H. and Gerard G. "An exploratory study of three-dimensional photothermoelasticity". (Applied Mechanics vol. 28. 1961, pp 35-40).

5. Oppel G.U. "Forschungs Gebiete ing. 7, 1936, p. 240-248 Naca T.M. 824, (1937).

6. Weller R. "Journal of applied physics" t. 10, 4, p. 226, (1939).

7. Cheng Y.F. "A dual observation method for determining photoelastic parameters in scattered light". Experimental Mechanics, Vol. 7, n° 3, p. 140-144, 1967.

8. Robert A. and Guillemet E. "Nouvelle méthode d'utilisation de la lumière diffusée en photoélasticimétrie à trois dimensions" R.F.M. n° 5/6 p. 147-157, (1963).

9. Gross-Petersen A. "Compensation method in scattered light photoelasticity" I.U.T.A.M. Symposium. "The photoelastic effect and its applications" Bruxelles. September 1975 (Springer Editeur).

10. Srinath and Sarna "Determination of optically equivalent model in three dimensional photoelasticity" Experimental mechanics vol. 14, n° 3, p. 113-122 (1974).

11. Brillaud J. and Lagarde A. "Construction et mise au point d'un appareil permettant la détermination des lignes neutres et de la biréfringence en tout point d'un feuillet plan d'un fluide biréfringent en écoulement laminaire permanent , ou d'un solide photoélastique". Compte rendu-final du contrat D.G.R.S.T. n° 71-7-2886, Janvier 1976.

12. Robert A. and Royer J. "Principe de mesure des biréfringences à l'intérieur d'un solide transparent en vue de son application à la photoélasticimétrie tridimensionnelle" C.R. Ac. Sc. série B, t. 281, 1975 p. 373-376.

13. Desailly R. and Lagarde A. "Sur une méthode de photoélasticimétrie tridimensionnelle à champ complet". "Journal de Mécanique Appliquée" - Vol. 4 n° 1 (1980).

14. Lagarde A. " Progrès dans la mesure de grandeurs optiques en milieu tridimensionnel, Solide ou Fluide, biréfringent" Conférence générale, Congrès canadien de mécanique appliquée, Moncton 7-12 Juin 1981.

15. Drucker D. and Mindlin R. "Stress analysis by three dimensional photoelastic methods". J. Appl. Phys. Vol. 11 (1940).

16. Mindlin R. and Goodman L. "The optical equations of three-dimensional photoelasticity" J. Appl. Phys., 20 (1949).

17. Lee L., "Effects of rotation of principal stresses on photoelastic retardation" Exper. Mechanics, Vol 4, n° 10 (1964).

18. Aben H. "Optical phenomena in photoelastic models by the rotation of principal axes" Exper. Mechanics. Vol. 6 n° 1 (1966).

19. Vandaele, Dossche M. and Van Geen R. "Cinq leçons de photoélasticité tridimensionnelle" Séminaire d'Analyse des contraintes de l'Université Libre de Bruxelles, n° 8, 1969.

20. Aben H. (see. n° 17).

21. Brillaud J. and Lagarde A. "Méthode ponctuelle de photoélasticimétrie tridimensionnelle" R.F.M. n° 84, (1982).

22. Brillaud J and Lagarde A. "Ellipsometry in scattered light and its application to the determination of optical characteristics of a thin

slice in tridimensional photoelasticity" Symposium I.U.T.A.M. "Optical méthods in Mechanics of solid" Poitiers, September 1979 Ed. A. Lagarde (Sijthoff Nordoff).

23. Desailly R. "Méthode non-destructive de découpage optique en photoélasticimétrie tridimensionnelle - Application à la mécanique de la rupture" Thèse d'Etat n° 336 Poitiers (1982).

24. Desailly R. and Lagarde A. "Méthode de découpage optique de photoélasticité tridimensionnelle" ; Journées d'extensométrie, Poitiers, Septembre 1981, R.F.M. n° 2, 1984.

25. Brillaud J. and Lagarde A. "Photoélasticimétrie ponctuelle automatique à détection linéaire - Application" VIIth international conference on experimental stress analysis. Haifa Israël 23-27 August 1982.

26. Lagarde A., Brillaud J., Desailly R. "Paramètres optiques en photoélasticité tridimensionnelle" Conférence aux journées françaises d'extensométrie 1981 - R.F.M. n° 4, 1983.

27. Brillaud J. "Mesure des paramètres caractéristiques en milieu photoélastique tridimensionnel. Réalisation d'un photoélasticimètre automatique - Applications". Thèse de doctorat d'Etat, Poitiers 1984.

28. Hickson V.M. "Errors in stress determination at the free boundaries of "Frozen stress" photoelastic model" J. Appl. Phys. Vol. 3, n° 6, p. 176-181 (1952).

CHAPTER 3

PUNCTUAL METHODS IN PLANE PHOTOELASTICITY

APPLICATION TO CONTACT MECHANICS

Let us outline the main steps in the progress of point methods which
have followed developments in photodetectors, electronics, modulation techniques with rotating polarisers, rotating compensators, and eventually the
use of a reference signal to overcome difficulties inherent in direct photometric measurement. The main fundamental ideas are outlined. An example
show the importance of taking rotating power into account when checking
whether a state of stress is plane, and Senarmont's method can be adapted
for this purpose. Duality considerations show that some seemingly different methods are based on a common principle. Finally, an application of
the determination of contact forces between two photoelastic solids illustrate the use of the new methods.

1. Point measurement methods in plane photoelasticity

Stress calculations by integration of the equilibrium equations require a rather dense fringe pattern. Hence complementary measurements of birefringence have been employed since the inception of photoelasticity,
when fringe patterns were sparse because the glasses used at that time had
low photoelastic constants. Amongst the compensation methods we cite those
of Tardy |1| and of Sénarmont |2|. Neither of these methods require a com-
pensator as they use the polariscope elements to establish the fractional
fringe order.

1.1 Tardy and Sénarmont methods

The Tardy method employs the elements of a circular polariscope. The
quarter-wave plate behind the model is aligned at $\pi/4$ to the principal
axes at the point considered in the model. In a dark field, light is again
extinguished by an analyser rotation through $\phi/2$ ($\phi = 2\pi\delta/\lambda$, with relative
retardation δ, and wavelength λ). The method gives an unambiguous value for

relative retardations smaller than half a wavelength in absolute magnitude. It presupposes knowledge of the principal directions.

The Sénarmont method uses a plano-circular polariscope. The quarter-
wave plate and polariser are aligned at $\pi/4$ to the model axes. The supple-
mentary condition to be imposed makes this method less popular than that
of Tardy.

Since Tardy's method dates from 1929, Sénarmont's from 1840, we shall
denote the special alignment, common to both methods, of the quarter-wave
plate at $\pi/4$ to the model axes as "Sénarmont configuration for the quarter
wave plate". If in addition the polariser axis λ is placed parallel to a
principal axis of the quarter-wave plate, the arrangement will be called
"Sénarmont configuration for polariser and quarter-wave plate".

The development of photodetectors has brought new perspectives into
partial fringe evaluation. It allows dynamic measurement since the transmitted light intensity can be readily recorded (See section 2). However,
calibration presents a major difficulty. The maximum light intensity that
can pass through the model is difficult to evaluate. The photomultiplier
gain may vary and is sensitive to thermal effects and tube "fatigue". The
light emitted by the lamp may also vary, and the unloaded model may trans-
mit light unevenly over the surface.

Photodetector calibration can be avoided by using the null methods
mentioned or by a modulation technique.

1.2 Sapaly method

As far as we know, Sapaly was the first to introduce, in 1960, the
rotating analyser |3| in the Tardy set up and to measure phase |4| in or-
der to obtain relative retardation, the principal directions obviously
being known. The transmitted energy has the expression

$$E = \frac{1}{2} E_o \left(1 - \cos \left(2 \, \Omega t + \phi \right) \right)$$

The reference signal, of form $\cos 2 \, \Omega t$ was obtained by a train of light pulses through two holes in a rotating disk driven by the same motor
as the analyser. Nowadays, the preferred method for the reference signal
is direct impingement on the rotating analyser of a suitably polarised
pencil of rays. The Sapaly solution is in a sense a Tardy method automa-

tised by conversion from a null method to a phase method. It permits measurement of the algebraic value of relative retardations smaller than $\lambda/2$
in absolute magnitude.

1.3 Robert-Ferré method

In 1969, Robert and Ferré |5| proposed an ellipsometer consisting of
a circular polariser unit and a rotating analyser. The transmitted light
intensity is here where α denotes the orientation of the fast axis at the

$$E = \frac{1}{2} E_o \left(1 - \sin \phi \sin (2 \Omega t - 2\alpha)\right)$$

point considered in the model.

A measurement of phase against a suitable reference signal supplies
the principal directions by indicating the fast axis. The amplitude ratio
of alternating to continuous component of E gives $\sin \phi$, that is the ellipticity, the light vibration emerging from the model. Hence the name "ellipsometer" has been given to the apparatus which completely defines the
form of light. The relative retardation is measured unambiguously only if
its absolute value remains below $\lambda/4$. However, alignment of the quarter-
wave plate in Sénarmont configuration permits an increase in this limit
to $\lambda/2$.

1.4 Allison-Nurse method |6|

In 1971, Allison and Nurse proposed an automatic polariscope free of
these limitations. It comprises three component systems :

- A classical plane polariscope showing isoclinics and isochromatic
fringes,

- A plane polariscope rotating at constant speed yield principal directions by phase measurement,

- A novel set-up of a stationary, adjustable plane polariscope with
a rotating compensator to give relative retardations.

The compensator is a birefringent disk of thickness varying linearly
with angle of revolution θ, (fig. 1), so that the retardation also varies
linearly with θ from a value δ_o to $(\delta_o + 2 \pi\Delta)$, where Δ is the retardation
gradient. If the compensator rotates at constant speed about an axis pa-
rallel but not collinear with the light path, the retardation on the light

path has a "saw-tooth" variation with a value $\delta' = \delta_o + \theta\Delta$ at a given position θ.

The compensator is arranged so that the directions of its principal axes remain constant at any point on the light path as the compensator is rotated about its own axis. A servomechanism brings the principal optical axes of the compensator and of the model at the point considered into coincidence by rotation about the polariscope axis and aligns polariser and analyser at $\pi/4$ to these principal axes.

Let ϕ_1, (ϕ_2) and $-\psi_1$, $(-\psi_2)$ be the retardations in model and compensator for the wavelength λ_1, (λ_2), and suppose that the photoelastic constants do not vary with wavelength. Then

$$\phi_1 = \frac{2\pi\delta}{\lambda_1} \quad , \quad \phi_2 = \frac{2\pi\delta}{\lambda_2} \quad , \quad \psi_1 = \frac{2\pi}{\lambda_1}(\delta_o + \theta\Delta) \quad , \quad \psi_2 = \frac{2\pi}{\lambda_2}(\delta_o + \theta\Delta)$$

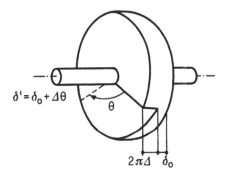

Fig. 1   Compensator

The analyser thus transmits at λ_1 the light intensity

$$E_1 = \frac{1}{2} E_{o1}\left(1 - \cos(\phi_1 - \psi_1)\right) = \frac{1}{2} E_{o1}\left(1 - \cos\frac{2\pi}{\lambda_1}(\delta - \delta_o - \theta\Delta)\right)$$

and at λ_2

$$E_2 = \frac{1}{2} E_{o2}\left(1 - \cos\frac{2\pi}{\lambda_2}(\delta - \delta_o - \theta\Delta)\right)$$

If the two wavelengths are interchanged in rapid succession and the light energy is equally distributed over both, the photomultiplier receives an intensity

$$E = \frac{1}{2} E_o \{1 - \cos \{\pi \left(\frac{1}{\lambda_1} + \frac{1}{\lambda_2}\right) (\delta-\delta_o-\theta\Delta)\} \cos \left(\pi \left(\frac{1}{\lambda_1} - \frac{1}{\lambda_2}\right) (\delta-\delta_o-\theta\Delta)\right)\}$$

Light is extinguished only when the rotating compensator is in a position characterised by

$$\theta = (\delta - \delta_o)/\Delta$$

Hence the retardation is measured by $\delta = \delta_o + \theta\Delta$.

The compensator executes a certain number of rotations. At the start of each cycle a counter is set to zero and then "digitises" the angle of rotation. An electric signal actuated by the light extinction stops the count in each cycle. After some twenty cycles, the mean value of the counts is formed and used to calculate the retardation.

This method requires complex equipment. To our knowledge it has been the first to

a) combine full-field observation with precise local measurement,

b) read directly the relative retardation, whatever the fringe order, up to a value restricted only by the compensator design. A further advantage is that its sensitivity is independent of the measured retardation, being linked to the retardation gradient of the compensator.

1.5 Lagarde-Oheix extension of the Tardy method |7-11|

In 1973 Lagarde and Oheix proposed a simpler solution which can be considered as an extension of the Tardy method and which uses the rotating analyser, introduced by Sapaly, to determine the principal directions and the relative retardation.

The set-up shown schematically in Fig. 2 includes

a) a circular polariser unit P_c,

b) the model with principal directions x_1, x_2 and retardation phase angle ϕ,

c) an alignable quarter-wave plate Q with its slow axis x taken as the reference axis so that $(x, x_1) = \beta$,

d) a rotating analyser $(x, A_t) = \Omega t$

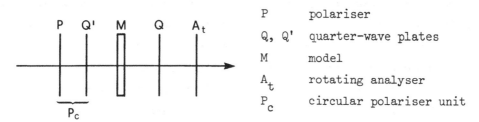

P   polariser

Q, Q'   quarter-wave plates

M   model

A_t rotating analyser

P_c circular polariser unit

Fig. 2   Optical arrangement along the light beam

The plane is orientated by the sense of propagation of the light.

The transmitted light intensity has the form

$$E = \frac{1}{2} E_o \left(1 - \varepsilon \mathscr{A} \sin (2 \Omega t + \psi)\right)$$

with $\mathscr{A}^2 = \sin^2 \phi \ \sin^2 2\beta + \cos^2 \phi$

$$\tan \psi = \tan \phi \ \sin 2\beta \quad , \quad \cos \phi \ \cos \psi > 0 \quad , \quad |\psi| < \pi$$

$\varepsilon = + 1$ for a right-handed circular polariser unit

$\varepsilon = - 1$ for a left-handed circular polariser unit

In the following assume $\varepsilon = + 1$, and introduce a reference signal of the form

$$e = e_o (1 - \sin 2 \Omega t)$$

The two measurable quantities, amplitude and phase, vary with the orientation of Q. In order to establish the Sénarmont configuration for Q, the principal axes of both Q and the model must first be set parallel. This position is characterised by zero phase difference and minimum amplitude. Note, only the second criterion remains usable when ϕ is near $(1/2 \ \pi \pm k\pi)$ for $(k = 0, 1, 2, \ldots)$, since the amplitude is then too small for accurate measurement of ψ.

Let x_1 denote the principal direction coinciding with the slow axis of Q.

A rotation of $- \pi/4$ will now place Q into the Sénarmont configuration. Phase measurement then gives the relative retardation as with the Sapaly method.

The procedure is easily carried out on a classical polariscope $|9\text{--}12|$. The ambiguity in fringe order can be eliminated by visualisation of the fringe pattern or by a technique of perturbation with a calibrated bire-fringent, used with a single wavelength as described in section 3. Recour-se to two wavelengths is also available and offers successive (indirect method) or simultaneous (direct method) measurement as alternatives.

Let the photoelastic constants C_1, C_2 for the two wavelengths λ_1, λ_2 have the same value so that the retardations are the same. If they differ we introduce a fictitious wavelength $\lambda_1^* = \lambda_1 C_1/C_2$.

We measure successively φ_1 and φ_2 characterised by

$$\delta = (\frac{\varphi_1}{2\pi} + k_1)\,\lambda_1 \quad \text{and} \quad \delta = (\frac{\varphi_2}{2\pi} + k_2)\,\lambda_2$$

$$\varphi_1\lambda_1 - \varphi_2\,\lambda_2 = 2\pi\,(k_2\lambda_2 - k_1\,\lambda_1)$$

with integers k_1, k_2. Several ways exist for evaluation of δ :

a) by a programmable calculation which presupposes $|\delta| < (\lambda_1\lambda_2/|\lambda_1\ \lambda_2|)$

b) by reading from a table of fringe order values : the preceding equation implies that $\varphi_1\lambda_1 - \varphi_2\lambda_2$ has only discrete values corresponding to integer fringe orders k_1, k_2. This procedure has been used by Srinath and Sarma $|13|$ for positive retardations and measurements of φ_1, φ_2 bet-ween 0 and $2\ \pi$. It can also be used for unknown sign of retardation and for measurements of φ_1, φ_2 between $-\ \pi$ and $+\ \pi$ $|9,\ 11,\ 14|$.

c) by nomogram $|10,\ 11|$: φ_2 varies linearly with φ_1 since

$$\varphi_2 = \frac{\lambda_1}{\lambda_2}\ \varphi_1 + 2\pi\ (k_1\ \frac{\lambda_1}{\lambda_2} - k_2)$$

Graphically, this variation leads to straight-line segments within a square field that can be graduated in relative retardations. With the mea-sured φ_1, φ_2 as co-ordinate axes, one can read directly for the point φ_1, φ_2 the relative retardation on one of these segments. The nomogram of Fig. 3 has been drawn for $\lambda_1^* = 406.6$ nm, $\lambda_2 = 546.1$ nm, and Araldite as material.

This procedure avoids the trial-and-error process of the preceding
method

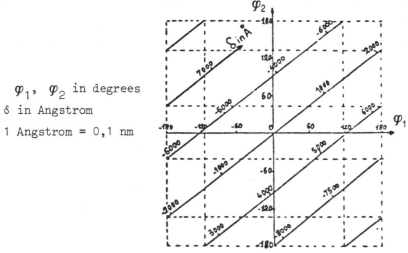

φ_1, φ_2 in degrees
δ in Angstrom
1 Angstrom = 0,1 nm

Fig. 3   Nomogram for birefringence measurements.

The direct method does not use a reference signal but, with quarter-
wave plates valid for both wavelengths, invokes the phase difference ψ,
which exists between the rays E_1 and E_2 of the two wavelengths traversing
the model,

$$\tan \psi = \frac{\sin 2\beta (\tan \phi_1 - \tan \phi_2}{1 + \sin^2 2\beta \tan \phi_1 \tan \phi_2}$$

with $\cos \psi \cos \phi_1 \cos \phi_2 > 0$.

Note that in the plate Q the fast axes for the two wavelengths must
coincide while in the first quarter-wave plate Q' they may be crossed.

As long as $|\phi_1 - \phi_2| < \pi$, that is $|\delta| < \frac{1}{2} \lambda_1\lambda_2/|\lambda_1 - \lambda_2|$, a zero va-
lue for $\tan \psi$ occurs only when $\sin 2\beta = 0$; that is when the principal
axes of model and Q are parallel. Hence the principal directions can be
found. The phase difference passes through an extremum in the Sénarmont
configuration $\phi_1 - \phi_2$ and its measured value then is proportional to the
relative retardation. By suitable calibration, a digital voltmeter can
be made to exhibit the algebraic value of the relative retardation, or
of the principal stress difference for the given model material.

This elegant solution for retardation measurement has been proposed
in June 1973 |7, 15| and published independently by Redner |16| shortly

shortly afterwards.

The direct method can be applied on a classical polariscope with quarter-wave plates sufficiently accurate for two wavelengths. It suffices to place in front of the image point for the considered point in the model the head of a split light-conducting glass fibre bundle (Fig. 4).

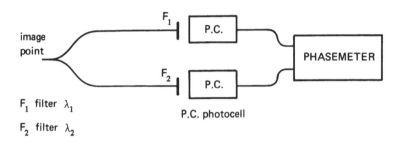

Fig. 4   Adaptation of the direct method on a whole-field polariscope.

In work |17| dating from 1967 |18|, Robert had developed a point method of linear detection of δ for |δ| < λ, using elements of a set-up with circular polariser unit and rotating analyser, which differs from the Sénarmont set-up by the orientation of the first quarter-wave plate. This complex method incorporates two slave mechanisms, one of which places the quarter-wave plate at π/4 to the axes of the birefringent. A reference signal is supplied by a splitter linked to the quarter-wave plate orientation.

Our indirect method is original in using only one slave mechanism and in creating simply and precisely the reference signal linked to the quarter-wave plate. In the arrangement for the classical polariscope, the reference signal is derived from the intensity measurement, at a point in the field, of a light ray by-passing the model. In the strictly localised point arrangement it is obtained by impingement on the rotating analyser of a ray plane-polarised at π/4 to the principal optical axes of the quarter-wave plate. To insure this polarisation, a sector of polaroid material traversed by the ray is rotated in unison with this quarter-wave plate.

We note that an analogous procedure has been adopted later by Robert and Royer |17|. A high-quality sinusoidal reference signal is in fact required for the synchronous detection as well as for the phase measurement.

Quarter-wave plate imperfections are particularly inconvenient in the direct method where differential measurements are taken, but such perturbations can be eliminated |10, 14| ;

a) either by taking measurements for two orthogonal positions of the circular polariser unit and averaging the results,

b) or by ensuring that the optical axes of the two quarter-wave plates always remain parallel.

Note : The indirect method is obviously more sensitive than the direct method, since full-scale phasemeter deflection corresponds to one wavelength. Hence the absolute error in retardation may also be smaller in the indirect method. However, when large retardations with steep gradients are measured, the errors caused by locating the measuring point and by its finite size may exceed the retardation error.

1.6 Redner method |16, 19|

This method uses Tardy's polariscope elements. Annular quarter-wave plates allow determination of the principal directions by the Nurse Allison method of synchronous rotation of analyser and polariser. The quarter-wave plates with parallel axes are placed at $\pi/4$ to the principal directions. The phase difference between two rays of different wavelengths (traversing the quarter-wave plates) yields the relative retardation. Redner was the first to use one quarter-wave plate for two slightly different wavelengths and hence was able to measure retardations up to about a dozen wavelengths.

1.7 Adaptations of the Sénarmont method (Lagarde-Oheix) |11, 20|

In 1975 Lagarde and Oheix proposed Sénarmont method adaptations similar to the Tardy method extensions. A uniform analyser rotation replaces the search for zero light intensity by a measurement of phase difference between two signals.

Let x_1, x_2 be the principal directions, x the fast axis of the quarter-wave plate Q, also $\alpha = (P, x_1)$, $\beta = (x, x_1)$, and Ωt the angle between

the rotating analyser and the direction P of the polariser. We introduce
the reference signal

$$e = \frac{1}{2} e_o (1 + \cos 2 \Omega t)$$

The transmitted light energy can be expressed as

$$E = \frac{1}{2} E_o \{1 + \mathcal{A} \cos (2 (\Omega t - \alpha + \beta) - \psi)\}$$

with $\mathcal{A}^2 = (\cos 2\alpha \cos 2\beta + \sin 2\alpha \sin 2\beta \cos \phi)^2 + \sin^2 2\alpha \sin^2 \phi$

$$\tan \psi = \frac{\sin 2\alpha \sin \phi}{\cos 2\alpha \cos 2\beta + \sin 2\alpha \sin 2\beta \cos \phi}$$

Let us extract a few relevant basic ideas.

If the light is assumed polarised at $\pi/4$ to the quarter-wave plate
axes, then $\alpha - \beta = \pi/4$ and hence

$$\mathcal{A}^2 = \sin^2 2\alpha (\cos^2 2\alpha (1 - \cos \phi)^2 + \sin^2 \phi)$$

$$\tan \psi = \frac{1}{\cos 2\alpha \tan \phi/2}$$

The polariser P and quarter-wave Q are rotated together and the changes in Amplitude of $\mathcal{A}$ are followed. In the cases when $\phi = 2 k\pi$ and
$\phi = (2 k+1) \pi$ characterized by the fact that $\mathcal{A}$ remains invariant (with values zero and maximum respectively for each of the ϕ values) can be avoided with the proper choice of the wave length. This condition is obviously
satisfied by chosing the sum of the two wavelength intensities $|30|$. Then
$\mathcal{A}$ vanishes if and only if the polarization direction is parallel to a
principal direction in the model, which can thus be found. This is true
for any wavelength.

Suppose now that the light is polarised along a principal optical
axis of Q so that $\alpha = \beta$. This leads us back to the relation introduced by
Sénarmont, except that we do not assume $|\alpha| = \pi/4$ and hence need not know
the principal directions.

If P and Q are rotated together, the phase difference ψ vanishes
when P is parallel to one of the principal directions (a method which
permits to determine it) if $\phi \neq \pi$, 3π,... This critical value of ϕ, characterized by the fact the phase remains zero when P and Q are rotated

together, can be avoided by chosing another wave lengt. Another possibility is to seek a zero amplitude corresponding to $\alpha = \pm\ \pi/8$. Once the
principal directions are thus determined, P and Q are placed in the Sénarmont configuration. The phase difference read is then ϕ, with periodicity
π. This is the equivalent of the indirect method. Measurement may be repeated for a second wavelength to remove the ambiguity.

A direct method can also be applied (maintaining $\alpha = \beta$). In this case
the phase difference between signals E_1 and E_2 obtained with wavelengths
λ_1 and λ_2 is considered. This presupposes one quarter-wave plate for both
wavelengths.

The principal directions follow from changes in ψ and $\mathcal{A}$ with α.
When $\psi = 0$ and the signal amplitude is a maximum, then $\alpha = 0$, $1/2\ \pi$, ...
Rotation by $\pi/4$ of polariser and quarter-wave now places them in Sénarmont
configuration.

One can also proceed, as described previously, by polarising the
light at $\pi/4$ to the optical axes of the quarter-wave, $\alpha-\beta = \pi/4$, and by
following the change of $\mathcal{A}$ with α ; the Sénarmont's configuration is
then found again when P is set parallel to x. This last procedure is generally preferable to the preceding one. In the Sénarmont's configuration
one measures $\psi = \phi_2 - \phi_1$ with periodicity 2π and $|\psi| < \pi$. The algebraic
value of δ is measured provided

$$|\delta| < \frac{1}{2}\ \lambda_1\lambda_2/|\lambda_2 - \lambda_1|$$

Results are comparable to those obtained by adaptation of the Tardy's
method wich, however, was more precise in determining the principal di-
rections, because near $\alpha = 0$ it had $|d\psi/d\ 2\alpha| = 2\ |\tan\ \phi|$ while here
$d\psi/d\alpha = 2 \sin\ \phi$, at the cost of more delicate manipulations. It had
$\psi = 0$ or π for $\alpha = 0$, according to the value of ϕ, and this led to fre-
quent changes of phasemeter scale. Moreover, the signal level dropped
near $\alpha = 0$, thus impeding phase measurement. The Sénarmont adaptations
avoid these inconveniences. Besides, for $\alpha - \beta = 0$, the determination of
the principal directions is not affected by quarter-wave plate imperfec-
tion, which have only second-order effects on δ measurement.

The main interest of this method lies in its ability to account for rotatory power, as will be pointed out in section 1.9 dealing with a specification for a modern photoelasticity meter. The proposed adaptations can obviously be used on a classical bench or on a strictly localised point measuring set-up. The reference signal is given in the first case by a ray traversing all the elements except the model, in the second by a ray traversing the rotating analyser and a polariser placed parallel to x and rigidly linked to Q.

1.8 Robert-Royer-Cadoret method |21|

This method uses the same components as the preceeding one, but interchanges their position along the light ray, giving the arrangement of Fig. 5. A sinusoïdal reference signal is formed as section 1.5, with phase origin aligned with respect ot the fast axis of the quarter-wave plate. The element A consists of two plane-polarising analysers, an annular one, A_a, having its optical axis displaced by $\pi/4$ from that of a central one, A_b.

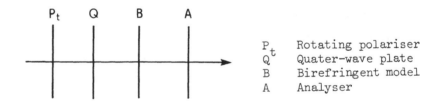

P_t	Rotating polariser
Q	Quater-wave plate
B	Birefringent model
A	Analyser

Fig. 5 Optical arrangement along the light beam

The elements A and Q are linked in rotation so as to give simultaneously arrangements analogous to those of para 1.7, fast axis of Q parallel to A_a axis and hence at $\pi/4$ to A_b axis. A synchronous detection procedure - corresponding to a zero phase method - uses the central ray and permits it to place A_a parallel to axis of B, and thus to determine, a principal direction in the model. Q and A_b are then in the preferred position leading to relative retardation by phase measurement between two signals of different wavelengths.

The method has the drawback that the signals disturbed by partial polarisation of the source introduce inaccuracies.

In order to obtain a high-quality rotating plane-polarised ray, the authors use a technique of a rotating birefringent similar to that proposed by Sapaly |22| in 1963 with a rotating half-wave plate.

1.9 A specification for a photoelasticity meter |9,11|

A few experimental results observed in a study of a semi-infinite plate loaded by a punch on its boundary may be indicated. An ordinary 6 mm thick rectangular, 150 . 200 mm, Araldite plate was used (Fig. 6).

All along a line parallel to the upper boundary, on either side of the load axis, the principal directions deviate significantly, up to 40 degrees, from theoretical values. The explanation is simple. An imperceptibly incorrect loading, due perhaps to an imperfectly plane plate, introduces a bending moment. The stress state then is three-dimensional and the material exhibits birefringence and rotatory power. This well known result was discussed in chapter 2. The method described below has enabled us to show up a rotatory power that is zero below the load, antisymmetric to either side, and increasing in magnitude possibly up to 40 degrees.

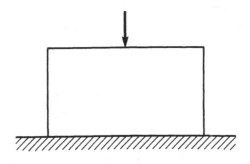

Fig. 6 Plane loading of an araldite plate

In the same type of experiment with correct loading (10 mm thick Araldite plate), isolated points are observed which we shall call "aberrant points". They correspond to local, hardly visible imperfections completely falsifying interpretation of apparently correct measurements. We have observed a rotation of the principal directions from their calculated values by up to 60 degrees without any effect on the value of relative retardation. The role of these aberrant points become more significant the more the measurements are localised. We have not observed them in Perspex.

Measurement of rotatory power appears to us as a precise criterion for ascertaining existence of a state of plane stress. It is an indispensable precaution because, even if the shear stresses are not noticeably disturbed, significant errors in the principal directions can lead to gross errors in stress calculation by integration of the equilibrium equations.

Remark

When a loading defect is the local cause of a bending state, it can be shown |31| that the measuring of the 3 physical parameters (those indicated in chapter 2) yield the classical values of photoelasticity.

1.10 Lagarde-Brillaud-Oheix methods |9,11,23,24|

These methods use the elements of the Sénarmont method.

The first one extends an adaptation of the Sénarmont method proposed in 1975 to the study of a retardation followed by a rotatory power. This extension can Be derived in a simple way by using Poincaré's representation.

The polariser axis and, for example, the fast axis of the quarter-wave plate are kept parallel. The preferred position of polarizer and quarter-wave plate is no longer the Sénarmont configuration (orientation at $\pi/4$ to the axes of the birefringent) but is displaced from it by the value of the rotatory power. The reference signal is obtained as indicated previously.

With the notation of section 1.7 and with R as the rotating power, the transmitted energy now is

$$E = \frac{1}{2} E_o \left(1 + \mathcal{A} \cos (2\Omega t - \psi)\right\}$$

$$\mathcal{A}^2 = \sin^2 2\alpha \cos^2 \phi + \cos^2 2\alpha \cos^2 (2R + 2\alpha)$$

$$+ \sin^2 2\alpha \cos^2 \phi \sin^2 (2R+2\alpha) + 2 \sin 4\alpha \cos \phi \sin 4 (R + \alpha)$$

$$\tan \psi = \frac{\sin 2\alpha \sin \phi}{\cos 2\alpha \cos(2R + 2\alpha) + \sin 2\alpha \cos \phi \sin(2R + 2\alpha)}$$

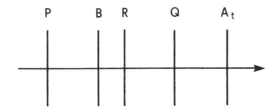

P	B	R	Q	A$_t$

P   Polariser
B   Birefringent
R   Rotating power
B+R Model
Q   Quater-wave plate
A$_t$ Rotating analyser

Fig. 7 Optical arrangement along the light beam

The changes of phase with orientation of P and A are considered.
For $\phi \neq k\pi$ (always satisfied for at least one of the two wavelengths
used) the phase difference vanishes with periodicity π when the quarter-
wave plate axes coincide with those of B, which can thus be determined.

If the quarter-wave plate is now removed, with P remaining in the
previously determined position, the transmitted energy is

$$E = \frac{1}{2} E_o \left(1 + \cos (2\Omega t + 2R)\right)$$

A phase measurement yields 2R.

After the quarter-wave plate is re-inserted and orientated in the
preferred position at $(\pi/4 + R)$ to the axes of the birefringent, we obtain

$$E = \frac{1}{2} E_o \left(1 + \cos 2R \cos (2\Omega t - \phi)\right)$$

A phase measurement supplies the relative retardation for $|\delta| < \lambda$.
For retardations exceeding a wavelength, one may have recourse to a second wavelength and employ an indirect method. Use of a direct method is
more problematic since the rotatory power may change with wavelength and
measuring point.

The preceding method resorts only to phase measurements. A slightly
different method $|23|$ requires neither parallel axes of quarter-wave pla-
te and polariser nor removal of the quarter-wave plate but is in fact ba-
sed on use of a new ellipsometer $|25|$ (see section 4.2) and utilises cri-
teria of zero phase or of sinusoidal amplitude. It leads to a linear detection of the optical characteristics of a slice in a three-dimensional
medium $|26,28|$. A contemporary method of linear detection proposed by
Robert, Cadoret and Royer $|29|$ may also be mentioned.

1.11 Duality by interchange of polariser and analyser

We have pointed out since 1973 $|15|$ that in adaptations of the
Tardy's method, the source and the photodetector can be interchanged
$|9,11,14|$ without any alteration in the expression for the light energy.
We have made the same remark about the extension of Sénarmont's method
$|11,23,24,26|$ to measurement of the optical characteristics of a birefringent and of a rotatory power. In other words, the components along
the light path can be interchanged so that for instance the arrangement
of Fig. 8 is obtained.

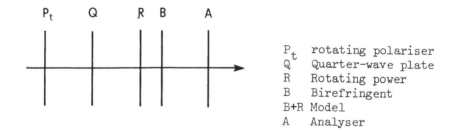

Fig. 8 Optical arrangement along the light beam

For R = 0, one thus arrives again at the method proposed in section
1.8 which can be deduced immediately from that proposed in section 1.7.

1.12 Method of Isodynes

This method, has been developed by J.T. Pindera and S.B. Mazurkie-wicz in 1977 |32,33|.   It deserves particular mention since it is based on the phenomenon of diffusion of light that is utilised as an analyzer, along two orthogonal directions as for the method of Y.F. Cheng |34|.

The method requires the translation of a laser beam in a plane pa-rallel to the surfaces of the model, which must be immersed in a liquid having the same index of refringence. If x, y are the orthogonal axes lying in the plane that is examined and if the beam is parallel to x, the phase angle due to a length dx in the loaded model will·be $d\phi = \frac{2\pi}{\lambda} C\sigma_y dx$. If the principal directions y and z do not vary along x one has

$$\phi = \frac{2\pi C}{\lambda} \int \sigma_y \, dx$$

The intensities diffused in the directions $\vec{\zeta}_1$· and ζ_2 normal to $\vec{x}$ and forming the angles $\pm \pi/4$ with $\vec{z}$ are modulated respectively by $(1 + \sin \phi)$ and $(1 - \sin \phi)$.

The incident light being circular, one can record the images of the beam observed in the directions $\vec{\zeta}_1$ and $\vec{\zeta}_2$. The fringes of the two pat-terns correspond to variations of $\lambda/2$.

The fringes are isodyne lines, i.e. lines along which $\int \sigma_y \, dx$ is constant. The authors also introduce a translation of the beam parallel to y in order to obtain the I_y isodynes along which $\int \sigma_x \, dy$ is constant. One has evidently

$$\sigma_y = \frac{\partial I_x}{\partial x} \quad \text{and} \quad \sigma_x = \frac{\partial I_y}{\partial y}$$

Since $\sigma_x + \sigma_y = \sigma_1 + \sigma_2$

it will be possible to separate the principal strains if the traditional field of the isochromatics is also known.

2. Application to contact mechanics

When two bodies are in contact and loaded, the stresses at the contact point are singular and often produce surface damage. This surface damage is directly influenced by the contact stresses and the capability to determine these stresses is very important.

This problem was approached for the first time by Hertz in the last century and has not yet been completely solved. The analytical solutions which have been proposed to date involve assumptions which permit only particular cases to be treated. Solutions are required to treat more general geometries of the contacting bodies and to deal with materials with more constitutive laws general.

The calculation of contact stresses for measurements obtained by photoelasticity or interferometry is an interesting method |35| |36| |37| but the results obtained are strongly dependent on the accuracy of experimental values. The integration of equilibrium equations from the classical data yielded by photoelasticity does not always give satisfactory results because of errors in the measured values of the direction of the principal stresses and of the birefringence in the vicinity of the contact area. The determination of the sum and difference of the principal stresses by an interferometric method |37| leads to a pressure distribution similar to the one obtained by the analytical method in the case of a plane contact |38| (for a plane stress problem). Again in the case of plane stress, starting from isocromatic fringes and considering the particular case of a lubricated Hertzien contact, Stupnicki |39| obtains the pressure distribution as foreseen by the elastohydrodynamic theory.

For the last fifteen years, the development of the finite element methods has permitted researchers |40 to 43| to take into account the real geometry of the contact but they require the definition of a friction law to describe the behaviour of contact surfaces.

We have studied the plane stress contact between two photoelastic solids : a rectangular plate fixed on one side with the opposite side loaded by a pad (fig. 9).

This study combines the numerical and experimental methods. A finite element method allows us to calculate the influence coefficients which

relate the unit values of discrete contact stresses to shear stresses in
certain parts of the materials in which the measurements are carried out.
The contact stresses are then obtained by numerical solution of the inverse problem

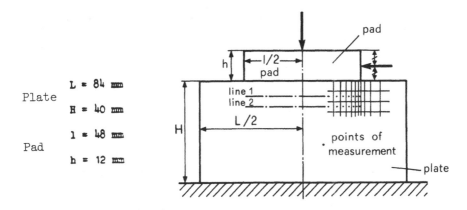

Plate   L = 84 mm

                     H = 40 mm

                     l = 48 mm
Pad
                     h = 12 mm

Fig. 9   Plate loaded with a pad

2.1   Photoelastic measurements |41| |42| |43|

The punctual measurement method for photoelastic determinations, adapted from Tardy's method (cf 1.5) was applied using a whole field circular
polariscope with a rotating analyser and flexible light conductors |8| |10|
|14|. The visualisation of isochromatics allows one to avoid the indeter-
mination of fringe orders without using a second wavelength.

As indicated previously, the experiment must be accurately performed with check's at different points of the model to insure that the rotatory power is very small and may be neglected (cf 1.10)

The influence of experimental errors was minimized by eliminating
the defects of quarter-waves (their neutral lines are maintained parallel, cf 1.5) and by loading the Araldite model sufficiently to reduce the
relatively small residual birefringence to a negligible quantity. In spite
of all the precautions, errors resulting from locating the point, in a
field where significant gradients in principal directions and in gradients
of angular birefringence occurred. For this reason the difference between
the real and the measured values may be several degrees. As a consequence the difference of principal stresses is much more reliable than

their orientation.

2.2 Calculation method |44| |45| |46|

Let us consider a two-dimensional elastic solid submitted to a pressure distribution p and whose contact boundary is divided into n elements
(Fig. 10). At a point i of the solid the elementary stresses created by
the pressure element p_j are :

$$(\sigma_x^p)_{ij} = (\sigma_x^p)_{ij}^u \, p_j$$

$$(\sigma_y^p)_{ij} = (\sigma_y^p)_{ij}^u \, p_j$$

$$(\sigma_{xy}^p)_{ij} = (\sigma_{xy}^p)_{ij}^u \, p_j$$

where $(\sigma_x^p)_{ij}^u$, $(\sigma_y^p)_{ij}^u$, $(\sigma_{xy}^p)_{ij}^u$ are the stresses created at point i by
a unit pressure applied to an element j. The stresses arising at point i
from the pressure distribution p can be expressed in the following way :

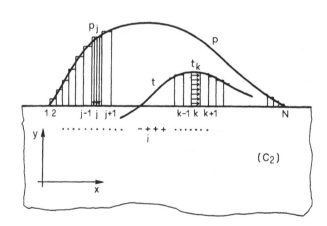

Fig. 10 Discretisation of the loading at the contact

$$(\sigma_x^p)_i = \sum_{j=1}^{N} (\sigma_x^p)_{ij} = (\sigma_x^p)_{ij}^u \, p_j$$

$$(\sigma_y^p)_i = \sum_{j=1}^{n} (\sigma_y^p)_{ij} = (\sigma_y^p)_{ij}^{u} \; P_j$$

$$(\sigma_{xy}^p)_i = \sum_{j=1}^{n} (\sigma_{xy}^p)_{ij} = (\sigma_{xy}^p)_{ij}^{u} \; P_j$$

In the same way the stresses created at the point i by the shear stress distribution may be expressed as follows :

$$(\sigma_x^t)_i = (\sigma_x^t)_{ik}^{u} \; t_k$$

$$(\sigma_y^t)_i = (\sigma_x^t)_{ik}^{u} \; t_k$$

$$(\sigma_{xy}^t)_i = (\sigma_{xy}^t)_{ik}^{u} \; t_k$$

One may then write, by using the principle of superposition :

$$a_{ij}^P \; P_j + a_{ij}^t \; t_j = (\sigma_2 - \sigma_1)_i \; \cos 2\alpha_i$$

$$(i = 1 \text{ to } n)$$

$$b_{ij}^P \; P_j + b_{ij}^t \; t_j = (\sigma_2 - \sigma_1)_i \; \sin 2\alpha_i$$

where the influence coefficients a_{ij}^P, a_{ij}^t, b_{ij}^P et b_{ij}^t are defined by :

$$a_{ij}^P = (\sigma_x^P)_{ij}^{u} - (\sigma_y^P)_{ij}^{u} \qquad\qquad a_{ij}^t = (\sigma_x^t)_{ij}^{u} - (\sigma_y^t)_{ij}^{u}$$

and

$$b_{ij}^P = 2 \, (\sigma_{xy}^P)_{ij}^{u} \qquad\qquad b_{ij}^t = 2 \, (\sigma_{xy}^t)_{ij}^{u}$$

The influence coefficients are determined in a numerical way by a finite element calculation using isoparametric elements with eight nodal
points.

In a similar way, the measurement at n points of the principal stress
orientation α_i and of the difference of principal stresses $(\sigma_2 - \sigma_1)_i$ allow the determination of the contact actions p_i and t_i as solutions of a
linear system of equations. In practice, some instabilities may appear
due to the errors of the experimental values and to the errors of the calculate values of the influence coefficients. These cause significant dif-

ficulty because the terms of the principal diagonal of a_{ij}^t are nearly zero. This approach involves the system of equations shown previously where
the trigonometric values have been eliminated

$$(a_{ij}^P \ p_j + a_{ij}^t \ t_j)^2 + (b_{ij}^P \ p_j + b_{ij}^t \ t_j)^2 = (\sigma_2 - \sigma_1)_i^2$$

or

$$F_i(X) = F_i \ (p_1 \ldots p_n, \ t_1 \ldots t_n) = 0 \qquad (i = 1 \text{ to } 2n)$$

These equations are non linear with the solution depending only on the
difference of principal stresses $(\sigma_2 - \sigma_1)$. However, $2n$ measurements are
required to determine the unknowns by a deterministic approach.

Though, the errors on the calculated or measured values are small (a
few per cent) some instabilities of the calculations appear and make it
impossible to obtain the solution of the non linear system $F_i \ (X) = 0$. The
solutions may be then found for

$$|F_i \ (X)| < \varepsilon \text{ where } \varepsilon \text{ is fixed a priori by using an iterative process.}$$

2.3 Some results |45| |46|

Point measurements of the difference of principal stresses are performed along two parallel lines at positions 6 and 10 mm from the contact
line.

One example of the results obtained is presented in Fig. 11 (normal load
1250 N) and Fig. 12 (normal load 1250 N and shear load 175 N).

Starting from the measurements of the difference of principal stresses, the contact stresses (normal and shear) were calculated by dividing
the contact region into 22 elements.

The normal stresses σ_{yy} and the shear stresses σ_{xy} at the contact are
presented (Fig. 13-14) for different values of the normal loading. The
results obtained confirm the symmetry of normal stresses about the loading axes and the antisymmetry of the shear stresses. The distribution of
the calculated pressure are in accordance with the results of Nisida and
Saïto |37| for a similar configuration. The ratio of the shear stresses
versus normal stresses, which correspond to the local friction is presented in Fig. 15. The coefficient of local friction is between - 0.9 and

+ 0.9 which confirm that there is no simple relation between normal and shear stresses |47|.

The influence of a shear loading on the stress distribution at the contact is given in Fig. 16 . Note that the pressure distribution becomes very nonsymmetric under shear loading. The antisymmetry of the shear stress distribution is lost but the maximum value of this stress does not change.

Starting from the distribution of the calculated normal and shear stresses, we have determined with a finite element analysis, the isochromatic pattern and the orientation of principal stresses in some points. The results are in accordance with the experimental observations and confirm the accuracy of the above approach.

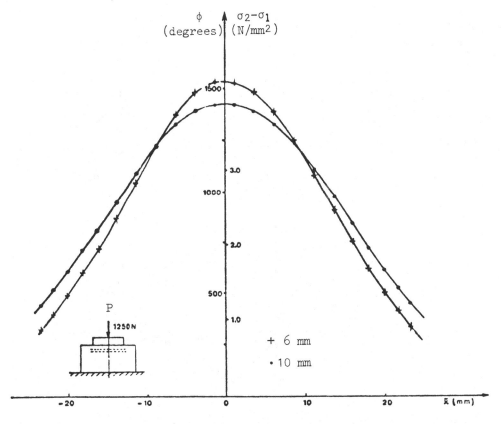

Fig. 11 Difference of principal stresses measured during
a normal load of 1250N

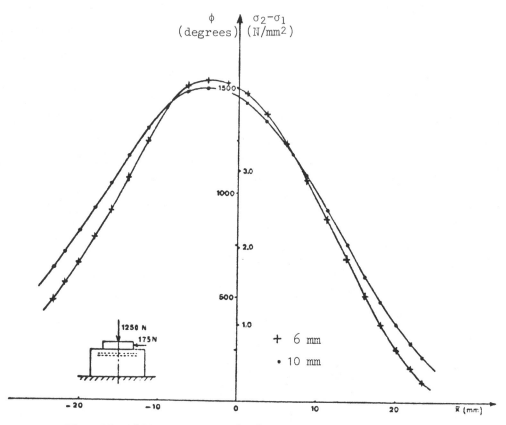

Fig. 12 Difference of principal stresses measured during
a normal load P = 1250 N and a shear load T = 175 N

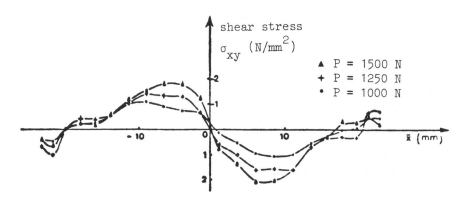

Fig. 13 Distributions of shear stresses at contact point
for different values of the normal load P.

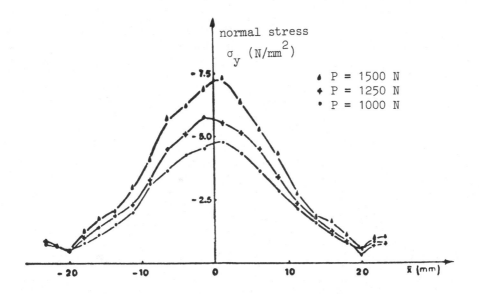

Fig. 14 Distributions of normal stresses at contact point
         for different values of the normal load P.

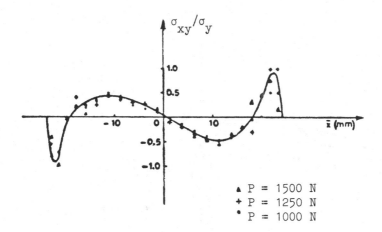

Fig. 15 Variation of local friction at contact point
         in the case of normal load P

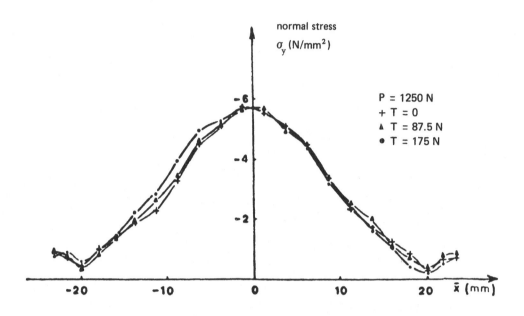

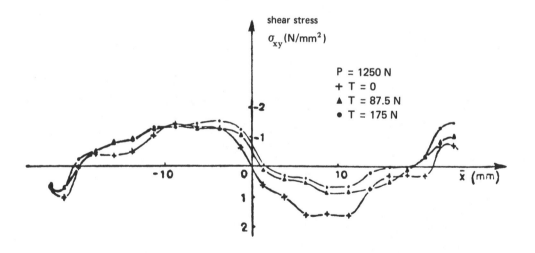

Fig. 16 Distribution of normal and shear stresses at contact point
for different values of the ratio T/P

CHAPTER 4

A THREE-DIMENSIONAL POINT PHOTOELASTIC METHOD OF OPTICAL SLICING
WITH APPLICATIONS

We have seen in the preceding chapter that any photoelastic medium,
thin or thick, can be represented by a model consisting of an birefringent plate and rototory power. The objective of this chapter is to measure three physical paremters in any photoelastic medium ; principal
directions, angular birefringence, and rotatory power. We also know
that such an operation for a thin slice yields the two parameters of
interest to the mechanician.

We present two versions of a method of linear determination of the
three physical parameters. We also give two examples indicating applications
of these methods.

We begin with the principle of ellipsometry with linear detection adapted to scattered-light measurements, because it is of interest and also
because it is used for introducing the first version of the point method.

We believe that the method we describe offers a satisfactory solution
to a difficult problem that has been the object of many investigations for
the past century. The proposed solution benefits from previous studies. Its
originality which its success is the application of an experimental procedure with precise criteria independant of inherent fluctuations in
scattered light measurements.

1   Principle of ellipsometry of linear detection and influence of
    depolarization of scattered radiation

We should mention that the depolarized part of scattered radiation
constitutes the principal obstacle in the precise measurement of the
characteristics of an elliptical vibration.

1.1 Principle of ellipsometry [1, 2]

The light vibration that we propose to characterize encounters successively a rotable quarter—wave plate and then an analyser rotating at an angular velocity Ω. A light beam passes through a polarizer whose axis is parallel to the fast axis of the quarter-wave plate and then through the rotating analyser to form reference signal (Fig. 1).

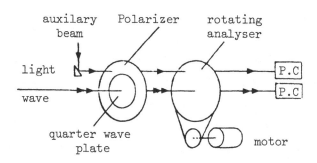

Fig. 1 : Shematic diagram of ellipsometer

The ellipticity of an elliptical vibration equal to $|tg\ I|$ the ratio of the minor to the major axis, is referred to these principal axes oX and oY with oX being the major axis. We designate by $\beta = (oX, oX_1)$ the angle between the fast axis oX_1 of the quarter—wave plate and the major axis of the ellipse, $\Omega t = (oX_1, A)$ the angle between the axis of the analyser $A(t)$ and oX_1 (Fig. 2). These angles are measured according to the direction of light propagation.

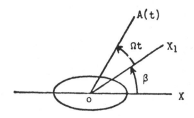

Fig. 2 : orientation of axes

The energy of the light vibration transmitted by the analyser is :

$$E = \frac{E_o}{2} \ (1 + \cos 2\beta \cos 2I \cos \Omega t - \sin 2I \sin \Omega t)$$

$$E = \frac{E_o}{2} \ (1 + \mathscr{A}\cos (2\Omega t + \psi))$$

where $\mathscr{A} \cos \psi = \cos 2\beta \cos 2I$ and $\mathscr{A} \sin \psi = \sin 2I$

$$tg \ \psi = \frac{tg \ 2I}{\cos 2\beta}$$

The reference signal has the form : $E_r = \frac{E_{ro}}{2} \ (1 + \cdot\cos 2\Omega t)$

ψ is the phase difference between the alternating components of the active and reference signals.

The determination of orientation and the ellipticity of the type of light is made in two steps :

- In the first step we orient the quarter-wave plate such that the absolute value of the phase difference is $\frac{\pi}{2}$, which corresponds to a value of $\beta = \pm \frac{\pi}{4}$; thus the axis of the ellipse are known.
- In the second step, starting from the previous setting we rotate the quarter-wave plate by $\pm \frac{\pi}{4}$; thus the measure of the phase difference obtained is $2I$ or $\pi - 2I$.

We note that the fast axis of the quarter-wave plate coincides with the major axis of the ellipse when $|\psi| < \frac{\pi}{2}$, and with the minor axis of the ellipse when $|\psi| > \frac{\pi}{2}$.

- For the special case of rectilinear light ($2I = 0$) the phase difference is zero when $|\beta| < \frac{\pi}{4}$ (mod π) and equal to π when $\frac{\pi}{4} < |\beta| < \frac{\pi}{2}$ (mod π).

The sinusoïdal amplitude is zero for $\beta = |\frac{\pi}{4}|$ (mod π) consequently, the direction of the rectilinear vibration is the bissector of the angle defined by the positions of the fast axis of the quarter-wave plate defined by $\beta = \frac{\pi}{4}$ and $\beta = -\frac{\pi}{4}$ (characterized by zero sinusoïdal amplitude), within a sector of value $\frac{\pi}{2}$ in which any position of the fast axis of the $\lambda/4$ plate leads to a zero phase difference. Thus the orientation of the

rectilinear vibration is known. We can show that the defects of the
quarter-wave plate do not have any influence on the determination of the
axes of the ellipse, and enter only as second order terms in the determination of the ellipticity (More complete details on the ellipticity are
given in |3, 4|)

1.2 Influence of depolarization of scattered radiation

Experience shows that even for a direction of observation normal to
the principal beam, the property of rectilinear polarization predicted by
Rayleigh's law is not satisfied.

The scattered light at a point in the model can therefore be represented more generally by two random orthogonal components of amplitudes, A,
and, a, respectively.

Under these conditions, starting with the two orthogonal components,
the thick medium transmits two elliptical vibrations of the same
ellipticity, which are in opposite directions and have their axes crossed.

The total energy emerging from the thick medium is the sum of the
energies of each vibration, since they are in phase.

$$E_a = \frac{a^2}{2} \left[1 - \mathcal{A} \cos (2 \Omega t + \psi) \right]$$

$$E_A = \frac{A^2}{2} \left[1 + \mathcal{A} \cos (2 \Omega t + \psi) \right]$$

$$E = E_a + E_A = \frac{A^2 + a^2}{2} + \frac{A^2 - a^2}{2} \mathcal{A} \cos (2 \Omega t + \psi)$$

where as before $\mathcal{A} \cos \psi = \cos 2\beta \cos 2I$ and $\mathcal{A} \sin \psi = \sin 2I$ or :

$$E = a^2 + \frac{A^2 - a^2}{2} \left[1 + \mathcal{A} \cos (2 \Omega t + \psi) \right]$$

The result is as if we had linearly polarized light of intensity
$(A^2 - a^2)$ superposed on a continuous background of natural or completely
depolarized light of intensity a^2.

The method of measurement of the form of light that we present is
completely independent of the depolarization of the scattered light, since
it makes use only of phase measurements or zero-sinusoïdal amplitude.

For the same reason, the method is not affected by any light intensity fluctuations that we encounter in general when using scattered light.

2   Point method with linear detection

2.1 Application of ellipsometer to the principle of a set-up of point measurement of the optical parameters of a slice |2, 5, 6, 7|

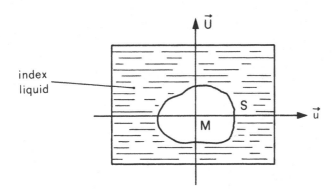

Fig. 3 : Solid in an immersion tank

To classify the point, let us consider a birefringent solid placed in a tank of liquid of the same index of refraction illuminated by a beam propagating in the $\vec{U}$-direction (Fig. 3). The observation of the light scattered from a point M of the model is made in the $\vec{u}$-direction normal to $\vec{U}$. Furthermore, the principal ray $\vec{U}$ while always passing through point M can be oriented on a plane orthogonal to $\vec{u}$.

According to Rayleigh's law the light scattered in the $\vec{u}$-direction from point M is linearly polarized along a direction P orthogonal to plane $(\vec{u}, \vec{U})$. Thus at point M of the model we set-up a rotable polarizer.

The ellipsometer described in section 1.1 is assembled outside the tank where the scattered light beam emerges from points of the model. The medium along path MS can be represented schematically by a birefringent plate B' followed by a rotatory power R'. We have then along the ray $\vec{Msu}$ the arrangement of Fig. 4.

Using calculations of the Poincaré sphere we can express angles β and $2I$, defined in section 1, as a function of angles, $α' = (X_1, b')$ and

$\theta = (b',P)$, angles oriented according to the direction of light propagation, b' being the fast axis of the birefringent plate B' and ϕ' its angular birefringence. We obtain :

$$E = \frac{E_o}{2} \left[1 + \mathscr{A} \cos (2 \, \Omega t + \psi) \right]$$

$$\mathscr{A} \sin \psi = \sin 2\theta \sin \phi'$$

$$\mathscr{A} \cos \psi = \cos 2\theta \cos 2(R' + \alpha') - \sin 2\theta \cos \phi' \sin 2(R' + \alpha')$$

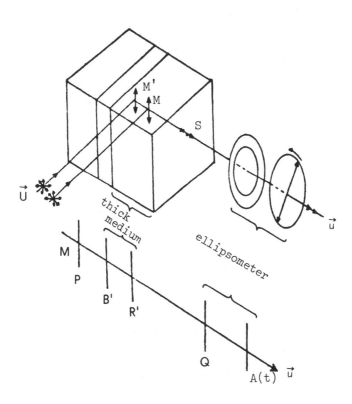

Fig. 4 : Representation of elements along ray $MS\vec{u}$

2.1.1 First step

The aim of this step is to place a compensator at the exit of the tank
in such a manner as to nullity the effects of the birefringent associated
with slice MS of the model.

We orient the direction of polarization until we obtain a phase difference of zero or π. Then the light emanating from the model is linearly
polarized, and the direction of polarization at the entrance of B' is one
of the axes of birefringent B' which are thus determined.

Having fixed the direction of polarization to a position corresponding
to a phase difference ψ of zero or π, we orient the quarter-wave plate until we obtain a sinusoïdal amplitude of zero. It is clear from the discussion in section 2.1 that we determine, without ambiguity, the orientation
of the linear vibration emerging from the thick medium MS. When the birefringence of B' is not equal to a half-wave we obtain the value of the
rotatory power R'. When this is not the case we can determine only the direction of the linear vibration emerging from the model with respect to the
direction of polarization, which is sufficient for defining the action of
the thick medium MS.

- Finally we set up, in front of the scattered beam at the exit of the
model, a compensator whose axes are at $\frac{\pi}{4}$ with those of the quarter-wave
plate, which is maintained in a position corresponding to zero sinusoïdal
amplitude, i.e., shifted from the value of the rotatory power with respect
to the axes of the birefringent. The direction of polarization P is shifted of
the value π/4.

It is easy to show, by means of the Poincaré sphere for example, that
it is now possible, by adjusting the birefringence introduced by the
compensator, to render the assembly of the thick medium and compensator
equivalent to the known rotatory power alone. We note that this approach
has already been used by Gross Petersen |8, 9| (Fig. 5).

The neccessary adjustement of the compensator is found easily when the
phase difference ψ is zero.

In the special case when the birefringence of B' is equal to one half-
wave, the compensator is adjusted to this same value, the assembly of

compensator and thick medium is then equivalent to a rotatory power, which
is easy to measure by the orientation of a linear vibration at the exit of
the model (cf. sect. 1).

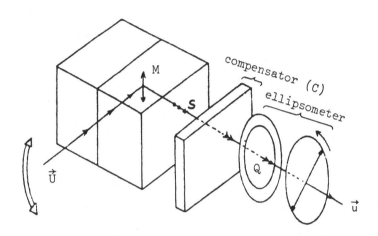

Fig. 5 : Utilization of compensator

. Orientation of the principal beam $\vec{U}$:

$$\left.\begin{array}{l} \text{Coincidence } \vec{U} \\ \text{with one axis of B'} \end{array}\right\} \Longleftrightarrow \psi = 0 \text{ (or } \pi)$$

axes B' known

. Orientation of Q :

$$\left.\begin{array}{l} \text{Axes of Q at } \overset{+}{-} \frac{\pi}{4} \text{ with the direction} \\ \text{of a rectilinear vibration issued} \\ \text{from (B' + R')} \end{array}\right\} \Longleftrightarrow \mathcal{A} = 0$$

R' known

. Introduction orientation and adjustement of C

Shift of $\vec{U}$-direction by $\pi/4$ - axes of compensator at $\pi/4$
with those of quater wave plate - we adjust C until the
obtaining $\psi = 0$ then B' + R' + C = R'.

2.1.2 Second step

We now displace the beam $\vec{U}$ so that it illuminates point M' near M located along the direction of observation $\vec{u}$ (Figs. 6 and 7). This beam translation has already been used by Robert and Guillemet |10|.

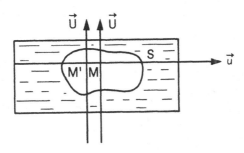

Fig. 6 : Translation of principal beam $\vec{U}$ in the immersion tank.

We thus add to the assembly of thick medium and compensator a thin sheet usually represented schematically by a birefringent plate B of angular birefringence ϕ^* and a fast axis b designated by α^*, $\alpha^* = (X_1,b)$. We can generalize this representation by attaching a rotatory power R.

We now find along the path of the scattered light the elements represented schematically in Fig. 7 .

As before, we can adjust the direction of polarization P until we obtain a zero phase difference, then orient the $\lambda/4$ plate until we obtain zero sinusoïdal amplitude. We have already shown that we obtain then the axes of B and the value of the rotatory power R + R' (R' is already known).

We can determine the birefringence by a simple phase measurement. In fact by designating $\theta^* = (b, P)$, the angle that axis b makes with the direction of polarization, the energy of the transmitted vibration is written as :

$$E = \frac{E_o}{2} \left[1 + \mathcal{A}\cos (2 \, \Omega t + \psi) \right]$$

with :

$$\mathcal{A}\sin \psi = \sin 2I = \sin 2\, \theta^* \sin \phi^*$$

$$\mathcal{A}\cos\psi = \cos 2\beta \cos 2I = [\cos 2\theta^* \cos 2 (R + R' + \alpha^*)$$

$$- \sin 2\theta^* \cos \phi^* \sin 2 (R + R' + \alpha^*)]$$

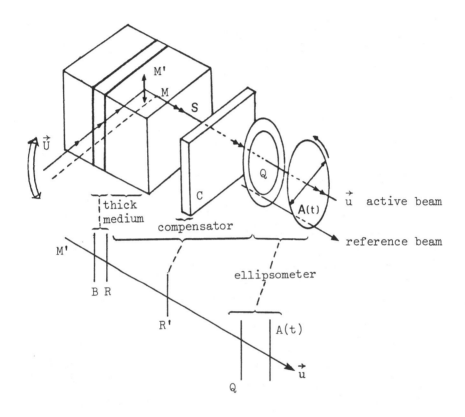

Fig. 7 : Representation of elements along path $MS\ \vec{u}$

    . Determination :

        axes B $\Big\}$ same as first step
        R + R'

    . measure of ϕ^*

        shift of $\vec{U}$-direction by $\pi/4$ and $\psi = \phi^*$

At the time of determination of the axes of the birefringent B and the rotatory power, when the direction of polarization coincided with one of the axes of the birefringent, the quarter-wave plate was positioned without ambiguity such that $\beta = \frac{\pi}{4}$ (cf. sect. 1).

Two cases can then be presented :

$$\theta^* = 0 \quad \text{then} \quad \alpha^* + R + R' = -\frac{\pi}{4}$$

$$\theta^* = \frac{\pi}{2} \quad \text{then} \quad \alpha^* + R + R' = -\frac{3\pi}{4}$$

Starting from each of these two positions; if we shift the direction of polarization by $+\frac{\pi}{4}$, $\psi = \pm \phi^*$ and the phase difference ψ gives directly the value of the angular birefringence ϕ^*.

The direction of polarization coincides with the fast or the slow axis of the birefringent, according to whether the measured phase difference is positive or negative.

Remark :

- The phase measurement gives ϕ^* within $2 k\pi$ however the fringe order k can be easily obtained by making the thickness of the thin sheet approach zero.

In practive we have followed this second method which has the advantage of being easily put into application.

- We could have presented this method also without making reference to ellipsometry (|11| to |15|).

2.2 Extension of the method to the case when the scattering phenomenon is used as an analyser

Let us take again the sequence of elements described in sect. 2.1. Let us invert the direction of light propagation considered in the schematic of fig. 7, to obtain Fig. 8.

A (t) (now a direction of the rotating polarization) and the fast axis of the birefringent b' are now referred to by the angles $-\Omega t$ and $-\alpha$ with respect to the fast axis x_1 of the quarter-wave plate, similarly the angle between P (now the direction of the analyser) and B', becomes $-\theta$ since we have reversed the direction of light propagation.

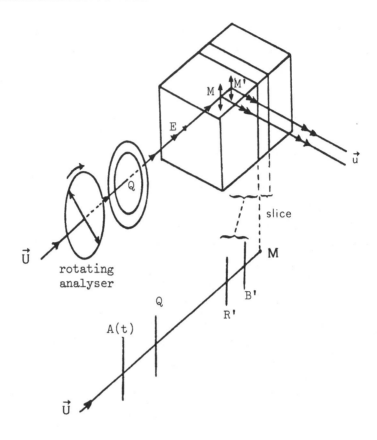

Fig. 8 : Representation of elements after inversion
of direction of light propagation

We note, by contrast, that the rotatory power remains equal to R'.

The calculation of light intensity emitted in the direction of the analyser leads to an expression which is identical to that related to the schematic of Fig. 7.

The schematic of Fig. 9 is interpreted easily : A principal beam consisting of rotating rectilinear light passes through the quarter-wave plate, then enters the model along the axis of observation of sect. 2.1. (traversing it in the reverse direction), encounters the rotatory power and the birefringent and finally at point M the observation is made in the direction of the principal beam as in sect. 2.1. It follows that in order

to measure the characterics of a sheet MM' masked by a thick medium MS
(Fig. 10) we set-up at the exit of the tank, in the direction A, a
removable compensator, a rotable quarter wave and a rotating analyser.

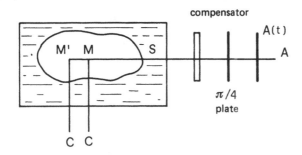

Fig. 9 : Schematic of the immersion tank and **external** elements.

2.3 Assessment of possibilities

Two possibilities exist :

a) Illuminate points M and M' along the movable direction C and collect
on a photocell the scattered radiation in the direction A. The scattering
phenomenon plays the role of the polarizer.

b) Illuminate in the direction A and collect on a photocell the light
scattered from points M and M' successively in the movable direction C.
The scattering phenomenon plays the role of the analyser.

In both cases, for an identical position of the elements, the signal
received by the photocell is the same (within a constant multiplier) ; the
experimental procedure is the same.

Remark :

We note the current method by Robert and Royer |17| which make use also
of the scattering phenomenon as an analyser and allows the determination of
the three physical parameters of a thick medium. The characteristics of the
thin slice, represented schematically by a birefringent, are deduced from
those of the thick medium which surrounds it.

The photoelasticimetry introduced here can be adapted to either case.

3   Design of three-dimensional photoelasticimeter |5| |16|

3.1 Case where scattering is used as polarizer (Fig. 10)

In order to set the direction of polarization, we have constructed a
plano-convex cylindrical lens of large dimensions fitting inside the
immersion tank. This lens contains a liquid identical to that of the tank.

A movable arm with mirrors positioned precisely allows the principal
beam to arrive at normal incidence on the convex surface of the lens. A
translation of this movable arm allows the illumination of any point along
the axis of observation.

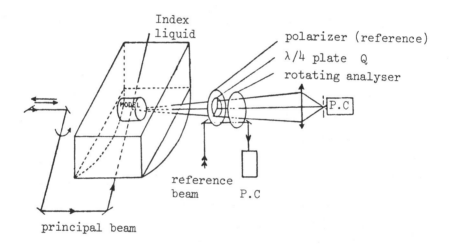

Fig. 10 Schematic of set-up using the scattering phenomenon as polarizer.

The assembly of compensator / quarter-wave plate / rotating analyser is
grouped on one stand. The compensator can be fixed with respect to the
quarter wave plate to assure an angle of $\frac{\pi}{4}$ between their axes. This
mounting facilitates the rapid positioning of the compensator.

Two photocells produce voltage outputs proportional to the light
intensities of the reference and active signal. A narrow-band filter,
matched to a frequency double that of the rotation of the analyser,
eliminates on one hand the DC component of the voltage, and on the other

hand the parasitic signals coming from the fluctuations of the laser, and
the background noise of the photomultipliers.

3.2 <u>Case where scattering is used as analyser</u> (Fig. 11)

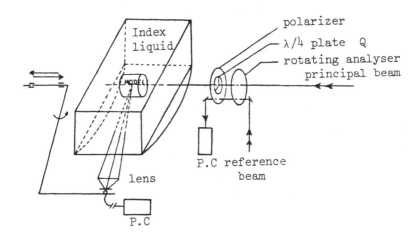

Fig. 11 : Schematic of set-up using the scattering phenomenon as analyser.

We use the same set-up, but we place a viewing system on the moving
arm. In order to avoid problems associated with poor quality of quarter-
wave plates, we have replaced them with compensator adjusted for a quarter
wave.

An automatic version of the appooaratus has been designed. The duration
of one measurement is approximalety 30 sec. We should note that the
sensitivity of the method is of the order of $\lambda/100$ for the birefringence
and one to two degrees for the principal directions and the rotatory power.

A silica powder is mixed in the epoxy resin to enhance its light
scattering properties, thus we can use a low power laser (15 mW).

4   <u>Example of determination of stress-tensor</u>
      <u>Prismatic bar under torsion</u> |16|

The ability of knowing with precision the mechanical parameters α and
$(\sigma' - \sigma'')$ at every point of any plane of the model affords the possibility
of determining the stress tensor.

The problem of a prismatic bar under torsion of equilateral-triangle
cross section, well known theoretically, present a most general case of

rotation of secondary principal directions along the incident and
scattered light beams.

Theoretical review : The cross section is an equilateral triangle of
height 3a (Fig. 12). Let o be the mass-center of the cross section oxy and M
the torsion along $\vec{z}$. The matrix form of the stress tensor is

$$T = \begin{vmatrix} 0 & 0 & \tau_{xz} \\ 0 & 0 & \tau_{yz} \\ \tau_{xz} & \tau_{yz} & 0 \end{vmatrix}$$

with $\tau_{xz} = ky \ (x + a)$

$\tau_{yz} = \dfrac{k}{2} \ (x^2 - 2ax - y^2)$ $k = \dfrac{5M}{9\sqrt{3}.a^5}$

We choose to determine the stress tensor along the line AM, $y = \dfrac{a}{2}$

$z = 0$

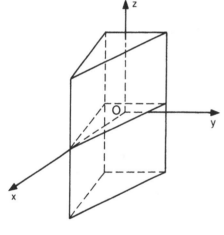

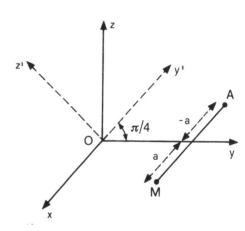

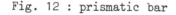

Fig. 12 : prismatic bar   Fig. 13 : the stress tensor is
                                                          determined along AM

We can easily see that the principal stresses are zero for the plane
parallel to xoy. On a plane parallel to xoz they correspond to a state of
pure shear but the principal directions are constant along a direction
parallel to $\vec{y}$. The situation is analogous for the plane parallel to yoz
and the direction parallel to ox.

In order not to limit the generality of the example chosen we use a new system oxy'z' obtained from system oxyz by a rotation of $\frac{\pi}{4}$ about axis ox (Fig. 13). The matrix form of the stress tensor becomes.

$$T' = \begin{vmatrix} 0 & \tau_{xy'} & \tau_{xz'} \\ \tau_{xy'} & \sigma_{y'} & 0 \\ \tau_{xz'} & 0 & \sigma_{z'} \end{vmatrix}$$

with :

$$\sigma_{y'} = -\sigma_{z'} = \frac{k}{2}\left[\, x^2 - 2ax - \frac{1}{2}(y' - z')^2 \,\right]$$

$$\tau_{xy'} = \tau_{xz'} = k.\frac{\sqrt{2}}{2}(y' - z')(x + a)$$

We can easily verify that the direction of the secondary principal stresses on a plane parallel to xoz' vary considerably along oy'. The same situation occurs with regard to plane oxy' along oz'.

We recall that the determination of the stress tensor in matrix form in a system oxy'z' at a point M along the line $y = \frac{a}{2}$ and $z = 0$ consists of two steps :

- The first step consists of determining the direction and difference of the secondary principal stresses at points P_i'', M_i, P_i' and Q_i'', M_i, Q_i' on plane AMz' and AMy', respectively. (Fig. 14). We compute for the two sets of points the couples of value $(\sigma_{z'} - \sigma_x, \tau_{xz'})$ and $(\sigma_x - \sigma_{y'}, \tau_{xy'})$, respectively. We follow the same procedure for planes y'M_iz'in order to obtain at points M_i the values $\sigma_{y'} - \sigma_{z'}, \tau_{y'z'}$

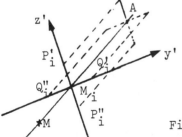

Fig. 14 : Measurement planes and mesh points

The second step consists of integrating the first equation of
equilibrium by the procedure already discussed in Ch. 3, sect. 1, which
provides σ_x (M) ; we have here σ_x (A) = 0 on the free boundary. We then
determine $\sigma_{y'}$ and $\sigma_{z'}$ from the values of $\sigma_{z'} - \sigma_x$ and $\sigma_{z'} - \sigma_{y'}$ obtained
previously. The stress tensor is then completely determined.

Experimentally we have used a mesh such that $P''_i M_i = M_i P'_i = Q''_i M_i = M_i Q'_i$
= 1,5 mm (Fig. 14). Points M_i are 2 mm apart. For each point the optical
characteristics are related to sheets of 2 mm thickness.

The couples of values $\sigma_x - \sigma_{y'}$, $\tau_{xy'}$ can be obtained without ambiguity
of sign by determining experimentally the orientation of the fast axis of
the corresponding birefringent in the slice considered.

The values of the stresses obtained were normalized by the maximum
value of the shear in the section equal to $3 \, k \, \dfrac{a^2}{2}$.

The principal stresses obtained by diagonalizing the previously
determined tensor are shown in Fig. 16. The agreement with the theorical
values is satisfactory.

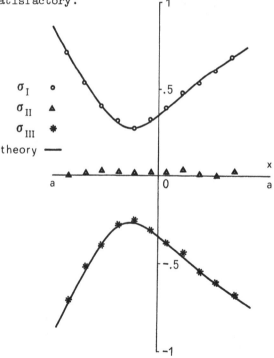

Fig. 15 : Values of principal stresses σ_I, σ_{II}, σ_{III}
(along line $y = \dfrac{a}{2}$, $z = 0$)

5   Stresses in the base of a turbine blade |16|

At the request of an industrial sponsor we undertook to study the
stress distribution in a well-defined zone in the vicinity of point M
(Fig. 16, where x(M) and z(M) are shown, the third being y(M) = 7 mm). This
heavily stressed zone is located below the joint of the blade and the
turbine shaft in the vicinity of the first tooth of the base of the "pine
tree" (dovetail joint) and near the free surface ; its volume is of the
order of several cubic centimeters.

The mechanical loading visualized is the centrifugal force acting on
the blade. The characteristics of the equivalent torsion are given.

We designed the photoelastic model and the loading fixture with a view
to approaching the actual conditions as much as possible.

The blade has a complex geometry. It consists of a thin wing and a
thicker base in the forme of a "pine tree" which allows the junction with
the engine shaft. A grooved housing on a circular arc is provided on the
carrying shaft in a shape matching that of the blade. Mounting is accomplished by sliding the "pine tree base" into its housing.

The photoelastic model made of Araldite F consists of two parts :

- The blade proper limited to its lower third (40 cm approximate
height).

- A massive anchoring containing the grooved housing.

The molding of this assembly of a complex geometrical shape was
conducted by adhering to two fundamental conditions :

- Geometrical precision for correct assembly

- Absence of residual stresses in the model

For reason of easier removal from the model the anchorage was made in
three parts : a support block and two "jaws" forming the sides of the
grooved housing.

The loading system consisted   the following :

- An anchoring base proper

- An upper aluminium plate (AU 4G) of 20 mm thickness connected to the
base by six columns of large section, on which is placed an adjusting
screw.

- A small tension rod equipped with electrical strain gages for

calibration purposes. This rod transmits the effect of the screw to a
metallic plate placed at the upper part of the blade.

This plate allows the distribution of the tensile force over the entire
blade to which it is connected by means of a sheet of glass cloth reinfor-
ced epoxy.

It should be noted that the lower part of the mounting made entirely
of araldite allows the illumination of any point in the zone of interest
along three orthogonal directions.

The assembly of model and loading fixture is mounted on a platform
with two micrometer slides at two orthogonal directions.

The precision of positioning is of the order of approximately 5/100 mm.
The connecting elements between the assembly and the micrometer slides are
designed to allow for easy orientation of the model along the three
orthogonal directions of study.

We have determined the optical characteristics along three orthogonal
directions for approximately 150 points within the zone of interest, i.e.,
approximately 450 measurements (Fig. 17).

The coordinates of the points of study were measured with respect to
the coordinates xyz of the reference point M. The mesh intervals were
2 mm along directions Ox and Oz and 4 mm along Oy.

The indexing is done experimentally in a simple fashion using the
abrupt discontinuity at the level of light scattering upon passage through
the boundary between the index liquid and the model.

At every point studied we determined the optical characteristics
associated with sheets of 2 mm thickness of normal directions $\vec{x}$, $\vec{y}$, $\vec{z}$. We
have computed $(\sigma_y - \sigma_z)$, τ_{yz}, $\sigma_x - \sigma_z$, τ_{xz}, $\sigma_x - \sigma_y$, τ_{xy}.
The results shown in Fig. 18 represent the value of the equivalent
stress σ_e in the Von Mises sense at various points of the grid for a
tensile load of 1470 daN.

The araldite F used has a photoelastic constant of 29.3 Brewster, a
young's modulus of 370 daN/mm^2, and a Poisson's ratio of 0.37.

We observe that in the entire field of study the equivalent stresses
varied in relative magnitude from one to three.

These results have allowed the verification of a computer program.

They can also be carried over to the prototype piece which is metallic
and has Poisson's ratio close to that of the model. The stress tensor was
not determined because it would have required a larger number of measurements all along a path of integration, starting from a point on the free
boundary.

It is important to emphasize that for the load applied, the deformation
of the base remain very low and are comparable to those obtained under the
actual conditions of encasement of the "pine tree" in the turbine disk.
This condition could not have been realized with the stress freezing
technique which could, among other things, disturb the loading conditions at
the level of the "teeth" at the base of the blade.

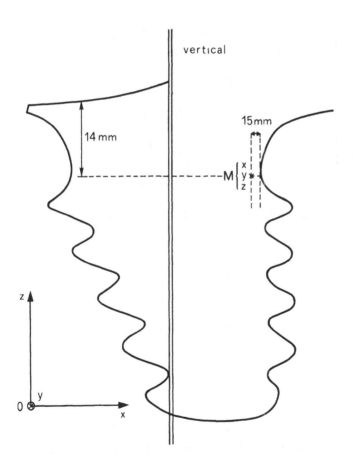

Fig. 16 : Reference system of the studied zone.

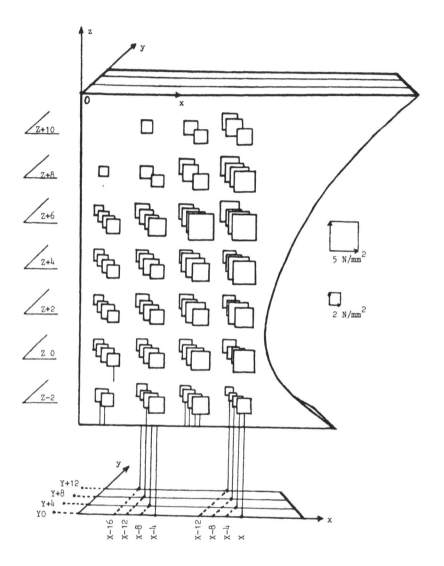

Fig. 17 : Values obtained for the equivalent stresses
at the nodes of the grid of the zone of
study.

References

1. Brillaud J., Lagarde A., "Mesure des formes de lumière en diffusion et applications" (C.R.Ac.Sc. Paris t. 281, 20.10.1975. série B).

2. Lagarde A., Brillaud J., "Construction et mise au point d'un appareil permettant la détermination des lignes neutres et de la biregringence en tout point d'un feuillet plan d'un fluide birefringent en écoulement laminaire permanent ou d'un solide photoélastique" (compte-rendu final de contrat D.G.R.S.T. n° 71 7-2886) janvier 1976.

3. Brillaud J., "Contribution au developpement des méthodes ponctuelles en photoélasticité tridimensionnelle" Thèse de 3ème cycle Université de Poitiers, 1978.

4. Lagarde A., "On some aspects in the development of photoelastic measurements" Symposium I.U.T.A.M. "Optical methods in mechanics of solids" Poitiers September, 1979, Ed. A. Lagarde (Sijthoff Noordhoff).

5. Brillaud J., Lagarde A., "Méthode ponctuelle de détection linéaire en photoélasticimétrie tridimensionnelle. Réalisation d'un photoélasticimètre et application" Communication à la 6ème conférence internationale" Experimental stress analysis", Munich, September 1973.

6. Brillaud J., Lagarde A., "Point by point methods of three-dimensional photoelasticity utilizing the phenomenon of light scattering as or polarizer or an analyser" VIII th. Symposium on Experimental Research in Mechanics of Solids, Varsovie 1978.

7. Brillaud J., Lagarde A., "Punctual determination of stress deviator in three-dimensional photoelasticity" 14th congres I.U.T.A.M. Delft, September 1976.

8. Gross-Petersen, "A compensation method in scattered light photoelasticity" I.U.T.A.M. Symposium. "The photoelastic effect and its applications" Bruxelles, September 1975, Springer verlag.

9. Lagarde A., "Progrès dans la mesure des grandeurs optiques en milieu tridimensionnel solide ou fluide birefringent". Conférence Générale - Congrès Canadien de Mécanique appliquée. Moncton 7-12 juin 1981.

10. Robert A., Guillemet E., "Nouvelle méthode d'utilisation de la lumière diffusée à trois dimensions" R.F.M. n° 5/6 p. 147-157 (1963).

11. Lagarde A., Brandt A.M., Brcic V., Laermann K.H., "Modern experimental methods of strain-stress and displacement evaluation in continous media" - conferences C.I.S.M. Udine, September 1975, Lagarde A. coordonnateur.

12. Lagarde A., Brillaud J., Oheix P. "Détermination de la birefringence et du pouvoir rotatoire équivalents à un milieu épais - Application à la photoélasticimétrie" - C.R.Ac.Sc., t. 282, (2 février 1976), série B.

13. Lagarde A., Oheix P., Brillaud J., "Determination of the birefringence and rotatory power characterizing a thick medium. Applications in photoelasticity" Mech. res. comm. vol. 3, p. 107-112, 1976.

14. Brillaud J., Lagarde A., "Méthode ponctuelle de photoélasticimétrie tridimensionnelle" "Application". R.F.M. n° 82. 1982.

15. Brillaud J., Lagarde A., "Photoélasticimétrie ponctuelle, automatique, à détection linéaire. Application". VIIth international conference on experimental stress analysis. Haifa - Israël - 23-27 avril 1982.

16. Brillaud J., "Mesure des paramètres caractéristiques en milieu photoélastique tridimensionnel. Réalisation d'un photoélasticimètre automatique. Application". Thèse de doctorat d'état - Poitiers. 1984.

17. Robert A., Royer J., "Principe de mesure des birefringences à l'intérieur d'un solide transparent en vue de son application à la photoélasticimétrie tridimensionnelle". C.R.Ac.Sc., Série B, t. 281 - 1975. p. 373-378.

CHAPTER 5

A THREE-DIMENSIONAL WHOLE-FIELD PHOTOELASTIC METHOD OF OPTICAL
SLICING WITH APPLICATION TO FRACTURE MECHANICS

The basis of this new method is the optical isolation of a slice by
two plane parallel beams emitted from a laser. The image of the slice can
be utilized to conduct an analysis analogous to that obtained in a linear
light field polariscope.

The properties of coherence and polarization of the diffused light are
used to show that the optical characteristics of the slice are taken into
account by the correlation factor of the scattered radiations from its two
faces. The use of polychromatic illumination allows the improvement of
fringe contrast.

After some illustrative examples, a study of a semi-elliptical surface
crack is given. Using our new method, to our knowledge, we have obtained
for the first time measures of K_I and σ_{on} for a material of Poisson's ratio of 0.37.

1. <u>Fundamental concept</u> | 1, 2, 3, 4 |

We isolate a slice of the photoelastic model by two plane parallel
sheets of light emitted from the same laser beam (fig. 1). The two-dimensional scattered light field is analysed in the direction z orthogonal to
the plane of the two illuminated sections. At the level of the point where it is emitted, the radiation is linearly polarized along a direction
perpendicular to the direction of incidence and analysis (Rayleigh's law).

To get an idea of the phenomenon, let us limit ourselves to the simple case where the slice is represented by a simple birefringent plate.

If the scattered radiations are coherent, the possibilities of interference of the radiations emitted by the first and second illuminated
sections of the model depend on the birefringence of the isolated slice.
If its birefringence is zero or if one of the principal directions is pa-

rallel to that of polarization of the scattered light, the polarization
of the light emitted by the first section is not modified by traversing
the slice ; the two radiations have the same polarization, therefore they
can interfere. By contrast, if the angular birefringence is an odd multiple of π and if one of its principal directions makes an angle of $\frac{\pi}{4}$ with
the direction of polarization, the radiation emitted by the first section
becomes orthogonal to that emitted by the second section and cannot interfere with it.

    In practice, the possibilities of interference of these two radia-
tions will be translated in the image plane of the middle plane of the
slice formed with the aid of an optical system of axis z, by a special
modulation of a speckle (granular) field.

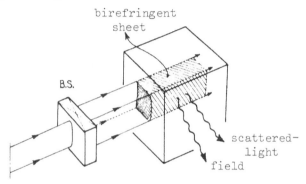

Fig. 1 Schematic of Optical slicing

2. Speckle monochromatic fields |4, 5, 6 |

    We will now describe the statistical properties of the speckle-field as
well as the optical filtering necessary for the application of the optical
slicing method.

2.1 Description of phenomenon

    The speckle phenomenon was observed mainly after the advent of the la-
ser. It is characterized by a random distribution of light spots.

    This phenomenon is produced by coherent illumination. For example, a
speckle pattern is formed if light transmitted from a scattering medium is
collected on a screen with the help of an optical system (which could be
the eye).

    This construction of light is the consequence of interference of wa-

ves coming from different points of the scattering medium and leading to
regions of maximum or minimum intensity. These interferences are possible
in the case of utilization of an optical system because the light rays
emitted from two neighbouring points interfere at the level of the common
part of their spots of diffraction. The random character of the phase of
light vibrations coming from a scatterer is found again at the level of
light distribution on the screen (fig. 2).

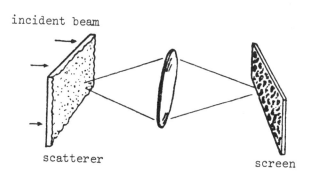

incident beam

scatterer   screen

Fig. 2   Formation of a speckle field

2.2 Statistical properties |4, 5|

The scattering medium behaves as a large spatially coherent source.
For the scattered light the phase relationship between the various
points is a definite function. The illumination received at every point
of the plate is also a definite function which could be determined by
knowing the transfer function of spatial frequencies of the optical system and the law of distribution of complex amplitudes at the level of
the scattering medium. In fact, one does not wish to know this complicated distribution because for applications we have recourse to a statistical treatment.

Let us denote by x_o, y_o and x, y the systems of orthogonal axes
which define the scattering plane and the screen respectively. Further,

let us designate by $a(x_o, y_o) = |a(x_o, y_o|e^{i\theta(x_o, y_o)}$ the complex ampli-
tude of the monochromatic wave transmitted from the scattering medium which
is assume to be linearly polarized. We considere the case where the
phase variations introduced by scattering correspond to a large number of
wave lengths such that the phase $\theta(x_o, y_o)$ is then a random variable of
equal probability over the range $(-\pi, +\pi)$. We assume the random process
to be stationary and ergodic, generating the random variables $|a(x_o, y_o)|$
and $\theta(x_o, y_o)$ which allows the equating of the spatial and statistical
means of the complex amplitude.

Let us designate by $< |(a(x,y)| >$ the spatial mean of the absolute
value of the amplitude :

$$< |a(x,y)| > = \lim_{L\to\infty} \frac{1}{L^2} \int_{-L/2}^{+L/2} \int_{-L/2}^{+L/2} |a(x,y)| \, dx \, dy$$

In practice the mean values are taken over a small domain which is large
in relation to the grain.

Let us designate by $E(|a(x,y)|)$ the statistical mean :

$$E(|a(x,y)|) = \int_{0}^{+\infty} |a(x,y)| . f(|a(x,y)|.d(|a(x,y)|)$$

where $f(|a(x,y)|$ is the probability density function of the amplitude
modulus.

The complex amplitude is apparently linear since no element that
could modify the polarization enters in the path. This amplitude,

$$A(x,y) = \sqrt{I(x,y)}.e^{j\psi(x,y)}$$

is obtained on the observation plane of the speckle field as the sum over
a domain of the complex amplitude existing at the exit of the scattering
medium. The extent of the domain depends on the mode of formation of the
speckle field.

We can show that the illumination $I(x,y)$ and the phase $\psi(x,y)$ of
the speckle field are random independent variables, that the probability
density of the phase is constant, and that the illumination obeys a negative exponential law.

The calculation of the variance σ_1^2 of the illumination of the speckle field shows that : $\sigma_I^2 = <I^2 (x,y)> - <I (x,y)>^2$

$$\sigma_I^2 = E\left(I^2\right) - \left(E\left(I\right)\right)^2$$

$\sigma_I^2 = <I (x,y)>^2$

we deduce that the contrast of the speckle field defined by the ratio of
the standard deviation over the mean value of the illumination is equal
to unity :

$$\rho = \frac{\sigma_1}{<I (x,y)>} = 1$$

We could easily apply the preceding results to the case where the
wave would be elliptically polarized provided it was nearly constant locally at the exit of the scatterer. In section 3 we demonstrate this
process for a polarization introduced along the path of the scattered
radiation.

2.3 Demonstration of contrast of a speckle field

We recall this technique of demonstrating the contrast of a granular field described by Eliasson and Dandliker. They take into account
in their article |6| the influence of gamma characteristic of the film
on which the field is recorded. We will limit ourselves here to the use
of a film acting as a quadratic receiver. The amplitude transmittance
$t(x,y)$ of this type of emulsion as a function of the illumination received is of the form (fig. 3) :

$t(x,y) = b - c\tau I(x,y)$

where b and c are constants, τ is the exposure time and $I(x,y)$ is the
illumination received.

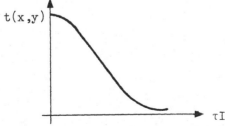

Fig. 3 Amplitude transmittance of film as a function of the received
        illumination

Let us express the received illumination as a function of the sum of a constant part and a centrally fluctuating part :

$$I (x,y) = <I (x,y)> + \tilde{I} (x,y)$$

$$<\tilde{I} (x,y)> = 0$$

We note that : $\sigma^2 = <|I - <I>|^2> = <|\tilde{I} (x,y)|^2>$

The film plane on which the speckle field is recorded is placed on an optical filtering stage (type Marechal) shown schematically in fig. 4

Photographic plate

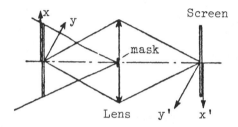

Fig. 4   Arrangement of optical Filtering

The film plane is illuminated by a convergent beam, and its image is formed by means of a lens whose center is found at the point of convergence of the incident beam.

The transmitted amplitude consists of the sum of a constant part and an oscillating part :

$$a (x,y) = a_o t(x,y) = a_o (b - c\tau<I (x,y)>- a_o c\tau \tilde{I} (x,y),$$

where a_o is the value of the incident amplitude. The Fourier transform of this amplitude is visualized on the plane of the lens forming the image. It follows that all the energy furnished by the continous component is concentrated in the central order, whereas the energy diffracted by the oscillating component of the transparency is distributed over the extent of the spectrum of the speckle field.

Let us place a mask at the center of the lens in a manner to filter out the central order. In this case only the oscillating component of the amplitude transmitted by the film reaches the image plane. Now express the mean value of the received illumination at the image plane as :

$$<\xi (x',y')> = a_o^2 \, c^2 \, \tau^2 \, <(\tilde{I} (x,y))^2>$$

$$<\xi (x',y')> = a_o^2 \, c^2 \, \tau^2 \, \sigma^2$$

$$<\xi (x',y')> = a_o^2 \, c^2 \, \tau^2 \, <I (x,y)>^2 \, \rho^2$$

The mean illumination received on the image plane is then proportional to
the square of the contrast of the speckle field recorded on the plate (a
condition that the mean illumination of this field be constant).

3. Calculation of the correlation factor of image fields of scattering
    sections |6, 8 |

Consider two points M_1 and M_2 lying on the first and second sections
illuminated such that their images form on the same point of the observation plane (fig. 5).

We have seen that in the very general case the thin slice and the thick
medium which follows it along the direction of observation $\vec{z}$, could, each
of them, be represented by a simple birefringent plate followed by a rotatory power.

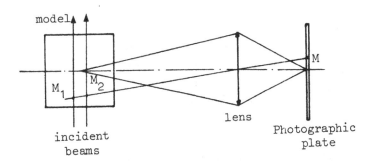

Fig. 5 Arrangement for collecting the characteristics of the sheet isola-
        ted by the two beams

Let us designate the axes x_1, y_1 of the simple birefringence
plate associated with the thin slice the angular birefringence by ϕ
and the rotatory power by R. Let us set $\alpha = (x, x_1)$ and $(x_1, x_2) = R$

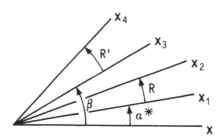

Fig. 6   Orientation of Axes of Elements Encountered Along the Scattering

Beam

In the same manner, for the birefringent medium located after the slice, let us indicate the axes of the birefringent plate as x_3, y_3 ; the angular birefringence of the plate by ϕ' and the rotatory power of the medium by R'. Let us set $\beta = (x, x_3)$ and $(x_3, x_4) = R'$.

The rays emitted from a point of a scattering section and converging at the image point are limited by a small solid angle and have for this reason traversed the same elements.

The complex vectorial amplitude scattered by point M_1 in the direction Z is put in the form :

$$a_1 (x,y) = |a_1 (x,y)| e^{j\theta_1 (x,y)} \vec{y}$$

After the action of the birefringent plate of the thin slice the complex amplitude is expressed a vector of Jones in   the x_1, y_1 axes

$$|a_1 (x,y)| e^{j\theta_1(x,y)} \begin{vmatrix} \sin \alpha^* \\ \cos \alpha^* e^{-j\phi^*} \end{vmatrix}$$

At the exit of the thin slice this expression represents the complex vectorial amplitude in the x_2, y_2 axes. The expression for the amplitude on the x_3, y_3 axes is

$$|a_1 (x,y)| e^{j\theta_1(x,y)} \begin{vmatrix} \sin \alpha^* \cos (\beta - \alpha^* - R) + \cos \alpha^* \sin (\beta - \alpha^* - R) e^{-j\phi^*} \\ - \sin \alpha^* \sin (\beta - \alpha^* - R) + \cos \alpha^* \cos (\beta - \alpha^* - R) e^{-j\phi^*} \end{vmatrix}$$

At the exit of the model the complex amplitude on the axes x_4, y_4

is written as

$$|a_1(x,y)| \cdot e^{j\theta_1(x,y)} \begin{vmatrix} \sin \overset{*}{\alpha} \cos (\beta - \overset{*}{\alpha} - R) + \cos \overset{*}{\alpha} \sin (\beta - \overset{*}{\alpha} - R) \, e^{-j\overset{*}{\phi}} \\ - \sin \overset{*}{\alpha} \sin (\beta - \overset{*}{\alpha} - R) e^{-j\phi'} + \cos \overset{*}{\alpha} \cos (\beta - \overset{*}{\alpha} - R) \, e^{-j(\overset{*}{\phi}+\phi')} \end{vmatrix}$$

We consider that the variation of birefringence on the (x,y) plane is produced slowly with respect to that of the phase term $\theta_1(x,y)$, i.e., the variables $\overset{*}{\alpha}$, β, $\overset{*}{\phi}$, ϕ' and R are constants in the small zone contributing to the formation of a grain. At the level of the latter the polarization is that of the corresponding object point since no element capable of altering the polarization intervenes along the path after the birefringent medium.

We can then write the complex vectorial amplitude on the x_4, y_4 axes on the image plane in the form

$$|A_1(x,y)| e^{j\psi_1(x,y)} \begin{vmatrix} \sin \overset{*}{\alpha} \cos (\beta - \overset{*}{\alpha} - R) + \cos \overset{*}{\alpha} \sin (\beta - \overset{*}{\alpha} - R) \, e^{-j\overset{*}{\phi}} \\ - \sin \overset{*}{\alpha} \sin (\beta - \overset{*}{\alpha} - R) e^{-j\phi'} + \cos \overset{*}{\alpha} \cos (\beta - \overset{*}{\alpha} - R) e^{-j(\overset{*}{\phi}+\phi')} \end{vmatrix}$$

where $A_1(x,y)$ and $\psi_1(x,y)$ are random variables.

In the same manner the vectorial amplitude of the second section is written on the (x_4, y_4) system as

$$|A_2(x,y)| e^{j\psi_2(x,y)} \begin{vmatrix} \sin \beta \\ \cos \beta \cdot e^{-j\phi'} \end{vmatrix}.$$

Let us designate by $A(x,y)$ the complex amplitude of the resulting field

$$A(x,y) = A_1(x,y) + A_2(x,y)$$

We calculate the illumination $I(x,y)$ of the speckle image field by noting

$$I_1(x,y) = |A_1(x,y)|^2 \quad \text{and} \quad I_2(x,y) = |A_2(x,y)|^2$$

$$I(x,y) = |A(x,y)|^2 = I_1(x,y) + I_2(x,y)$$

$$+ 2\sqrt{I_1(x,y)}\sqrt{I_2(x,y)} \sin \alpha^* \sin (\alpha^*+R) \cos (\psi_2-\psi_1)$$

$$+ 2\sqrt{I_1(x,y)}\sqrt{I_2(x,y)} \cos \alpha^* \cos (\alpha^*+R) \cos (\psi_2-\psi_1+\phi^*)$$

or also $I(x,y) = I_1(x,y) + I_2(x,y) + 2\left(\sqrt{I_1}\sqrt{I_2}\sqrt{\cos^2 R - \sin 2\alpha^* \sin 2(\alpha^*+R)}\right.$

$$\left. \sin^2 \frac{\phi^*}{2} \cos (\psi_1-\psi_2-\eta)\right\}$$

with η a function of α^*, ϕ^*, R ; ψ_1 and ψ_2 functions of x and y.

The correlation factor γ of these two fields in then expressed as

$$\gamma = \sqrt{\cos^2 R - \sin 2\alpha^* \sin 2(\alpha^*+R)} \sin \frac{\phi^*}{2}$$

It is remarkable that the factor γ is a function only of the characteristics of the sheet defined by the two illuminated sections independent of perturbations introduced by the medium between the sheet and the
boundary model.

4. Consideration of correlation factor of the Field-Images of the
Scattering Sections $|4,6, 7|$

Let us calculate the mean value of the illumination because of the
assumptions of ergodicity and the independence of the random variables defining the complex amplitudes we can write :

$$<I(x,y)> = E(I) = E(I_1) + E(I_2)$$

$$+ 2\gamma E(\sqrt{I_1}) E(\sqrt{I_2}) E(\cos (\psi_1-\psi_2-\eta))$$

Now $E(\cos (\psi_1-\psi_2-\eta)) = 0$ since the phases are independent and of
equal probability over $(-\pi, +\pi)$; we deduce :

$$<I(x,y)> = <I_1(x,y)> + <I_2(x,y)>$$

Let us calculate the variance of the illumination using the same
assumptions we write :

$$\sigma^2 = <(I-<I>)^2> = E((I-<I>)^2)$$

$$\sigma^2 = E\left((I_1 + I_2 + 2\gamma \sqrt{I_1} \sqrt{I_2} \cos(\psi_1 - \psi_2) - E(I_1) - E(I_2))^2\right)$$

$$\sigma^2 = E\left((I_1 - E(I_1))^2\right) + E\left((I_2 - E(I_2))^2\right) + 4\gamma^2 E\{I_1 I_2 \cos^2(\psi_1 - \psi_2)\}$$

$$+ 2E\,(I_1 - E(I_1))\,E(I_2 - E(I_2)) + 4\gamma\,E(\sqrt{I_1} \sqrt{I_2}\,(I_1 - E(I_1) + I_2 - E(I_2))$$

$$E\,(\cos(\psi_1 - \psi_2 - \eta))\,.$$

then, $E\left(\cos^2(\psi_1 - \psi_2)\right) = \dfrac{1}{2}$

It remains therefore to go back to the spatial means

$$\sigma^2 = \sigma_1^2 + \sigma_2^2 + 2 <I_1> <I_2> \gamma^2$$

utilizing the fact that the contrast of the speckle field, which could be obtained starting from each scattering plane, is equal to unity, we obtain

$$\sigma^2 = <I_1>^2 + <I_2>^2 + 2\gamma^2 <I_1> <I_2>$$

We can easily calculate the contrast of the resulting speckle field by assuming the mean illuminations to be equal :

$$<I_1> = <I_2>$$

$$\rho^2 = \frac{1+\gamma^2}{2}$$

The contrast of the speckle field then takes into account the correlation factor, i.e., the optical characteristics of the sheet.

5. Principle of the Method and Expected Results |4, 8|

From what we have said the method consists of isolating a thin sheet by two plane beams scattering the same intensity at every point. The film on which the speckle image will be recorded after filtering of the central order gives a mean illumination $<\xi>$ proportional to the square of the contrast

$$<\xi> = k\,\{1 + \left(\cos^2 R - \sin 2\alpha^* \sin 2(\alpha^* + R) \sin^2 \frac{\phi^*}{2}\right)\}$$

This expression is, within a constant term, identical to the illumination given by a linear polariscope with light field for the study of the sheet (cf. sect. 1.5).

In the case where the sheet is reduced to a birefrigent plate   (R=0)
we observe then a pattern of isochromatics and zero isoclinic, $\alpha=0$.

According to the results of Chap. 2, we know that if $R < \frac{\pi}{6}$ we must
observe an isoclinic zone which gives the locus of points where the secondary principal directions are, on the median plane of the sheet, parallel to x. The peaks of the isochromatics represent the loci of points
where $\phi = k\pi$.

Let us evaluate the contrast $\mathscr{C}$ of the fringes :

$$\mathscr{C} = \frac{<\xi>_{max} - <\xi>_{min}}{<\xi>_{max} + <\xi>_{min}} = \frac{\rho^2_{max} - \rho^2_{min}}{\rho^2_{max} + \rho^2_{min}}$$

The peaks of ρ^2 corresponding of 0 and 1 of γ

$$\rho_{max} = 1 \; ; \; \rho_{min} = \frac{1}{\sqrt{2}}$$

Therefore $\mathscr{C} = \frac{1}{3}$

The fringe contrast is weaker than in a classical polariscope (1/3
instead of 1).

6. Application and Results |4, 6, 10|

6.1 Set-up for obtaining a constant intensity from the scattering sections

We have seen that the information sought is provided by the contrast
of the speckle field when the illuminated sections scatter the same intensity at all points.

This condition is not realized along a polarized laser beam traversing a loaded model. In fact, the polarization of the existing light at
a point in the beam depends on the birefringence encountered along the
beam. When the observation is done normal to the direction of incidence,
the scattering being treated by linear analysis, the incident beam appears
in a discontinuous manner. When we illuminate a plane section of a loaded
photoelastic model, we observe in the same manner a modulation of the scattered field. This phenomenon which forms the basis of Weller's method is
an inconvenience in the framework of our study. For this reason we have ap-

plied a technique of spatial synthesis of the natural light which makes use of a Babinet compensator |4| and allows the determination of a constant scattered intensity for whatever birefringence encountered along the incident beam.

6.2 Photographic Setup

The source of coherent radiation is a Krypton laser of 600 mW luminous power. The beam splitting apparatus used to form two parallel sheets of light consists of (Fig. 7) :

- A half wave plate allowing the rotation of linear polarization of the incident beam
- A lens of 10 mm focal length
- A wollaston prism which provides two orthogonal polarized beams (vertical and horizontal)
- A half wave plate used to bring the directions of polarization of the two beams at $\pi/4$ with the vertical
- A Babinet compensator placed horizontally and introducing a path difference of only one wavelength in order to prevent a folding of the beam
- A cylindrical lens which allows the focusing of the two beams

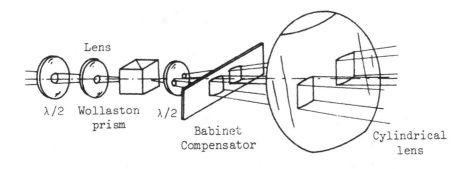

Fig. 7   Schematic of beam splitting setup for obtaining two parallel plane beams

The intensities of the two beams are equalized by means of rotation of the half-wave plate. The distance between the two light sheets is ad-

justed by the translation of the wollaston prism. The translation of the
cylindrical lens allows the focusing of the beams at the level of the model. At the point of the cylindrical lens these beams have a width of
5 mm ; their focal lines are located at 1.3 m from the lens. Under these
conditions the light sheets have a theoretical thickness of 0.15 mm over
80 mm. In practice, because of aberrations, the slice is much thicker and
extends over a bigger depth.

The images of the two illuminated sections are formed in the direction normal to their planes by means of a lens of 150 mm focal length
opened at f/4.5 on a kodak photographic plate 50253. The exposure time
is 3 min.

6.3 Filtering Setup

The film after developing and fixing is placed on a filtering setup
by reflection. This type of setup is more effective than those operating
by transmission (Fig. 8).

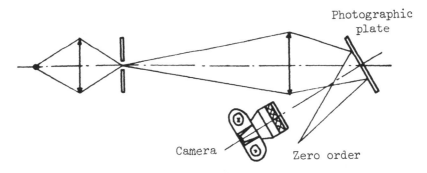

Fig. 8   Spatial Filtering Setup using reflected light

We form a slightly converging beam by means of a mercury vapour lamp
and a system of lenses. The image of the film plane on which the beam is
reflected in part, is formed by means of a camera whose front lens is located at the point of convergence of the beams.

The orientation of the axis of the film plane and camera with respect to that of the incident light, is adjusted in such a manner as to
obtain maximum visibility of the fringes.

6.4 Results

We present the results obtained in a study of a cylindrical specimen loaded in torsion.

The torsion fixture is of a simple design such that when we study a sheet at $\pi/4$ with the axis it permits a variation of the secondary principal directions along the incident beams, (which was taken into account in the theory) as well as along the scattered beams. This is the most general situation in a three-dimensional study.

The araldite D specimen and its loading fixture are placed in an immersion tank containing a fluid with matched index of refraction to the model such that its axis is orthogonal to the direction of incidence and inclined at $\pi/4$ with the vertical (Fig. 9)

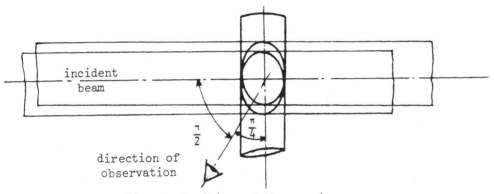

Fig. 9 Position of the specimen

Let us recall that we must obtain a fringe pattern similar to that obtained by means of stress freezing and slicing where the slice is placed between parallel and vertical polarizers. For an identical slice in our case the polarization of the light emitted at the location of the illuminated sections is linear and vertical. Figure 11 shows the corresponding fringe pattern. For a cylindrical bar under torsion and an infinitely thin slice inclined at $\pi/4$ with the axis, the theory predicts regularly spaced circular isochromatics and radial isoclinics (Fig. 10).

Figure 11 is a combination of the $\alpha = \frac{\pi}{2}$ isoclinic, and the isochromatic fringe pattern. However, the isoclinic has a width which appears abnormal, due to the relatively large rotation of the secondary principal directions through the thickness of the slice studied.

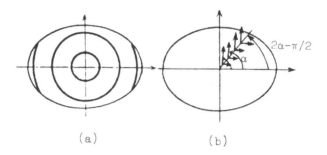

(a)   (b)

Fig. 10 Isochromatics (a) and Isoclinics (b) of an infinitely thin slice
inclined at π/4 with tha axis for a cylindrical bar under torsion

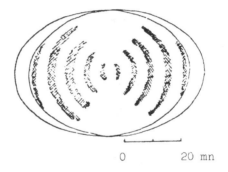

0   20 mn

Fig. 11 Fringes obtained for an optically isolated slice in a specimen
under torsion (Drawing from Photograph)

7. Improvement of Contrast |6, 9|

In order to increase the fringe contrast we make use of a technique
of improving the signal to noise ratio. This relies on polychromatic illumination leading to a replacement of filtering of the central order by
band filtering. We will limit ourselves in this report to clarify the basic ideas and show the results obtained.

7.1 Modulation of information

Let us consider two points P_1 and P_2 located on the first and second section, illuminated by polychromatic radiation.

Under the assumption of maximum interference where the correlation
factor γ is 1, the polarization of the light emitted from point P_1 re-

mains linear after traversing the slice and parallel to that emitted from
point P_2. When the observation is made along $\vec{z}$ we have two sources of polychromatic light displaced by the thickness of the slice. The analysis
of the resultant radiation in the direction $\vec{z}$ by means of a spectroscopic
arrangement, which produces a Fourier transformation in the space of temporal frequencies, produces a striated spectrum (Fig. 12). In the case
where $\gamma = 0$ the radiations emitted by the two points P_1 and P_2 cannot interfere, and the analysis of the resulting radiation by means of a spectroscope does not give rise to the striated spectrum (Fig. 13).

A(σ) and B'(σ) represent respectively, the spectral densities of the
incident radiation and the diffracted radiation energy by the spectroscope.

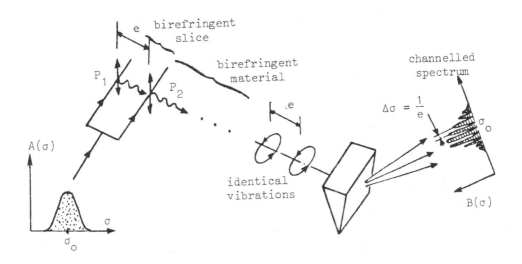

Fig. 12   Superposition of Polychromatic Radiations in the case where
          $\gamma = 1$

In order, to obtain simultaneously information over the entire sheet
studied, we form the image of the two sections, illuminated by polychromatic light, by means of the spectroscope (Fig. 14). For the regions of
the image where the correlation factor is maximum we obtain a random structure whose grains contain a sinusoïdal modulation. For the regions where
this factor is zero we obtain a random structure whose grains are simply
elongated in the direction of analysis by the spectroscope.

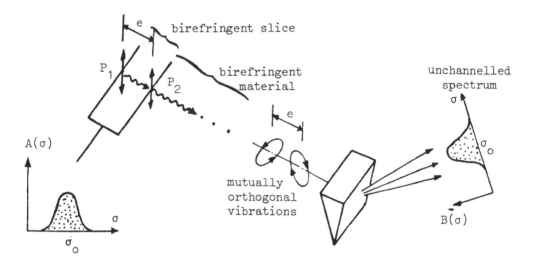

Fig. 13 Superposition of polychromatic Radiations in the case where
         γ = 0

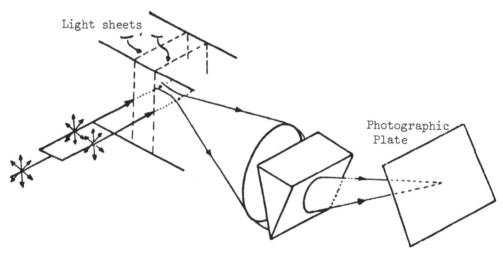

Fig. 14 Photographic recording of the spectrally Resolved Image

7.2 <u>Demonstration of the correlation Factor</u> |6, 9|

The demonstration of the image zones which show a modulation, that
is the visualization of the isochromatics and isoclinics of the sheet,
is effected by optical band pass filtering of the negative (Fig. 15). The
striated regions which diffract the light through the openings produce

light fringes corresponding to isochromatics of integer order ϕ = 2kπ and
to the isoclinic

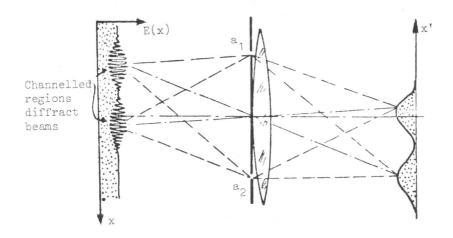

Fig. 15 Band pass filtering of the negative

The technique using polychromatic radiation shows clearly its supe-
riority at the stage of filtering of the image of the birefringent sheet.
After band pass filtering the energy collected in the filtering band
is zero when γ = 0, that is $<\xi>_{min}$ = 0. It follows that the contrast of
the filtered image is then theoretically equal to unity in the polychromatic case instead of 1/3 in the monochromatic case (Cf Sect. 5). In
practice the noise of the film tempers somewhat this result which remains
in any case better than before.

7.3 Experimental set-up [6, 10]

We described here the necessary additioned means for the application
of this method.

7.3.1 Fixture for illuminating the model

The source of polychromatic illumination consists of a dye laser
(Rhodamine 6G) pumped by an Argon ion laser. The light power is of the
order of 600 mW. The optimal spectral width is obtained by scanning with
a Lyot filter controlled by a fixture with an excentric cam.

The beam splitting fixture used for the formation of two parallel sheets of light was described in sect. 6.2.

7.3.2 Analysis fixture

The diffraction grating for the spectral analysis of the scattered
radiation is setup in a manner allowing autocollimation ; for this reason the grooves are oriented horizontally and the grating is inclined in
a manner such that the incident and diffracted rays are also on a horizontal plane.

The value of the angle of reflection on the horizontal plane was op-
timized in a manner such that the partial polarization introduced by the
diffraction in the grating is compensated by that corresponding to the
reflection from it.

The images is formed by means of a lens of 1 m focal length and an
objective lens of 300 mm focal length after reflection from a grating of
70 mm width containing 600 lines per millimeter.

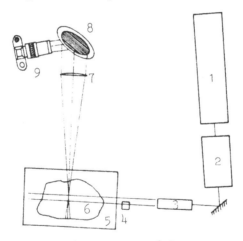

Fig. 16 Schematic of recording setup : (1) Argon laser, (2) dye laser,
        (3) Beam separation fixture, (4) Additional glass plate, (5)
        Index tank, (6) model, (7) lens, (8) diffraction grating (9)
        camera.

7.3.3 Filtering setup

The filtering setup was conceived with the following requirements in
mind :

    - to allow a good selectivity of band-pass filtering, which requires a point source

    - to provide a filtered image with the minimum possible speckle density, which necessitates a source of large spectral width

    - to produce an image without geometric distorsion.

The source selected is a xenon lamp whose spectrum is practically continuous over the entire visible range. It is spatially filtered before illuminating the negative. The front lens of a photographic objective system is placed at the point of convergence of the incident beam. This lens is masked to allow selection of the order of diffraction. In order to minimize the "noise" of the negative, the latter is immersed in glycerine between two glass plates

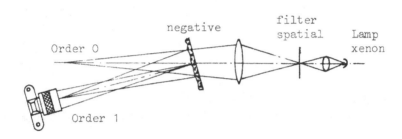

Fig. 17 Schema of Filtering setup

7.4 Example : Cylinder of square-cross section loaded in torsion about its Axis

We examine slice inclined at $\pi/4$ with the axis of torsion for different orientations of polarization of the scattered radiation and for two different thicknesses. The fringe contrast is satisfactory (Fig. 18).

    We should note that the analysis of these photographs follows the study conducted in chapter 2, $(6, 9, 10)$. For a large rotation of the secondary principal directions the isoclinic disappears (this is the case for the central zone of the photograph where the rotation is predicted by the theory).

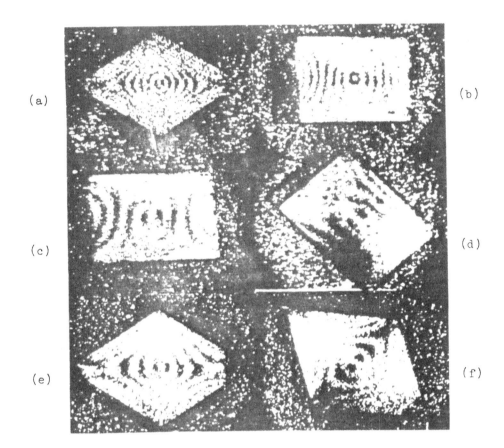

Fig. 18 Photographs of fringe pattern obtained for different slices sin-
gled out at $\pi/4$ from the axis of square beam in torsion (the thickness of
the sheet is 4.6 mm for the patterns a, b, c, d and 9 mm for the patterns
e and f). The direction of polarization of the scattered radiation is parallel to the axis of the slices (a, c, e, f) or inclined by $\pi/4$ with
these axes (b, d).

8. Study of a surface Crack [6, 11]

Among the various experimental methods, photoelasticity appears pree-
minent for the study of cracking in a three-dimensional medium. To our
knowledge, the classical method used to date is that of stress freezing
and slicing (12), which has the following major disadvantages :

   - the need to fabricate a new model for every case of loading and
above all for every crack length studied

    - a Poisson's ratio (nearly) equal to 0.5, which poses particularly delicate problems for surface cracks near a free boundary

    - the need to apply large deformations to obtain measurable birefringence.

    To avoid these difficulties we have used the method of optical slicing for the study of the semi-elliptical crack.

8.1 Characterization of photoelastic information

    We recall that photoelasticity provides a map of maximum shear stresses around the crack tip and not of the stress components. It therefore does not allow for a direct determination of the characteristic parameters K_I and σ_{on}.

    The problem then is one of establishing relations between the maximum shear and the values of the stress intensity factor and the normal stress σ_{on}. To this end we use the theoretical works of Benthem (14) and Kassir and Sih (13), for the regions near and remote from the boundary, respectively

8.1.1 Regions Remote from Free Boundary

    We assume that for the regions remote from the free boundary the stress distribution around the base of the crack is that given by Kassir and Sih for an enclosed elliptical crack (13). The maximum shear referred to the system $(\vec{n}, \vec{t}, \vec{z})$ at the base of the crack is expressed as a function of parameters K_I and σ_{on}.

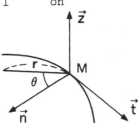

Fig. 19

$$(2\tau_m)^2 = \frac{1}{2\pi r}(K_I \sin\theta)^2 + 2\frac{\sigma_{on}K_I}{\sqrt{2\pi r}}(\sin\theta\,\sin\frac{3\theta}{2}) + \sigma_{on}^2$$

8.1.2 Regions near the Free Boundary

In order to analyse the information contained in the isochromatic patterns obtained for a sheet near the boundary we used the result by Benthem (14) dealing with a quarter infinite crack in a half-space. We have shown (16) that although the stress distributions around the crack tip were different from those predicted by Kassir and Sih, the shear distributions were remarkably close to the preceding analytical expression.

It stands to reason that, for the regions close to the boundary as well as those remote from it, we use the same expression to describe the maximum shear as a function of the two parameters K_I and σ_{on}.

8.2 Analysis of Photoelastic Data

First we observe that the analytical expression of Kassir and Sih describing the maximum shear as a function of parameters K_I and σ_{on}, in the normal plane at the botton of the crack is identical to that obtained by Irwin in two dimensions.

It follows that the techniques of analysis of the isochromatic fringe patterns used for a two-dimensional model are perfectly applicable to the plane sheet normal to the base of the crack.

We used a technique similar to that of Sanford and Dally |15|. The advantage of their method is that it uses a large number of measurements remote from the base of the crack and, in theory, minimizes errors. In using two parameters (K_I, σ_{on}) and Q measurement points we redure the problem to the solution of a rectangular system of Q equations with two unknows.

$$f_q = \frac{K_I^2}{2\pi r_q} \sin^2 \theta_q + \frac{2K_I \sigma_{on}}{\sqrt{2\pi r_q}} \sin \theta_q \sin \frac{3\theta_q}{2} + \sigma_{on}^2 - (2\tau_n)^2 = 0$$

where τ_n is the value of the shear stress obtained from the isochromatic pattern at a point q of coordinates (r_q, θ_q).

This system can only be solved numerically. The algorithm of computation uses the Newton-Raphson iterative method and least squares minimization procedure. The alogrithm, written in Basic, is solved by a microcomputer (HP 9835). The convergence of the numerical procedure is rapid and a few iterations are sufficient to obtain K_I with an error of only 1 % between two successive iterations.

8.3   Experimental study

8.3.1   Specimen dimensions

We selected a geometry which allows a comparison with the numerical results of Raju and Newman.

The rectangular cross section of the cylindrical specimen used had dimensions of 58 mm x 110 mm, the semi-elliptical crack had a length of a = 34 mm, a width of 2c = 34 mm (hence an ellipticity c/a = 1/2), and a depth defined by a/t = 0.6 where t is the thickness of the plate

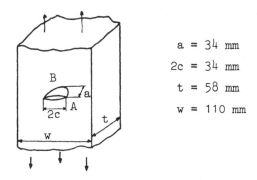

a = 34 mm

2c = 34 mm

t = 58 mm

w = 110 mm

Fig. 20   Specimen dimensions

The specimen was made using an epoxy photoelastic model material. The crack was produced during casting by including a metallic blade sharpened in the shape of a semi-ellipse. The blade was easily extracted after polymerization by cooling the blade quickly.

We were thus assured of obtaining a strictly semi-elliptical shape and were able to compare our measurements with the numerical results and, besides, a microscopic observation of the base of the crack revealed a radius smaller than 0.005 mm.

8.3.2   Fixture for translation and rotation of the model

Recall that a complete study of a photoelastic model requires a fixture for positioning and reference with six degrees of freedom (3 translations and 3 rotations). We resolved this problem by using a spherical immersion tank. A glass balloon of 500 mm in diameter is supported at its lower part by a hydrostatic bearing of water allowing an effort less orientation of the assembly of tank and model. An assembly of three translation slides is interposed at the junction of the model loading fixture and the tank. The ball-joint connection with the spherical tank allows a choice of the system of axes best suited for the study. The precision of positioning with respect to the radius of the sphere was in the order of one degree.

8.3.3 Photographic recording

The experimental procedure of referencing any sheet perpendicular to the base of the crack is a particularly simple operation thanks to the use of the orientable spherical tank. With the aid of the translation slides we bring to the center of the sphere the point of the ellipse where we wish to make a measurement. Next we orient the tank in a manner such that the viewing axis is tangent at the bottom of the crack bringing the plane of the crack into coincidence with the viewing axis (referenced with a HeNe laser).

A rotation of the sphere about the viewing axis does not modify the isolated sheet optically, but it simply changes the orientation of the model with respect to the direction of propagation of the incident beams. In this manner we can choose the isoclinic that we wish to visualize simultaneously with the isochromatic pattern. For a crack problem where only the isochromatic pattern is necessary for the analysis, we set the loading axis at $\frac{\pi}{4}$ with the direction of propagation of the incident beams in such a manner as to avoid most of the perturbation of the isoclinic zone.

We studied in detail two regions on the crack boundary which are designated as points A and B in Fig. 20.

We loaded the model with different loads corresponding to stresses at infinity of 2.46 MPa, 3.32 MPa and 3.77 MPa.

The image of the sheet studied is recorded by means of the setup described previously (Sect. 2.3) on a Kodak photographic plate 50/253 film. The exposure time used was 10 seconds for an output power of the dye laser of 400 mW.

8.3.4 Filtering and display of isochromatic patterns

The negatives obtained were filtered by means of the setup described previously (Sect. 2.4). The filtered images were recorded on a high contrast film and then printed with a magnification of ten with respect to the model. The isochromatic patterns shown correspond to sheets of 4 mm thickness isolated optically in the vicinity of points A and B for a loading of 3.77 MPa. We remark that the fringes were resolved starting at a distance from the base of the crack (taking into account the magnification) of the order of a millimeter.

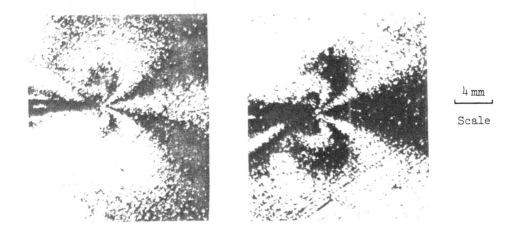

Fig. 21   Isochromatic pattern obtained for the point A for a load of
          3.33 MPa and 3.77 MPa.

We selected some thirty points of measurement in each isochromatic pattern in an annular region determined by a ratio of r/a between 0.02 and 0.2.

These data are treated following the procedure described before and yield the values of K_I and σ_{on}.

We present the results of the analysis of fringe patterns obtained at points A and B for loadings corresponding to stresses at infinity of 2.46 MPa, 3.32 MPa, and 3.77 MPa Fig. 22.

We maintain that the value of stress σ_{on} is close to that of the applied stress and near the value of the mean stress σ_x in the plane of the crack. In fact $\bar{\sigma}_x$ equals 2.86 MPa 3.86 MPa and 4.37 MPa for the three loading cases respectively ($\bar{\sigma}_x$ is calculating by subtracting the area of the crack from the specimen cross section).

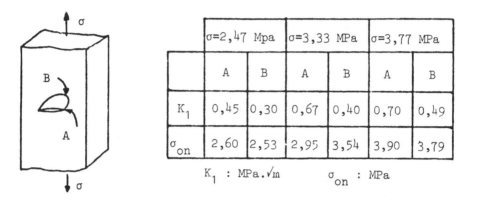

	σ=2,47 Mpa		σ=3,33 MPa		σ=3,77 MPa	
	A	B	A	B	A	B
K_1	0,45	0,30	0,67	0,40	0,70	0,49
σ_{on}	2,60	2,53	2,95	3,54	3,90	3,79

K_1 : MPa.$\sqrt{m}$ σ_{on} : MPa

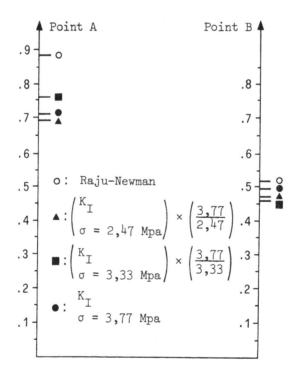

Fig. 22 : Results

To our knowledge, this is the first time that values of σ_{on} are presented from data obtained experimentally for a semi-elliptical crack.

We compared our results with those obtained numerically by Raju and Newman. First in order to normalize the measurements to only one loading of 3.77 MPa, the values obtained for 2.46 MPa and σ = 3.32 Mpa were multiplied by 3.77/2.46 and 3.77/3.32 respectively. We maintain that the relative precision between measurements made at a point of the specimen under the same experimental conditions but for different loads, is of the order of 5 %.

We note a good agreement between the experimental and numerical results for the farthest region (point B) and a difference higher than 10 % for the region close to the boundary (point A).

The importance of the observed discrepancy in our opinion, raises the question of a problem in fracture mechanic rather than a problem of measurement and computation. More precisely it is possible that these two methods take into account the influence of the boundary in a different manner and for the passage from a state of plane strain to one of plane stress. Also, Raju and Newman obtain K_I by interpolation of nodal forces and not by displacements.

References

1. Desailly R., "Visualisation of isoclinics and isochromatics in a bire-
    fringent slice optically singled out in a three dimensional model"
    Publication in the optics communications revue, vol. 19, n° 1, oct.76.

2. Desailly R., Lagarde A., "Application des propriétés des champs de gra-
    nularité à la photoélasticimétrie tridimensionnelle" C.R.Ac.Sc. t. 284
    p. 13-16, série B, 1976.

3. Desailly R., Lagarde A., "Rectilinear and circular analysis of a plane
    slice optically isolated in a three-dimensional photoelastic model"
    Mech. Research. commun., vol. 4 (2), 99-107, 1977.

4. Desailly R., "Application des propriétés des champs de granularité à la
    détermination sans capteur, des déplacements, des déformations et des
    contraintes au sein des milieux tridimensionnels" Thèse de docteur ingé-
    nieur, Poitiers 1978.

5. Dainty J.C., "Laser speckle and related phenomena" "Topics in applied
    physics", vol. 9, p. 9-26, Springer Verlag.

6. Desailly R., "Méthode non destructive de découpage optique en photoé-
    lasticimétrie tridimensionnelle - Application à la mécanique de la rup-
    ture", Thèse d'état, n° 336, Poitiers 1981.

7. Lagarde A., "Progrès dans la mesure de grandeurs optiques en milieu
    tridimensionnel, solide ou fluide, biréfringent", Conférence Générale
    CANCAM Monction 7-12, 5 juin 1981.

8. Desailly R., Lagarde A., "Sur une méthode de photoélasticimétrie tri-
    dimensionnelle non destructive à champ complet", Journal de Mécanique
    Appliquée, vol. 4, n° 1, 1980.

9. Desailly R., Froehly C., "Whole-field method in three-dimensional pho-
    toelasticity : Improvement in contrast fringes", Symposium I.U.T.A.M.
    "Optical methods in Mechanics of Solids" Poitiers september, 1979, Ed
    A. LAGARDE (Sijthoff Noordhoff).

10. Desailly R., Lagarde A., "Méthode de découpage optique en photoélasti-
    cité tridimensionnelle - Application", R.F.M. n° 1, 1984.

11. Desailly R., Lagarde A., "Etude d'une fissure débouchante par une mé-
    thode photoélastique tridimensionnelle de découpage optique", VIIth
    International Conference on Experimental Stress Analysis, Haïfa, Israël,
    23-27 avril 1982.

12. Smith C.W., "Use of three-dimensional photoelasticity in Fracture Méchanics", Exp. Mechanics, vol. 13, n° 12, 1973, p. 539-544.

13. Kassir M.K., SIH G.C., "Three dimensional stress distribution around an elliptical crack under arbitrary loadings", Journ. Applied Mechanics, t. 33, n° 3, 1966, p. 601-611.

14. Benthem J.P., Douma Th., "Graphs of three dimensional state of stress at the vertex of a quarter infinite crack in a half space", Report 677, Laboratory of Engineering Mechanics, Mekelweg 2. 2628 CD, Delft.

15. Sanford R.J., Dally J.W., "A general method for determining mixed mode stress intensity factors from isochromatic fringe pattern", Eng. Fract. Mechanics, vol. 11, 1979, p. 621-633.

16. Raju I.S., Newman J.C., "Stress intensity factors for a wide range of semi-elliptical surface cracks in finite thickness plates", Engineering fracture mechanics, t. 11, 1979, p. 817-830.

CHAPTER 6

DYNAMIC INTERFEROMETRIC PHOTOELASTICITY

We present the basis of interferometric photoelasticity before discus-
sing some applications in dynamics.

The intent of this chapter is to facilitate the understanding of the
following chapter which will concern holographic photoelasticity but also
to give an easy to use whole field method for determining stresses, for the
study of repetitive transient phenomena.

To obtain the algebraic values of the principal stresses at any point,
the photoelastic data giving information on the difference of the principal stresses are complemented by interferometric measurements giving information on the sum of these stresses. This local determination of the stress
tensor that does not require the knowledge of the boundary conditions gives
an elegant solution in dynamics ; in general, it is very attractive when
the boundary conditions are difficult to specifiy exactly (multi-layered
materials |1|, frictional contacts, one-sided connection).

Two features characterize classical interferometry :
- the use of light sources so-called thermic light sources which requires
operation near the zero path difference,
- the interference of two nearly parallel beams.

A two-wave interferometric setup permits us to divide a beam into two
different geometric paths ; in one of them, we introduce the disturbance to
be measured, the other one being used as a reference. Then we combine these
two beams so that interference records the information. The fringe formation
requires a coherence relation between the interfering beams.

To determine the disturbance produced by a given state of stress, the
method must be used in two steps. The photoelastic model is introduced in

one of the beams and the interference patterns before and after loading
are compared. The information is extracted either from point data |2| to
|5| or from the whole field |6| and |7| to |9|.

In what follows, we will assume that each beams have the same energy.

1. Favre's point method |2|

Favre's experiment performed in 1928, is based on the following principle. He used a two-beam Mach Zender interferometer and interfered with
the reference polarized beam successively the two plane polarized beams
propagating through the model (made of photoelastic glass), along the principal axis at the point in question.

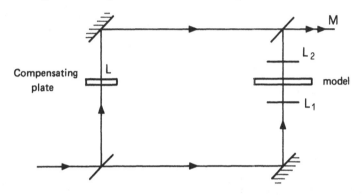

Fig. 1 Mach Zender Interferometer

To this end, the polarization direction of the incident beam is choosen
normal to the drawing plane : under these conditions, reflections and refractions do not alter the state of polarization. A half wave plate L_1 suitably oriented makes the linear incident vibration parallel to a principal
axis of the model ; the half wave plate L_2 has the inverse effect ; it
brings back the emerging linear vibration parallel to its initial direction.
These operations require the prior knowledge of the principal axis at the
point in question.

In the diagram, the reference beam propagates through the plate L which
compensates for the path difference introduced on the object beam by the
unloaded model. Because of the short coherence length of the light source
used, this condition is essential for obtaining interference. At point M,
the complex amplitudes have the same value : $e^{j\phi}$ (the interferometer adjus-

tement is said to be "at uniform field").

When the model is loadel for a direction of vibration parallel to the stress direction σ_1, the complex amplitude of the beam studied is $e^{j(\phi_1 + \varphi_1)}$ with

$$\varphi_1 = \frac{2\pi\delta_1}{\lambda} \qquad\qquad \delta_1 = e(n_1 - n_o) + \Delta e\,(n_1 - n)$$

n : index of refraction in air

n_o : index of the unloaded material

e. : thickness of the unloaded material

Δe : variation of the thickness due to the loading

The complex amplitude at point M is written as $e^{j\phi} + e^{j(\phi + \varphi_1)}$. The illumination is therefore proportional to $1 + \cos \frac{2\pi}{\lambda} \delta_1$.

δ_1 is expressed by $\delta_1 = (e + \Delta e)\,(n_1 - n) - e\,(n_o - n)$

For a problem of plane stress, according to Maxwells law, one has

$$n_1 - n_o = c_1\,\sigma_1 + c_2\,\sigma_2$$

(1)

$$n_2 - n_o = c_1\,\sigma_2 + c_2\,\sigma_1$$

and the thickness variation is :

(2) $\dfrac{\Delta e}{e} = -\dfrac{\nu}{E}\,(\sigma_1 + \sigma_2)$

From the first relation (1) it follows that :

$$\delta_1 = (e + \Delta e)\,[\,c_1\,\sigma_1 + c_2\,\sigma_2 + n_o - n\,] + e\,(n - n_o)$$

$$\delta_1 = \Delta e\,(n_o - n) + (e + \Delta e)\,[\,c_1\,\sigma_1 + c_2\,\sigma_2\,]$$

As $\dfrac{\Delta e}{e} \ll 1$ and $c_1\,\sigma_1 + c_2\,\sigma_2 \ll 1$, one has $\delta_1 \simeq \Delta e\,(n_o - n) + e[\,c_2\,\sigma_2 + c_2\,\sigma_2\,]$

Form (2) we can vrite :

$$\delta_1 \simeq e\,(c'_1\,\sigma_1 + c'_2\,\sigma_2)$$

with $c'_1 = c_1 - \dfrac{\nu}{E}\,(n_o - n)$; $\qquad c'_2 = c_2 - \dfrac{\nu}{E}\,(n_o - n)$

The illumination corresponding to the vibration parallel to σ_2 is proproportional to $1 + \cos \frac{2\pi}{\lambda} \delta_2$ with $\delta_2 = e\,(c'_1\,\sigma_2 + c'_2\,\sigma_1)$.

The measurement of absolute retardations δ_1 and δ_2 allow the determination of σ_1 and σ_2. They can be performed by a compensation method. When the path difference is larger than the wavelength, it is convenient to use white light (the retardation is compensated when the central fringe again appears bright). In such a zero method, a small coherence length is sufficient to cause the fringes to be displaced. However, accurate parallelism of the two beams is required to observe the model under normal incidence and to produce a reference wave having the same mean shape. So the lateral dimensions of the source must be smaller than the limit for which the interference fringes lost their contrast. It is a condition of spatial coherence.

Remark : The usual relation $\delta = \delta_2 - \delta_1 = Ce\ (\sigma_1 - \sigma_2)$ with $C = c_1 - c_2$ allows for verification.

2. Whole field methods

It must be stated that relatively few whole field investigations have been performed in interferometry. Various reasons can be given :
- often, the determination of isochromatics is sufficient
- the laboratory devices are rather difficult to make
- a large field interferometer is very expensive
- classical interferometry often uses a compensating plate of optical quality : namely, transparency, flatness and smoothness must be sufficient ; now the usual photoelastic materials have low hardness so their accurate machining is difficult,
- the reflections and the refractions introduce disturbances which are not often discussed even if they are well compensated for.

All these considerations perhaps explain why it was not until in 1964 that a whole field of isochromatics and isopachics (lines along which the sum of the principal stresses is equal) for a photoelastic model was obtained by Nisida and Saito |7|.

2.1 Nisida and Saito method

These authors use a Mach-Zender interferometer for a static problem whith an unpolarized monochromatic radiation source. This radiation can be resolved into two orthogonal components, each being parallel to one of the principal axes of the model at the point in question, for the measurement

beam coming from this point and for the homologous beam with which it in-
terfers at M.

If we assume the interferometer to be adjusted for an uniform field
when the model of optical quality is unloaded the complex amplitude at M is,
after loading,

$$\vec{A}_\ell = a \left(e^{j(\phi_1 + \varphi_1)} \vec{x}_1 + e^{j(\phi_2 + \varphi_2)} \vec{x}_2 \right).$$

The reference beam is defined by

$$\vec{A}_R = a \left(e^{j\phi_1} \vec{x}_1 + e^{j\phi_2} \vec{x}_2 \right).$$

These expressions show that classical interferometry gives the interference between two waves corresponding to the loaded and unloaded model,
respectively.

The illumination E at M is given by :

$$(\vec{A}_\ell + \vec{A}_R)(\vec{A}_c^* + \vec{A}_R^*) \text{ that is } 4a^2 (1 + \cos \frac{\varphi_1 + \varphi_2}{2} \cos \frac{\varphi_1 - \varphi_2}{2})$$

or else $E = 4a^2 [1 + \cos \frac{\pi}{\lambda} \delta \cos \frac{\pi \delta'}{\lambda}]$

$$\delta = C e (\sigma_1 - \sigma_2)$$

$$\delta' = (c'_1 + c'_2)(\sigma_1 + \sigma_2) e$$

Let us remark that the introduction of the reference wave makes possible
to record information relative to the phases φ_1 and φ_2 by means of a quadratic receiver (eye or photographic plate) measuring energy.

The transmitted energy depends on the sum and the difference of the
principal stresses. It can be distinguished as :
- half tone fringes, $E = 4a^2$, which are the isochromatics of a circular polariscope with light field :

$$\frac{\pi C e}{\lambda} (\sigma_1 - \sigma_2) = (2q + 1) \frac{\pi}{2} , \ q \text{ integer}$$

- contrasting fringes defined by :

$$\frac{\pi}{\lambda} (c'_1 + c'_2)(\sigma_1 + \sigma_2) = (2q' + 1) \pi , \ q' \text{ integer}$$

which are the isopachics on which the energy takes the value

$$4a^2 [1 - \cos \frac{\pi}{\lambda} C e (\sigma_1 - \sigma_2)]$$

The isopachics exhibit a contrast inversion at every crossing with the isochromatics. A dark fringe becomes light and conversely. A half fringe shift occurs. This phenomenon was noticed by Post in 1953 |6|.

It is worth noting a difficulty we have encountered and which is not discussed in the literature to our knowledge. The refractions and the reflections disturb the form of light and alter the expected results : the phenomenon of half fringe shift does not appear |9||10|. With some coating of the mirrors, in order to remove these disturbances it is necessary to introduce precise compensation at the entrance and at the exit of the interferometer |9||11| (See Addendum).

Under theoretical conditions of observation, the isochromatic pattern disturbs the isopachic one. Their interaction is stronger |12| as the two systems of fringes become parallel and their spatial frequencies are nearly equal. On the other hand, it is possible to draw precisely both patterns if they are normal and the ratio of their spatial frequencies is large. Generally, the isopachic fringes are denser than the isochromatics. To satisfy this condition, Nisida and Saïto assumed $C'_1 + C'_2 \gg C$.

The order of the isopachic fringes generally can be determined as well as the order of the isochromatic one |13|. At a point, the principal stresses are obtained from their sum and difference by interpolation between two neighbouring curves of each pattern.

Although this method is less accurate than the point method, nevertheless it represents an advancement because of its easier application. A disadvantage is that it cannot be easily in regions where isochromatics and isopachics are nearly parallel.

In the usual models, the thickness varies from point to point and it is not possible to adjust the interferometer to a uniform field. After compensating for the mean thickness, fringes corresponding to lines of equal thickness appear (the variation of thickness along orthogonal paths can be quite arbitrary). By using moiré of both interferograms obtained before and after loading, Nisida and Saïto observed that the same fringes were obtained as with the model of optical quality.

2.2 Garnault and Lagarde Method |8, 9, 10|

This method avoids the previous difficulties and improves the precision

in the entire field. Like the previous one, it consists of the comparison of two states, namely, the fringe patterns before and after loading.
First of all its originality lies in the introduction of some distribution of optical path differences. To this end, we make use of the
possibilities of easy adjustment of a Mach-Zender interferometer : translation and rotation of the second separating plate, rotation of the mirror
of the reference beam. All this is performed by means of accurate remote
commands. We will describe the successive operations without justifying the
path differences introduced as a result of translations and rotations. The
reader is referred to a more detailed paper |10|.

The adjustment is made with the model in position but unloaded.

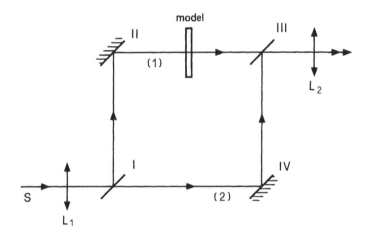

Fig. 2   Ajustable Mach Zender Interferometer

The first operation consists of the compensation for the mean thickness
$\bar{e}$ of the model (e = $\bar{e}$ + $\Delta\bar{e}$), without using any compensating plate but simply by introducing a uniform path difference by means of :
- a translation and a rotation of plate III (Fig. 2)
- a rotation of plate IV (Fig. 2).

Translation and rotation of plate III allow the localization of fringes
in the plane of the model : this leads to the sharpness of both fringes and
model edge on the photographes.

Finally, by means of a rotation of mirror IV, we introduce a path difference $\tilde{\delta}$ = bx + cy (related to a normal axis to the model plane) that is
a "coin d'air" in French optics language.

In the absence of loading, the transmitted energy is

$$E = \frac{a^2}{2} \left[1 + \cos \frac{2\pi}{\lambda} \{ \Delta\bar{e} (n_o - n) + \tilde{\delta} \} \right]$$

Then we observe the appearance of dark and light fringes, the first
ones being defined by

$$\frac{2\pi}{\lambda}\left(f (x, y) + \delta_2\right) = (2p_o + 1) \pi$$

with $f (x, y) = (n - n_o) \Delta\bar{e}$

This is the pattern of interference fringes which are lines of equal
path difference.

By introducing $\tilde{\delta}_2$ we obtain a dense pattern of interference fringes p_o
(about 7 fringes per cm at the scale of the model) ; this pattern displays
a known monotonic variation of the path difference along any line normal
to the fringes ; we can choose the mean direction of the pattern at will
(Fig. 3).

Let us point out the test procedure of a time translation $m \lambda$ of plate
III leading to a translation of the whole system of fringes (allowing us to
be sure of the monotonic nature of the variation) and bringing the fringe
of order p_o to the position previously occupied by the fringe of order
$p_o + m$ (which indicates the sense of increasing fringe order).

Under these conditions, the transmitted energy through the loaded model
(static loading or dynamic loading at an instant t) is expressed by

$$E = \frac{a^2}{2} \{ 1 + \cos \frac{\pi}{\lambda} [e(C_1' + C_2') (\sigma_1 + \sigma_2) + 2 \frac{\Delta\bar{e}}{\lambda} (n_o - n) + \delta_2]$$

$$\cos \frac{\pi e}{\lambda} C [\sigma_1 - \sigma_2] \}$$

As before, we obtain a double fringe pattern on the photographic plate
such that :
- half-tone fringes, which are the isochromatics of a circular polariscope
with field, given by :

$$\frac{\pi}{\lambda} Ce (\sigma_1 - \sigma_2) = (2q + 1) \pi/2$$

- a new pattern p_1 of dark interferometric fringes, given at a point M where
the unloaded fringe order was p_o, by :

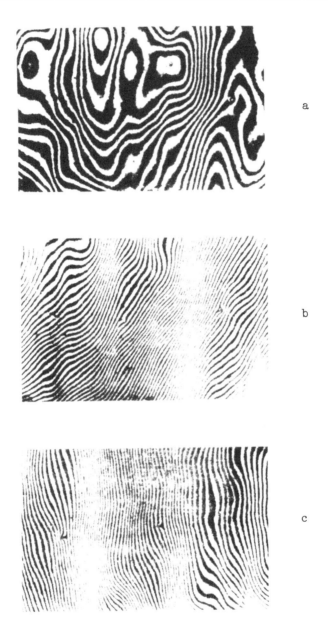

Fig. 3 : Photographs of the prior pattern for various adjustments of the
         mirrors (model made of P.S.M. 1)
         a – ordinary fringes
         b – dense fringes obliquely oriented
         c – dense fringes vertically oriented

$$\frac{\pi}{\lambda} \left[e(C_1' + C_2') \, (\sigma_1 + \sigma_2) + \frac{2\Delta\overline{e}}{\lambda} \, (n_o - n) + 2\tilde{\delta} \right] = (2p_1 - 1) \, \pi$$

The interference fringes p_1 result primarily from the displacement field of the fringes p_o. On the whole they conserve the same spatial frequency which is large with respect to that of the isochromatics without any a priori hypothesis concerning $C_1' + C_2'$. By taking into account the possibility of orientation of the initial pattern, it is always possible - at least sectionally - that the pattern of isochromatics does not disturb the pattern of fringes p_1. The shape of these latter fringes is therefore quite sharp ; moreover their order can be determined at every point. Indeed, the fringes p_1 are obtained in the major part of the field of the model by a displacement of the fringes p_o.

We can determine the fringe order p_1 :
- either with spatial continuity from an area where p_1 is already known, for instance, the areas not reached by the wave, where $p_1 \equiv p_o$,
- or with time continuity following the movement of fringe order at a point between initial time t_o and time t. It is possible in the given examples.

The variation of the fringes is therefore monotonic. The moiré of both patterns yields then the isopachics directly. When new fringes appear, their order can be determined by means of a new experiment from the initial pattern modified by an additional rotation of mirror IV $|10|$.

The sum of the principal stresses is proportional to the difference of the orders of the patterns

$$\sigma_1 + \sigma_2 = \frac{2 \, (p_o - p_1) \, \lambda}{e(C_1' + C_2')}$$

In as much as the patterns are dense, the interpolation between neighbouring fringes is quite satisfactory and in the case where constants C'_1 and C_2' are known, the sum $\sigma_1 + \sigma_2$ is therefore known everywhere without any qualitative considerations.

The knowledge of the isochromatic pattern

$$\sigma_1 - \sigma_2 = \frac{\lambda}{Ce} \, q$$

gives the stresses such that :

$$\sigma_1 = \frac{\lambda}{e \, (C_1' + C_2')} \, p + \frac{\lambda}{2Ce} \, q$$

$$\sigma_2 = \frac{\lambda}{e \ (c_1' + c_2')} \ p - \frac{\lambda}{2Ce} \ q$$

Let us return to the conditions of temporal and spatial coherence.

The width of the spectrum line must be smaller than $\frac{1}{n}$ times the mean wavelength where n is the total number of fringes desired. For 100 fringes in the field and λ = 5000 $\overset{o}{A}$, we have $\Delta\lambda$ = 50 $\overset{o}{A}$ and the coherence length is 50 microns.

The angular size of the source acts as before. This size must be as small as possible in order to ensure the contrast of the fringes and the parallelism of the beams. To some extent this last condition is not imperative if we localize carefully the fringes on the plane of the model.

3   Experimental study of a ring in stationnary vibrations

3.1 Experimental set-up

The model is a ring of internal radius 38 mm, external radius 58 mm and thickness 4,8 mm. It is made of CR. 39. On its exterior part, the ring contains two diametrically opposite bosses, by which the ring can be fitted in a heavy frame and where the moving coil of an electrodynamic dynamo can be attached. This latter system is working at its resonance frequency (535 hz). We perfected a set-up to measure the displacements of the coil in order to deduce the amplitude (accuracy of micron order) and the shift between the transmitted vibration and the reference vibration (coil vibration).

The light source of the interferometer and that of the photoelasticimeter are flashes emitted by a Bruel and Kjaer stroboscope. A generator furnishes current both to the stroboscope and to the excitation intensifier, so that a shift, varying from 0 to 2π, can be introduced between them : this allows observation at every instant during the period of the phenomenon.

3.2 Determination of constants C_1' and C_2'

Because of the viscoelastic behaviour of this material, these constants must be determined under condition very similar to that of the model to be analysed.

It was shown, for a prismatic rod built-in at one end and submitted to longitudinal vibrations, at the other one, that the Kelvin-Voigt model was very well suited to it near a resonance frequency |14|.

If ℓ is the length, m the mass of the rod, M the mass of the coil, the eigen-frequencies are given by

$$N_i = \frac{\beta_i}{2\pi\ell} \sqrt{\frac{E}{\rho}}$$

where β is given by the solution of $tg\beta = \frac{1}{\beta} \frac{m}{M}$

If we impose a vibration of frequency N_1 to the rod, the longitudinal displacement at point x is given by

$$U(x, t) = U(\ell, t) \frac{\sin \beta x/\ell}{\sin \beta}$$

It follows : $\sigma_1(x, t) = \frac{E\beta \sin \beta x/\ell}{\ell \sin \beta} U(\ell, t)$

$$\sigma_2 \equiv 0$$

The maximum stress is :

$$\sigma_{1max} = \frac{\beta}{\ell} \frac{E U}{\sin \beta} \sin \beta \frac{x}{\ell}$$

U being the amplitude of $U(\ell, t)$

Let us place in the interferometer (with the same apparatus which will be used for the ring) a prismatic rod excited at the resonance frequency in longitudinal vibrations.

If we polarize the light in a direction parallel to the axis of the rod, the light intensity is given by.

$$I = \frac{I_o}{2} [1 + \cos (\frac{2\pi e}{\lambda} c_1' \sigma_1 + \delta_o)]$$

We can deduce the total variation Φ of the argument of the cosine during a period

$$\Phi = \frac{2\pi e}{\lambda} c_1' \cdot 2 \sigma_{1max}$$

If we trace the variations of Φ versus U, we obtain a straight line, the slope of which gives c_1' ; the constant c_2' is obtained in a similar way by polarizing the light orthogonaly to the axis of the rod.

We find $c_1' = -90$ Bw   and $c_2' = -120$ Bw.

3.3 Determination of $\sigma_1 + \sigma_2$

Let us first photograph the isopachics of order p_o on the unloaded ring, and second the fringes of order p_1, at a time of the period, for instance, at time where the stretch of the vertical diameter is maximum (Fig.4 and Fig. 5).

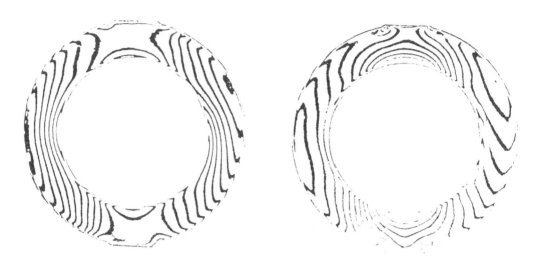

Fig. 4 : Interferometric fringes p_o.   Fig. 5 : Interferometric fringes p_1 at the instant when the diametral stretch is maximum

The determination of p_1 presents no difficulty because we can follow the variation of fringes during a period by shifting slowly the stroboscope.

To know the order $p = p_1 - p_o$, we traced the variations of p_1 and p_o along numerous radii and then we deduced the variations of p and consequently the position of the integer p orders on these radii.

3.4 Determination of $\sigma_1 - \sigma_2$

Instead of using the isochromatic fringes given by the interferometer it is better to use those given by a circular polariscope (we have only to mask the reference beam of the interferometer). To increase the density of information, from the photographic plate, we determined with a microphotometer the fractional isochromatic fringes (one-eight of order).

Knowing $\sigma_1 - \sigma_2$ and $\sigma_1 + \sigma_2$ on the entire field, it is easy to compute σ_1 and σ_2, then we can trace easily the pattern of σ_1 and σ_2 (Fig. 6 and Fig. 7).

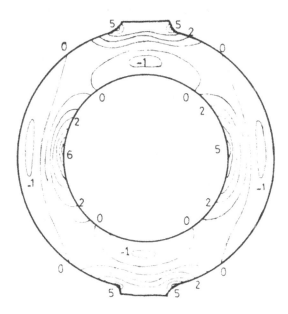

Fig. 6 : For an isobar of order n, the stress σ_1 is 0,60 n daN/mm^2.

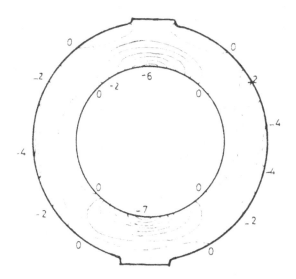

Fig. 7 : For an isobar of order n, the stress σ_2 is 0,60 n daN/mm^2.

4   Study of repetitive transient phenomena

We studied the stress state in a plane rectangular birefringent model, clamped on one edge and submitted at the middle of the opposite edge to re-petitive shocks given by the loading system described in chapter 8.

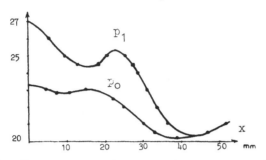

Fig. 8 : Order p_o (unloaded)   Order p_1 (loaded)

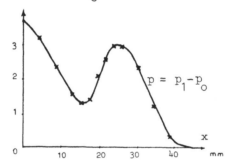

Fig. 9 : Isopachic fringe order $p = p_1 - p_o$

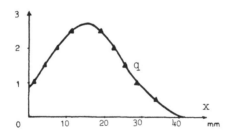

Fig. 10 : Isochromatic fringe order q

We recorded the fringes obtained by interferometry and the isochromatics given by photoelasticity at a time t = 23 µs after the beginning of the
shock. At this time, the stress wave created by the shock has not yet reached the edges of the model. We have two distinct areas : a disturbed area
which corresponds to the position of wave at instant t, and an area
where the stresses are zéro.

Figure 8 shows the variations of p_o and p_1 orders along the axis ox of
the shock. Let us remember that the p_o order is determined only to within
a constant. This does not matter because we only need the difference
$p = p_1 - p_o$, given by Figure 9. Figure 10 shows the variations of the q
order of isochromatics. From p and q we obtain the variation of σ_1 and σ_2
along the axis ox (Fig. 11).

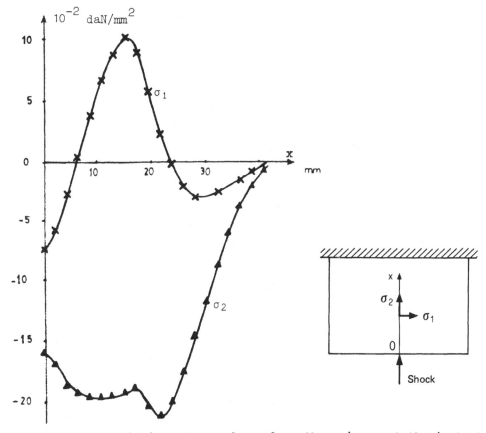

Fig. 11 : Variation of σ_1 and σ_2 along the axis ox at the instant
        t = 23 µs.

We have made the plot of p_o, p_1, p and q along lines parallel to the edge of the rectangle. Figures 12 and 13 show the isopachics and the iso-chromatics. We then deduced the graph of σ_1 and σ_2 in the disturbed area of the model (Fig. 14, 15).

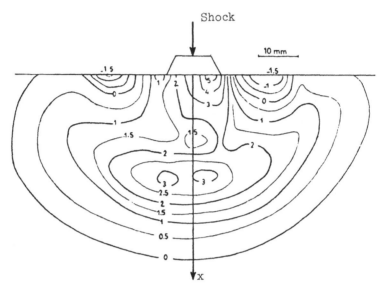

Fig. 12 : Isopachics of order p.

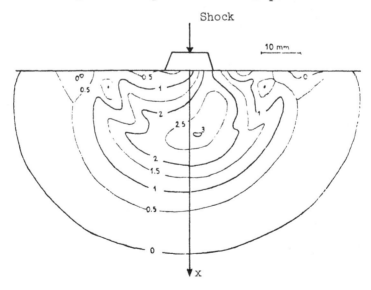

Fig. 13 : Isochromatics of order q.

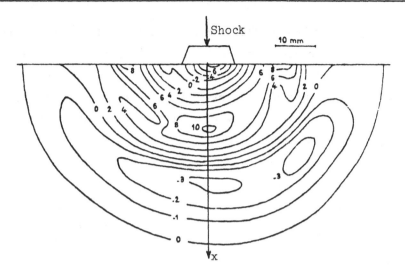

Fig. 14 : Isobars σ_1 expressed as 10^{-2} daN/mm^2.

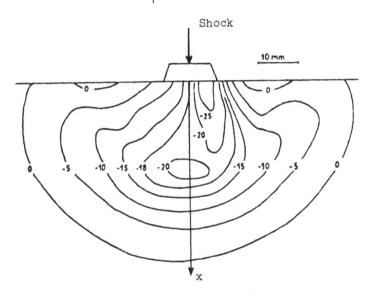

Fig. 15 : Isobars σ_2 expressed as 10^{-2} daN/mm^2.

Remarks :

Let us notice that, using recordings on rectilinear or circular polariscopes, it is possible to determine the orientation of principal stresses
by, densitometric measurements.

Practically, it is not neccesary to use a classical interferometer. The

coherence length of the laser allows interferometry with very inexpensive
equipment. Then we introduce a large "coin d'air" in the setup the consequence of which being that the interferometric fringes in the unloaded
or loaded states have a monotonic variation along each orthogonal path. So
we obtain directly the isopachics by the moiré method. Practically because
the complexity of the superimposed fringes, it is often easier to count the
fringes p_o and p_1. Thus, if we want to be in the best conditions, we only
have to orient the mirrors as we said before.

In areas of dense stresses, the intricacy of isochromatics and inter-
ferometrics fringes makes their determination very hard and often impossible. In such cases we must use the holographic photoelasticity which allows
the elimination of the isochromatics.

References

1. Maury V. "Mécanique des milieux stratifiés" 1970 - Dunod

2. Favre H. "Sur une nouvelle méthode optique de détermination des tensions intérieures" - Revue d'optique théorique et expérimentale 8,
   193-213, 241-261, 289-307 (1929).

3. Fabry C. "Sur une nouvelle méthode pour l'étude expérimentale des tensions élastiques" - C.R.Ac.Sc. 190, 457-460 (1930).

4. Favre H., Schumann W., Stromer E. "Ein photoelektrisch-interferometri-
   sches Verfahren Zur vollständigen Bestimmung von eleven Spannungszus-
   tänden" - Bulletin suisse de la Construction. Septembre 1960.

5. Bowler B. and W. Schumann "On the complete determination of dynamic
   states of stress" - Exp. Mech., March 68.

6. Post D. "A new photoelastic interferometer suitable for static and
   dynamic measurement" - Proc. SESA 12-1, 1954.

7. Nisida M. and Saito H. "A new method of experimental stress analysis".
   Exp. Mech., 1964.

8. Lagarde A. et Garnault J. "Résolution expérimentale de problèmes dynamiques plans pour un matériau élastique biréfringent" - C.R.Ac.Sc.,
   T. 273, p. 194-196, Juillet 1971.

9.   Grellier J.P., Garnault J., Lagarde A. "Méthode interférométrique de détermination des contraintes pour des problèmes plans statiques et dynamiques et influence des polarisations partielles". - Revue Française de Mécanique n° 70, 1979.

10. Garnault J. et Lagarde A. - Rapport D.G.R.S.T. n° 6700549.

11. Derouet J. "Effet des polarisations partielles en interférométrie" - J. Optics, Paris, 1982, vol. 13, n° 6, p. 359-366.

12   Stanford R.J. and Durelli A.J. "Interpretation of fringes in stress-holo interferometry" - Exp. Mech. 11(4), p. 161-166, 1971.

13. Durelli A.J. and Riley W.F. "Introduction to photomechanics" - Prentice-Hall, 1965.

14. Lagarde A. "Etude du comportement mécanique et photoélastique d'un barreau biréfringent soumis à des vibrations". - Thèse, Paris, 1962. N.T. 127, Ministère de l'Air.

15. Grellier J.P. "Méthodes de photoélasticité et d'interférométrie ponctuelle et à champ complet. Application à la détermination des contraintes planes en régime dynamique transitoire". - Thèse de Docteur Ingénieur, Poitiers, 1982.

ADDENDUM

COMPENSATION OF REFRACTION AND REFLECTION IN THE INTERFEROMETRY

Let us point out a difficulty which has not been noticed by other authors.

Reflections and refractions disturb the isopachic fringe pattern. The shipt of half-a-fringe when crossing an isochromatic is not respected at every point of the model (fig. 16b).

Reflections and refractions disturb the fringe patter, i.e., a fringe is translated by a distance equivalent to about half a fringe.

For natural incident light, let us call a' the component in the plane of incidence and b' the component in the normal plane of the beam at the entrance of the model.

An interferometer introduces a polarization such that $\dfrac{b'^2}{a'^2} = 2.5$.

To obtain natural light at the entrance of the model we polarize the incident light with three plane glass plates, , inclined at an angle of $\pi/3$ to the axis of the beam. This gives us $b'^2/a'^2 = 1/2.5$ so that we obtain b' = a' at the entrance to the model.

The effect of the divergence of the incident beam (about ± 3°) can be significantly diminished by using two systems of two plates at an incidence of 50° with axis of the beam

At the exit from the interferometer a similar system counterbalances the effect of plate III.

This compensation is valid – see Fig. 16 –
More complete theoretical studies are reported in |9, 11, 15|.

Figure 17 gives some results obtain by J. Derouet |11| using electromagnetic theory of light and shows the importance of coating choice (Aluminium fig. 17c) is well suitable.

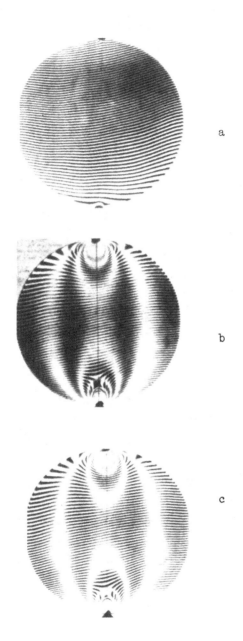

Fig. 16 : Compensation of refraction and reflection in
interferometry
a - unloaded model
b - loaded model without compensation system
c - loaded model with compensation system

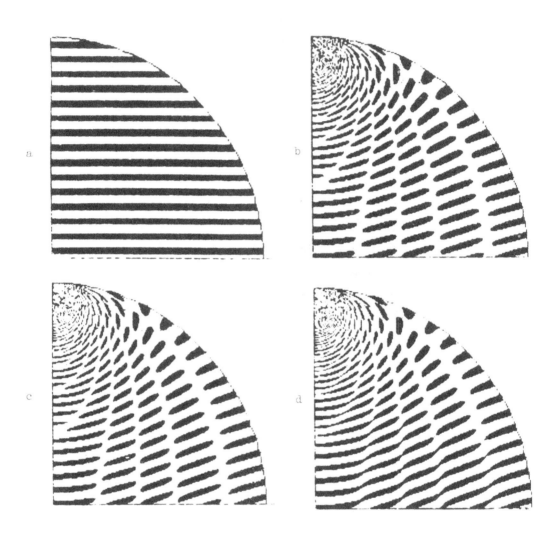

Fig. 17 Disturbing of fringe pattern for diametrally loaded disk
        a - unloaded
        b - loaded-perfect interferometer
        c - loaded aluminium coating
        d - loaded-steel coating

CHAPTER 7

DYNAMIC HOLOGRAPHIC PHOTOELASTICITY. APPLICATION

First we will describe briefly this method before developing its
dynamic aspects.

It is well known that holography is a technique of recording the
amplitude and phase of a light wave. As early as 1946 D. Gabor |1| gave the
basis of the technique. But it has been necessary to wait for the advent of
highly coherent sources (lasers) and a positive improvement |2| (the sepa-
ration of measurement and reference beams) in order for holography to reach
the spectacular developments of the last few years.

Here we deal with its application to photoelastic media which, to our
knowledge, is presently limited to two dimensional problems. In our presen-
tation we shall use a directed measurement beam (also known as an object
beam) and we shall not deal with some aspects related to scattered light
holography.

Let us recal briefly the holographic process.

1   The holographic process |3, 4|

To store the information related to the phase and the amplitude of a
light wave, like interferometry, holography uses a reference beam and a
photosensitive emulsion. The hologram of an object is a negative on which
is recorded the modulation, in the form of interference, between a reference
wave and a wave coming from the object (Fig. 1).

1.1   Recording

Whereas in interferometry the two interfering beams were quite parallel, in holography the coherence conditions allow large angles between
the two beams. Optical path differences varying from one point to another

are then systematically introduced. The interferogram then has dense
fringes the usefulness of which we shall see later

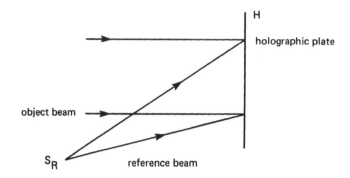

Fig. 1 : holographic recording

As an example, it is not rare to record a pattern with one thousand to ten
thousand fringes per centimeter. This requires the use of high resolution
photographic plates, therefore low sensitivity, which requires very stable
devices and very powerful light sources. This high resolution power allows
the recording of several interferograms on the same plate.

There is another difference between holography and interferometry :
the recording must involve a linear photographic process. The complex
amplitude of transmittance t_r of the hologram (ratio of the transmitted
amplitude and the received amplitude) must be limited to its linear domain
as a function of the illumination E. In holography this transmittance has
real value (Fig. 2).

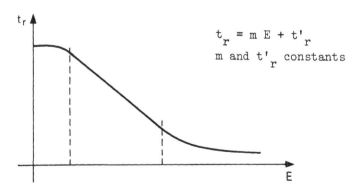

$$t_r = m E + t'_r$$
m and t'_r constants

Fig. 2 : Transmittance versus illimination

The transmittance varies as the illumination received by the hologram, that is, sinusoïdally. Locally, at a point M the incident waves can be regarded as plane waves with the angle β (Fig. 3) and thus the hologram will be regarded as a sine pattern of parallel fringes with interfringe spacing $i(M) = \dfrac{\lambda}{2\sin \beta/2}$

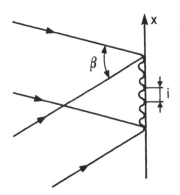

Fig. 3 : holographic recording with two parallel beams

1.2 Reconstruction

The second step of the holographic process is the reconstruction, i.e, the reading of the hologram. Whereas interferometry deals with the displacements of fringes on the record itself (where the visibility requires a system of relatively wide fringes), holography extracts the information by diffracting a monochromatic radiation through the hologram. The denser the fringes the more satisfactory are the conditions of the operation.

For the reconstruction, we use a reference beam under the same conditions as for the recording. At point M the hologram acts as a sinusoïdal grating which can diffract the monochromatic reference wave of wavelength λ along three directions (Fig. 4) given by

$$\sin \theta - \sin \theta_k = \frac{k \lambda}{i(M)}$$

For k = 0, zero order, θ_o = 0, we have the directly transmitted wave, it is the restored reference wave.

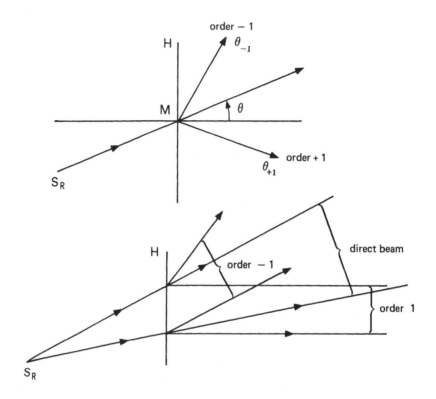

Fig. 4 : Diffraction of reference beam by hologram

For k = 1, we have a wave of which the direction of propagation and the amplitude are the same as those of the object wave $|5|$. It is the restored object wave.

For k = - 1, we have the conjugate wave the direction of which is nearly symmetric to that of previous one with respect to the reference wave.

In order to facilitate the description we consider linearly polarized waves along the direction $\vec{x}$.

We calculate the complex vectorial amplitude $\vec{A}_T$ of the diffracted radiation at M by the hologram. Let the complex vectorial amplitudes be $\vec{A}_O = a_O \, e^{j\phi} \, \vec{x}$ and $\vec{A}_R = a_R \, e^{j\phi_R} \, \vec{x}$ for the object and reference waves, respectively. The energy which is recorded on the plate at M is

$$E = (\vec{A}_o + \vec{A}_R)(\vec{A}_o^{**} + \vec{A}_R^{*})$$

The transmitted complex vectorial amplitude $\vec{A}_T$ of the hologram illuminated by $\vec{A}_R$ is expressed as

$$\vec{A}_T = (t_{r'} + m\ E)\ \vec{A}_R$$

$$\vec{A}_T = |t_{r'} + m\ \vec{A}_o\ \vec{A}_o^{*} + m\ \vec{A}_R\ \vec{A}_R^{*}|\ \vec{A}_R + m\ (\vec{A}_o\ \vec{A}_R^{**})\ \vec{A}_R + m\ (\vec{A}_R\ \vec{A}_o^{**})\ \vec{A}_R$$

The second term of the second group is written as $(\vec{A}_R\ \vec{A}_R^{**})\ \vec{A}_o$.

The terms of the second member represent in order, the complex amplitudes of the direct beam, the reconstructed object beam, and the conjugate beam. We note that we can also find, from the shape of the constant phase surfaces (wave fronts), the direction of propagation of the three orders |5|.

We remark that if we want to restore the exact phase of the object wave, the reference source and the hologram must be in the exact position that they occupied during recording.

We shall see that in holographic photoelasticity the conditions of reconstruction are more or less strict according to the application.

If angle θ is large enough, the three beams are separated and an observer looking in the direction 1 only discerns the restored object wave with its same phase and its amplitude (within a constant factor, which means the wave shape is the same).

In the case of a polarized measurement beam, let us specify that it is possible to record and to restore its polarization state with two orthogonally polarized reference beams |6, 7|.

2 Coherence conditions

The coherence conditions are more strict here than in interferometry. As an example, we can make a holographic interferogram on a 9 cm x 12 cm plate with 10,000 fringes per centimeter, only if the (spectrum) line width (necessarily laser) is $\lambda/120{,}000$). For a He-Ne laser, $\lambda = 6328\ \overset{o}{A}$ (red orange) we have $\Delta\lambda = \frac{1}{20}\ \overset{o}{A}$. For some setups using a Faraday cell (cf. 4.3.3, 6.1) the coherence length must be about 1 m ; the line width is then $\Delta\lambda = 6 \times 10^{-3}\ \overset{o}{A}$. With regard to spatial coherence, the conditions are less

critical if the fringes are localized on the hologram plane. Normally we
use laser sources which emit wave surfaces with a well defined phase, so
we need not be concerned about this condition. Otherwise we must use
sources of light such that the transverse size is smaller than the
diffraction mark of the optical systems used in the setup |8, 9|.

3 The Holographic Setup and its Advantages

The laser cavity emits a parallel beam of several millimeters in
diameter. It is convenient to use a short focal length objective, diver-
ging or converging, such as a microscope objective. This allows the beam
to be focused on a diaphragm aperture placed at the focal point of a converging lens such that the parallel object beam is obtained. A classical set
up is shown below (Fig. 5).

It is easy to modify the geometry of the beams and to rearrange the
setups on a small vibration-free table.

Compared to interferometry, holography offers many advantages.

The ability to work with large angles between the object and reference

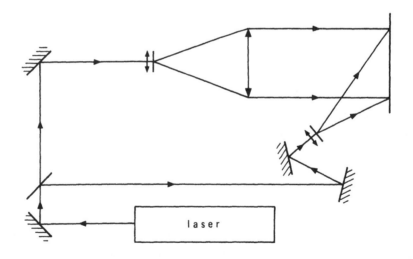

Fig. 5 : Typical holographic set-up

waves, allows us to suppress the second separating plate (beam splitter)
and keep a beam with a very small diameter throughout its path before the
model. The influence of flaws in the band is thus considerably reduced. We

can present linear polarization preserved by the reflection and use quarter-wave plates of small size and thus of good quality. The field of study is much less limited than in interferometry because of the price and the difficulties in designing a large field interferometer.

As we can obtain and record a great number of fringes it is possible to record several interferograms on one hologram. Frequently, the double exposure technique is used and the corresponding fringes act as two independant patterns. Upon reconstruction, all perturbations due to surface flaws are eliminated |10|.

Among other things it is of interest that a circular or rectilinear beam can be used. In the latter case, we only retain the component of the measurement beam which is parallel to the direction of polarization.

4 Holographic photoelasticity

We examine several methods performed first using the single exposure method.

4.1 Obtaining Isoclinics and Isochromatics

The first works were published in 1967 and 1968 |11, 12|. A classical holographic setup with a circularly polarized measurement beam and a linearly polarized reference beam gives upon reconstruction the isochromatic pattern and the isoclinic associated with the direction of polarization of the incident beam.

It is possible to obtain only isochromatics |12, 15|. Let us point out the procedure in order to classify the concepts and to show that the reconstruction conditions are not critical. The object and reference beams are unpolarized or circularly polarized. The complex amplitudes of the object beam and reference beam at the same point of the holographic plate are, respectively

$$\vec{A}_\ell = a\,[\,e^{j(\phi_1 + \varphi_o + \varphi_1)}\,\vec{x_1} + e^{j(\phi_2 + \varphi_o + \varphi_2)}\,\vec{x_2}\,]$$

$$\vec{A}_R = a\,[\,e^{j(\phi_1 + \phi_R)}\,\vec{x_1} + e^{j(\phi_2 + \phi_R)}\,\vec{x_2}\,]$$

ϕ_1 and ϕ_2 are the phases in the hologram plane for the active beam without a model ; ϕ_1 and ϕ_2 are two random functions when the beam is unpolarized and $\phi_1 - \phi_2 = \pm \dfrac{\pi}{2}$ for a circularly polarized beam.

φ_0 is the phase shift introduced by the unloaded model

$$\varphi_0 = \frac{2\pi}{\lambda} (n_0 - n)\, e$$

φ_1 is the phase shift introduced by the loading

$$\varphi_1 = \frac{2\pi}{\lambda} [\, (n_1 - n_0)\, e + \Delta_e\, (n_1 - n)\,]$$

The energy received by the photographic plate is

$$E = (\vec{A}_\ell + \vec{A}_R)\, (\vec{A}_\ell^{\,*} + \vec{A}_R^{\,*})$$

For the reconstruction let us use a beam with a slightly different orientation with a complex vectorial amplitude

$$\vec{A}_{R'} = a\, [\, e^{\,j(\phi'_1 + \phi'_R)}\, \vec{x}_1 + e^{\,j\,(\phi'_2 + \phi'_R)}\, \vec{x}_2\,]$$

where ϕ'_1 and ϕ'_2 are defined in the same way as ϕ_1 and ϕ_2.

The complex vectorial amplitude $\vec{A}_T$ restored in the hologram plate is

$$\vec{A}_T = (t'_r + m\, E)\, \vec{A}_{R'}$$

The complex vectorial amplitude in the first diffraction order is proportional to

$$(\vec{A}_R^{\,*}\, \vec{A}_\ell)\, \vec{A}_{R'}$$

that is to

$$[\, e^{\,j(\varphi_0 + \varphi_1 - \phi_R)} + e^{\,j(\varphi_0 + \varphi_2 - \phi_R)}\,]\, [\, e^{\,j(\phi'_1 + \phi'_R)}\, \vec{x}_1 + e^{\,j(\phi'_2 + \phi'_R)}\, \vec{x}_2\,]$$

This complex amplitude depends on the shape of the reference wave, whereas the associated energy does not, since it is proportional to $\cos^2 \dfrac{\varphi_2 - \varphi_1}{2} = \cos^2 \dfrac{\pi\, Ce}{\lambda}\, (\sigma_2 - \sigma_1)$. This expression determines the isochromatics of a circular polariscope with light field.

It is possible to obtain the isochromatics and the complete family of isoclinics |16|. Two reference beams with different polarization directions enter simultaneously during recording and reconstruction |16|.

4.2 Obtaining Isohromatics and Isopachics

Let us investigate now different double exposure methods.

In 1967 it was shown $|17|$ that double exposure holography produces interference phenomena "similar" to those given by interferometry. The first applications to photoelasticity were performed in 1968 $|12, 13|$.

We superimpose on the same hologram the two interferograms made before and after loading the model. Let us note that if the hologram is developed after the first exposure and held strictly in place during loading we observe directly in real time the interferences between the restored wave and the wave coming from the loaded model. The object and reference beams are always circularly polarized or unpolarized and have the same intensity. The complex amplitude of the object beam before loading is :

$$\vec{A}_{u\ell} = a \ (e^{j(\phi_1 + \varphi_o)} \ \vec{x}_1 + e^{j(\phi_2 + \varphi_o)} \ \vec{x}_2)$$

If we assume that the exposure times are equal and the conditions of linearity are satisfied during recording, the diffracted complex amplitude of order 1 is proportional to :

$$\vec{A}^*_R \cdot (\vec{A}_{u\ell} + \vec{A}_\ell) \ \vec{A}_R$$

i.e. $[2 + e^{j\varphi_1} + e^{j\varphi_2}] \ \vec{A}_R$

The diffracted energy of order 1 is then proportional to

$$1 + \cos \frac{\pi}{\lambda} \ Ce \ (\sigma_1 - \sigma_2) \ [\ 2 \cos \frac{\pi}{\lambda} \ e \ (C'_1 + C'_2) \ (\sigma_1 + \sigma_2) + \cos \frac{\pi}{\lambda} \ Ce \ (\sigma_1 - \sigma_2)]$$

Let us note that this double exposure technique requires the same position for the model before and after loading.

The above expression is different from that given by interferometry ; there is a complementary term : the one obtained by single exposure. However, in tracing the isochromatics and isopachics we encounter the same difficulties mentioned in Ch. 6. This fact explains the work performed to eliminate the isochromatics. Many solutions have been found.

4.3 Obtaining of Isopachics only

4.3.1 Stress-Freezing Technique

One method takes advantage of the sign change of the photoelastic constant with temperature |18, 19|. The loaded model is frozen at an elevated temperature. At ambient temperature a similar loading cancels the birefringence. This method is time consuming and of course cannot be applied under dynamic conditions.

4.3.2 Use of Polarizing Diffuser

Another method consists of introducing a diffuser such as tracing paper or glass ground on both sides immediately after the traverse of the model. The object and reference beams are linearly polarized and the diffuser can be represented schematically by a random birefringent and a rotatory power. We can then predict and verify |20| that we obtain two families of "pseudo-isopachics" defined by

$$C'_1 \sigma_1 + C'_2 \sigma_2 = (k + \frac{1}{2}) \frac{\lambda}{e} \quad \text{and}$$

$$C'_1 \sigma_1 + C'_2 \sigma_2 = (k' + \frac{1}{2}) \frac{\lambda}{e}$$

(k and k' are integers)
corresponding, respectively, to two orthogonal directions of polarization (0 and $\pi/2$ isoclinic). The method is simple only in regions of the model sufficiently removed from the $\pi/4$ isoclinic.

4.3.3 Use of Rotatory Power

Another method consists of traversing the model twice by the measurement beam at near normal incidence (a fact which requires a sufficiently complex setup and a large coherence length). Before the second traverse through the model, the beam passes through a rotatory power which rotates by $\pi/2$ the plane of polarization, and thus the resulting birefringence is zero |15|. If we use a reciprocal rotatory power |21 à 24|, e.g., two half wave plates at an angle $\pi/4$, we cannot be limited to near normal reflections from the plates used (see Fig. 6). The use of a Faraday cell (non reciprocal rotatory power) avoids this inconvenience |22,23| (Fig. 7).

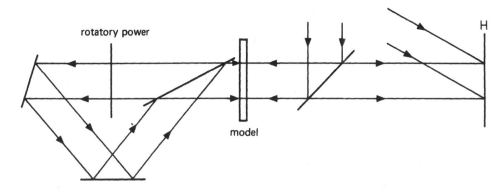

Fig. 6 : Single crossing of beam in the rotatory power

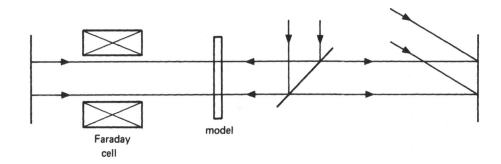

Fig. 7 : Use of Faraday cell

4.3.4. Direct Measurement of Thickness Variation

More recently |26| it was proposed to make the face at the entry of
the model semi-reflective (the transmitted beam gives the isochromatics by
photoelasticimetry). Only the reflected part is recorded holographically.
In double exposure, we obtain a pattern which is related to the normal displacement of the plate, i. e., thickness variations of the model, provided
these displacements do not include rigid body motions of the assembly during
loading.

5 Brief survey of Dynamic Applications

To our knowledge very few studies have been performed under dynamic
conditions. Only the advent of pulsed lasers has made it possible to apply
holographic techniques to dynamic problems. The power and the short dura-

tion of the pulse allow the study of the phenomenon which is assumed to be reproducible at different steps of its development by a method of repetition Let us note that a ruby laser with an amplifier and a Pockel's cell gives a pulse of 1 joule in 20 nanoseconds and a coherence length of several meters. During this time, in epoxy, a dilatational wave covers a distance of about 0,01 mm.

The first investigations were reported in 1971 by Holloway and Johnson |27|. They used active and reference beams which were circularly polarized. They took a first exposure with the unloaded model covering half of the holographic plate. A second exposure for the entire plate was made during the event. Thus we obtain in a single exposure the isochromatics, and in double exposure the isochromatics and the isopachics. As in static problems, the model must be exactly at the same place for both exposures.

In 1972 Holloway, Ranson and Taylor used the method described § 4.3.4

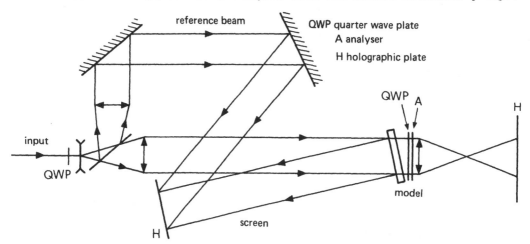

Fig. 8 : Set-up using the measurement of thickness variation of the model

6 Separating Isochromatics and Isopachics in Dynamic conditons [28, 29, 30]

6.1 General Principles of Separation by a Faraday Rotator in Dynamics

The procedure to be described is a whole-field method allowing simultaneous but separate recordings of isochromatic and isopachic-fringe patterns for transient-plane stress problems. This is achieved at any given time during the propagation of the disturbance through the model. The method

requires good reproducibility of the propagation phenomena (boundary conditions, stable material properties, etc.), as well as a perfect synchroni-
zing of the recording apparatus and the shock generator. Successive experiments thus may yield the pattern of schock propagation for several instants of time.

The experimental setup is composed of a light source (a Q-switched
ruby laser giving dual short pulses), a shock generator (air gun) and an
optical bench with a magnetic-optical rotator (Faraday's cell).

The optical bench, at least the elements giving the isopachic fringes,
has been used for static cases, but to date has apparently never been
applied to dynamic problems. The system uses equipment necessary for holographic interferometry with a Faraday's cell to eliminate the isochromatic
fringes from the interferometric recording.

The need to provide a large field for studying wave propagation requires an optical bench with large dimensions. Therefore, it was not possible for the optical elements, including the laser and the model, to be
located on the same rigid structure isolated from the surrounding vibrations. A very short double-exposure time was necessary for the isopachic
fringes which in turn required the position of the elements to be absolu-
tely unchanged. This severe constraint prohibits the use of a single pulse
laser for which the flashing rate is measured in tens of seconds. Instead,
we use a laser giving two light pulses separated by an adjustable delay of
5 to 220 microseconds. The first pulse lights up the model just before
being shot by the projectile and allows the first exposure of the unloaded
model. A second light pulse (the equality in intensity of the two light
pulses is not critical |31|) is emitted while the disturbance wave is tra-
veling through the model and allows a second exposure of the loaded model.
The procedure does not allow visualizing the wave more than 220 microseconds after the beginning of the shock. This is not a significant limitation in studying wave propagation in common materials.

As usual, the isochromatic fringes are obtained in a circular-light
polariscope. However it should be noted that if a light-field polariscope
is used, the first light pulse will evenly expose the film while the isochromatic fringes exposed during the second pulse will have little con-

trast (ratio 1/3). For this reason it is preferable to use a dark-field arrangement so that the first pulse going through the unloaded model will not expose the negative plate. The fringes of whole-order obtained from the loaded model with the second pulse will then have the best contrast.

In order to increase the precision of the information provided by the isochromatics, we have developed a technique of recording the isochromatics in light field (half-order fringes) with maximum constrast. An electro optic shutter, consisting of a Pockels cell placed in the path of the beam recording the light-field isochromatics, blocks the first laser pulse but lets the second pulse pass through. We thus eliminate the uniform illumination of the plate due to the first pulse and the half-order fringes have maximum visibility.

6.2 Description of the Experimental Device

The apparatus is shown on the diagram in Fig. 9.

6.2.1 Light Source

The short-duration light pulses of the ruby laser (RL), about 20 nanoseconds, stops the dynamic event such that large fringe gradients are captured without loss of contrast due to the movement of the wave during the exposure time. The ruby laser has its own oscillator and an amplifying stage in order to induce an energy of 1 joule for both pulses. The pulses are triggered by a Pockel's cell.

6.2.2 Shock Generator

The method requires a reproducible shock-generation procedure. Explosive loading methods such as lead azide or PETN on the model's edge were determined to be unsuitable for several reasons. The shock might last longer than 10 microseconds, is destructive in the area of the explosion, and is not accurately reproducible. For these reasons, explosives are best used with a cinematographic recording system (an ultrahigh-speed framing camera or a multiple spark-gap camera).

A 4.5 - mm bore air gun (AG) was developed and constructed which used compressed air up to 10 bars in pressure that could propel steel projectiles (balls or cylinders) at speeds ranging from 20 to 100 meters per second with a reproducibility of about 1 or 2 percent. The shock duration in the photoelastic material (Araldite B or Hysol 4290) varied from 20 to 50 microseconds.

6.2.3 Optical-measurement Apparatus

The main incident beam emitted by the ruby laser (RL, Fig. 9) lights up the model (M) in parallel rays 210 mm in diameter at normal incidence. The beam is reduced to a 25 - mm diameter going through the Faraday's cell (FC). A beam splitter (BS2) is set up exactly at the focal point of the image of the model M formed by the lenses L3 and L4 and the cell (FC). The latter condition is a requirement because for a non interferometric quality model, an incident beam emerging from the model should return to exactly the same point on the way back after reflection from BS2. The dioptric effect of the model (the thickness of which may vary no more than about 1/10 mm) is then well balanced. The beam having gone back through the model is reflected from beam splitter BS1 and arrives at the ground glass (GG) set in the model's image plane defined by lenses L1, L2 and L5. The model's image is thus collected on the ground glass which in turn diffuses light towards the holographic plate (H). In this way every portion of the hologram includes information for the image of the whole model. This is a widely used procedure which allows for better holographic reconstruction compared with a recording made in direct light (without ground glass).

A reference beam (RB), following a path (not shown in Fig. 9) of equal length to the working beam, impinges on the holographic plate (H). The holographic reconstruction is achieved by a 2 mW Helium-Neon laser.

A portion of the working beam is circularly polarized before passing through the model ; the beam then proceeds through the beam splitter BS2, a quarter-wave plate ($\lambda/4$), a rectilinear analyzer (A) and produces the whole-order isochromatic fringes on a photographic plate (PP); of course (PP) stands in the model's image plane given by the optical elements within (M) and (PP).

The photosensitive receivers are :

To increase the precision, light field isochromatics were obtained by introducing a Pockel's cell in an auxiliary beam after beamsplitter BS3 (fig. 9). The cell POC blocks the first pulse but let's the second pulse through.

The photosensitive receivers are :

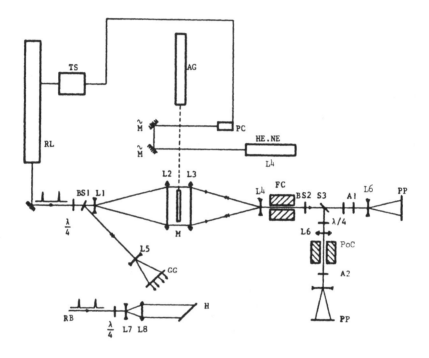

λ/4	– QUARTER WAVE PLATES	M	– MODEL
A	– ANALYSER	FC	– FARADAY CELL
BS	– BEAM-SPLITTERS	RL	– RUBY LASER
GG	– GROUND GLASS	AG	– AIR GUN
H	– HOLOGRAM	HE-NE	– HELIUM-NEON LASER
RB	– REFERENCE BEAM	PC	– PHOTOCELL
L	– LENS	TS	– TRIGGERING SYSTEM
PP	– PHOTOGRAPHIC PLATE	PoC	– POCKELS CELL
PC	– PHOTOCELL	M̃	– MIRRORS

Fig. 9 : Experimental set up to simultaneously obtain
separate isochromatic and isopachic patterns
for dynamic loading using a Faraday rotator

- holographic plates type AGFA 10E 75 for holographic recordings
- sheet film type AGFA PAN 400 for recording of isochromatics and
reconstructed holographic images.

The mirrors and partial mirrors are glass plates of 10 mm thickness
coated with multi-dielectric coatings, of interferometric quality matched
for a wavelength λ = 6943 $\overset{\circ}{A}$. The mirrors for normal incidence have a diameter of 30 mm and the mirrors and partial mirrors for 45° incidence have
a diameter of 50 mm. The importance of the diameters is compatible with
the condition of the coating resistance (20 MW/cm^2). In fact, it is impossible to receive on these mirrors a beam coming directly from the ruby
laser having a diameter of less than 20 mm without damaging them.

6.2.4   Faraday Rotator

The rotator is a solenoid supplied by a d-c generator of 80 V and
100 amperes with a tolerance of one percent in which is positioned a glass
rod SF 57 of 35 mm diameter and 180 mm length. In order to obtain 90 deg
of rotation, a magnetic field of 3200 Gauss is produced by a current intensity of 70 amperes. Even though the system is water cooled, the working time must be limited to one minute.

The adjustment of current intensity is achieved by placing a statically loaded model in the field of the experimental set up. However, the
optical elements are somewhat rearranged for that purpose. A quarter-wave
plate and an analyzer are located between BS1 and L5 on the return beam
and a flat film is used instead of the ground glass GG. By this procedure
a circular-light field polariscope is produced with a beam going twice
through the model. The solenoid current intensity is adjusted so that the
half-order isochromatic fringes disappear from the model's image when recorded on the film. The schematic diagram of this arrangement is shown in
Fig. 10.

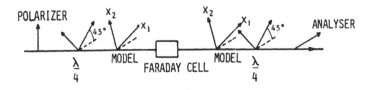

Fig. 10 : Schematic diagram to adjust the current intensity of the solenoid

6.2.5 Synchronization of the Air-Gun with the Ruby Laser

The recording of transient phenomena traveling at a high speed presents a difficult problem of synchronization.

The projectile of the air gun triggers the action of the ruby laser. Also, the projectile interrupts the light beam of a Helium-Neon laser twice (Fig. 9) and allows the velocity of the projectile to be measured. The signal given by the photocell (PC) is transmitted to a triggering system (TS) which is a high-speed computer made with integrated circuits in T.T.L. logic. In this manner the trigger impulse is controlled by the projectile velocity. This set up allows the wave to be accurately captured at a single position in the model at the moment of film exposure and gives good reproducibility in spite of the fluctuation of the projectile velocity from one shot to the other.

About the synchronization problem, precise details are given in |28, 31|.

6.3 Results

Experiments were conducted to visualize the isochromatic and isopachic-fringe patterns in a dynamic problem. The model was a disk of araldite B (Hysol 4290), 180 mm diameter and 5 mm thick, tested by radial impact. The projectile was a ball 4.5 mm diameter with a velocity of 54 m/s. In order to avoid destruction of the model, a 1 mm thick piece of metal made from an aluminium alloy was glued to the edge of the model. A new piece of metal was used for each shot. This operation did not significantly alter the boundary conditions. Figure 11 shows a vertical section of the set up.

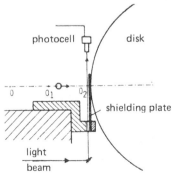

Fig. 11 Projectile shock on the disk

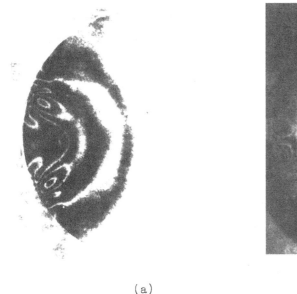

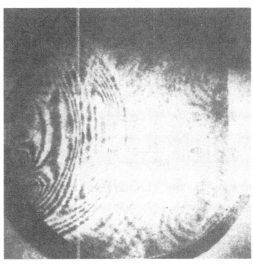

                    (a)   (b)

Fig. 12 Simultaneous isochromatic (a) and isopachic (b) fringe patterns
in a disk subjected to a radial impact for a delay of 44 microseconds,
recorded by the Faraday rotator technique.

    Figures 12 a and 12 b are examples of exposures showing isopachic
fringes and whole-order isochromatic fringes recorded 44 microseconds
after impact. The resulting isochromatic and isopachic-fringe patterns are
shown together in Fig. 13. The principal stresses σ_1 and σ_2 may be computed using eqs (2) and (3) providing that the optic-mechanical constants
C and C' are known.
    The phase shifts are related to the principal stresses σ_1 and σ_2 by
the well known relations

$$\varphi_1 + \varphi_2 = \frac{2\pi e}{\lambda} \, C' \, (\sigma_1 + \sigma_2) \qquad\qquad (2)$$

$$\varphi_1 - \varphi_2 = \frac{2\pi e}{\lambda} \, C \, (\sigma_1 - \sigma_2) \qquad\qquad (3)$$

where   e = model's thickness

           λ = light source wavelength

 C and C' = optic-mechanical constants of the photoelastic material

    The ratio C'/C can easily be obtained by observing the isochromatic
and isopachic - fringe patterns on the free edge of the model where the
stress state is one dimensional. At the point where the isochromatics are
of k-order and the isopachics are of k' - order we have,

$$\varphi_1 - \varphi_2 = \frac{2\pi e}{\lambda} C \, \sigma_s = 2k\pi$$

$$\varphi_1 + \varphi_2 = \frac{2\pi e}{\lambda} C' \, \sigma_s = 2k'\pi$$

where σ_s is the circumferential stress. The ratio of the above two equations leads to k'/k = C'/C. Study of Fig. 13 indicates the ratio C'/C is

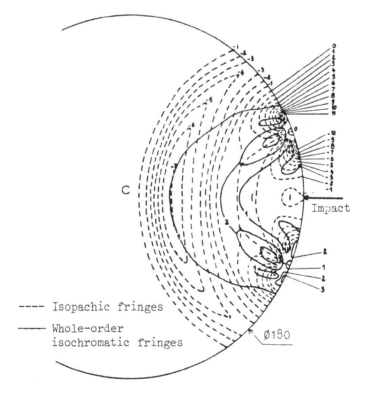

Fig. 13 Isochromatic-isopachic fringe patterns derived from Figs. 12 (a)
         and (b)

about 4. With C = 50 x 10^{-12} Pa^{-1} (value determined by others and in a-
greement with those found in the literature for Araldite B) we obtain :

C' = 200 x 10^{-12} Pa^{-1}

As a first application, we consider the calculation of principal
stresses along the axis of symmetry of the pulse 44 microseconds after
impact. The results are shown in Fig. 14.

The second application we present deals with the determination of
principal stresses at a point M of the disk as a function of time. The
point M selected is located on the axis of symmetry of the pulse at a distance x_M = 50 mm from the point of impact. Contrary to the first application, this calculation requires the conduct of many identical experiments,
which assumes the reproducibility of the phenomenon.

It has been shown that the real behaviour of the material is slightly
photoviscoelastic and introduces a variation of about 10 % from the stres-
ses obtained with the hypothesis of a pure photoelastic behaviour |28|.

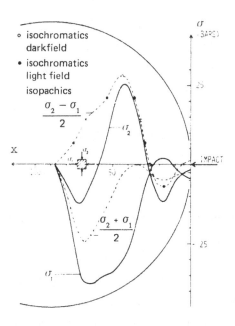

Fig. 14 : Variation of principal stresses along the symmetric
axis of the shock 44 µs after impact

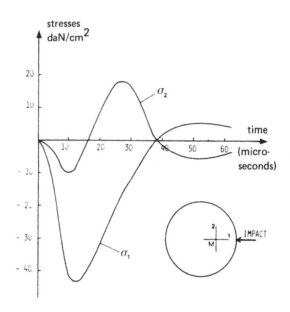

Fig. 15 : Variation of principal stresses σ_1 and σ_2 versus time
at a point of the symmetric axis of the shock distant
from 50 mm of the impact point

References

|1| Gabor D. A new microscopic Nature, 161, p. 777-778 (1948)

|2| Leith E.N. and Upatnieks J. Reconstructed wavefronts and commu-
       nication theory J.O.S.A. 52, p. 1123 (1962)

|3| Françon M. Holographie. Masson Cie Ed. (1969)

|4| Vienot J.Ch., Smigielski P., Royer H. Holographie Optique, développe-
       ments, applications. Dunod Ed. (1971)

|5| Froehly C. Un moyen de l'optique moderne : cours sur l'holographie.
       Ecole d'été internationale sur l'optique moderne en métrologie.
       Orsay Juillet 1971.

|6| Lohmann A.W. Reconstruction of vectorial Wavefronts. Appl. Opt. 4,
       (12), p. 1667, c. 99 (1965)

|7| Bryngdahl O. Polarizing holography J.O.S.A. 57, p. 545-546 (1967)

|8| Lurie M. Effects of partial coherence on holography with diffuse
       illumination J.O.S.A. Octobre 1966 vol. 56 n° 10, p. 1369

|9| De Velis J.B., Raso D.J., Reynolds G.O. Effects of source size on
       the resolution in Fourier transform holography J.O.S.A. vol. 57 n° 6
       Juin 1967 p. 843

|10| Surget J. Application de l'interférométrie holographique à l'étude
des déformations des corps transparents. Recherche aérospatiale, 3
(1970)

|11| Vienot J.Ch. et Monneret J. Interférométrie et photoélasticimétrie
holographique. Rev. Opt. 46(2) p. 75-79 (1967)

|12| Fourney M.E. Application of holography to photoelasticity Exp. Mech.,
8, (1), p. 37-38 (1968)

|13| Hovanesian J.D., Brüc V. and Powel R.L. A new experimental stress
Optic method : stress-holo-interferometry Exp. Mech. 8, (8), p. 362-
368

|14| De Bazelaire E., Prades B. Application de l'holographie et des for-
mes de lumière à la séparation des contraintes. Sym. Appl. de l'holographie. Besançon Juillet 1970.

|15| O'Regan R. and Dudderar T.D. A new holographic interferometer for
stress analysis. Exp. Mech., 11, (6), p. 241-247 (1971)

|16| Fourney M.E., Waggoner A.P. and Mate K.V. Recording polarization
effects via holography Exp. Mech., 10 (5), p. 177-186 (1970)

|17| Setson K.A. and Powell R.L. Hologram interferometry J.O.S.A. 56, (9),
p. 1161-1166 (1966)

|18| Nicolas J. Contribution à la détermination de la somme des contrain-
tes principales au moyen de l'holographie C.R.Ac.Sc. Paris 267-A,
p. 371-373 (1968)

|19| Nicolas J. Contribution à la détermination de la somme des contrain-
tes principales au moyen de l'holographie. Revue Française de Mécanique 28, p. 47-56 (1969)

|20| Ebbeni J., Coenen J. and Hermanne A. New Analysis of Holophotoelas-
tic Patterns and their Application. The Journal of Strain Analysis,
11, (1), p. 11-17 (1976)

|21| Hovanesian J.D. Elimination of ischromatics in photoelasticity.
Strain, 7, 151 (1971)

|22| Chatelain B. Etude expérimentale de deux méthodes d'observation si-
multanée ou indépendante des réseaux d'isochromes et d'isopaches re-
latifs à un modèle unique soumis à une seule sollicitation. Thèse
de 3e Cycle, Besançon, 1972.

|23| Cadoret G. Contribution à l'utilisation de méthodes de l'optique cohérente dans l'analyse expérimentale des contraintes. Thèse de Docteur Ingénieur, Paris VI, 1976.

|24| Gasvik K. Separation of the isochromatics-isopachics patterns by use of retarders in holographic photoelasticity. Experimental Mechanics, 16, (4), p. 146-150 (1976)

|25| Assa A. and Betser A.A. The application of holographic multiplexing to record separate isochromatic and isopachic fringe patterns. Experimental Mechanics, 14, (12), p. 502-504 (1974)

|26| Holloway D.C., Ranson W.F. and Taylor C.E. A neoteric interferometer for use in holographic photoelasticity. Experimental Mechanics, 12, (10), p. 461-465 (1972)

|27| Holloway D.C. and Johnson R.H. Advancements in holographic photoelasticity Exp. Mech., 11 (2), 57-63 (1971)

|28| Lallemand J.P. Interférométrie holographique et détermination des lois de comportement mécanique et optico-mécanique en régime transitoire. Thèse de Doctorat ès Sciences Poitiers 1980

|29| Lallemand J.P. et Lagarde A. Applications de l'holographie interférométrique à la détermination des contraintes principales planes dans un modèle photoélastique soumis à un choc. Revue Française de Mécanique n° 69, 1979

|30| Lallemand J.P. et Lagarde A. Separation of isochromatics and isopachics using a Faraday Rotator in Dynamic-holographic Photoelasticity. Experimental Mechanics May 1982

|31| Lallemand J.P. Synchronization Problems in Dynamic Holographic Photoelasticity. Experimental Mechanics Décembre 1981.

CHAPTER 8

PUNCTUAL POINT MEASUREMENT METHODS IN DYNAMIC PLANE PHOTOELASTICITY

We shall now discuss punctual point measurement methods for dynamic
plane-stress problems in the case of rapid loading. Such dynamic measurements deserve a special chapter, both because of their significance and because of the fact that, using   very simple instrumentation, one may obtain the stress tensor at every moment.

1. General considerations

As it is known, the punctual point measurement-methods are rarely used
due to the fact that the initial conditions cannot be accurately controlled and also because the loading is a time consuming operation (even if
reproducible). Nevertheless these methods have the advantage of providing
continuous accurate information useful, for example, to study a constitutive optico-mechanical law or to make possible a follow-up of the pattern
evolution between successive frames when a Cranz-Chardin camera is used.
Moreover, some present developments allow the separation of stresses for
cases (Cf § 5) where interferometric or holographic methods are difficult
to apply.

Using a repetitive technique in order to make the phenomenon reproducible is decisive for the application as well as for the conception of
a method to reduce exploration time at a model. Thus it may be important
to collect all the data at one point in the course of a single test. Like-
wise, many interesting problems especially in mechanical and civil engi-
neering may benefit from our application of an idea of Becker |1| : rapid
repetition of reproducible shocks. In order to simplify the set-up, several successive measurements can then be rapidly taken for the same point.

Point methods consist of measuring the intensities of several rays

traversing the model with normal, nearly normal, or oblique incidence.
Two types of photometric methods can be distinguished, those with static
components based on classical photoelastic elements only, and those addi-
tionally introducing light modulators.

Amongst the latter are the high-speed rotating analyzer used by Sapa-
ly |2| in 1961, and the Pockels cell applied by Robert and Dore |3| in
1974. Their pass band of about a tenth of the modulation frequency at best
hardly exceeds 10 kHz, which is insufficient for the study of transients
produced by short shocks, as will be seen on time records given in the following. The improved performance of Ker cells with rotating fields permits
extension of the pass band, as suggested by Royer |4|, doubtlessly at the
cost of complex equipment.

In view of the actual performance of highly sensitive photodetectors
and the significant reduction of pass bands by introduction of a modulator,
the direct photometric methods with static components appear to us best
adapted to a study of rapid dynamic phenomena.

Our quick survey of the various methods will present first the problem of determining the principal directions and the relative retardation.

The essential difficulty is the measurement of the maximum level of
transmissible light intensity E_o for signal calibration. Zero level obviously corresponds to the photodetector voltage for black. The level E_o
varies slowly due to the photodetector drift and slight fluctuations of
the light source. Modulation techniques overcome this drawback exactly as
in static applications.

If the principal directions are constant in time, a circular polariscope readily shows up the characteristic aspects of intensity extremes
and permits us to ascertain whether retardation has reached values exceeding λ. The maximum level then certainly represents E_o. If, moreover, the
principal directions are known for instance by symmetry considerations, a
dark-field polariscope plane-polarised at $\pi/4$ can be used and will eviden-
tly give the value of E_o as just described |5|.

2. Literature survey

Schwieger and Reimann |6| developed in 1964 a method using reflection
from a silvered layer in the middle plane of a plate under reproducible

transverse impacts. Though this method is not related to this chapter concerning two-dimensional problems, we mention this study because it makes use of the same photoelastic measuring concept, the circular-plane polariscope. The method has two consecutive stages :

- use of a dark-field polariscope and a quarter-wave plate with the same orientation (Hence inactive). The polariscope is directed at $\pi/4$ with respect to the principal directions, considered known and fixed during one phenomenon (this hypothesis was proved experimentally). A first recording of the intensity of the reflected beam is made.

- use of a circular-plane polariscope obtained after rotation of the $\pi/4$ quarter-wave (the two polariscopes have the same level E_o) A recording of the intensity is again made.

The level $E_o/2$, which was the same for both polariscopes was measured in the second arrangement with the model unstressed and free of residual birefringence, as was assured by the glass materials used.

We recently discovered that this method was the first to use a circular-plane polariscope which allows easy calibration of the signal and which was also employed by us in 1973 $|7|$.

The authors determined tan $\Phi/2$ with $\Phi = 2\pi\delta/\lambda$ for values $|\delta| < \lambda$. This sufficed in view of the low photoelastic constant of their glasses.

In 1968, Bohler and Schumann $|8|$ recorded simultaneous signals obtained in nearly normal incidence with a plane dark-field and with a circular polariscope. The levels E_o of the two signals were determined by compensator before each test. They found absolute values for the angle of the principal directions and relative retardations for values below λ.

Lagarde and Oheix $|7|$ in 1973 recorded intensities of four rays of two wavelengths initially polarised circularly and then plane-analysed in two directions differing by $\pi/4$. They were the first, we believe, to obtain unambiguously the orientation of the principal directions and the relative retardation for values up to several wavelengths. We shall revert to this method in view of its very general character.

For reproducible phenomena, Kuske and Robertson $|9|$ in 1974 proposed successive recording of the intensities transmitted by a plane dark field polariscope in two polariscope orientations differing by $\pi/4$. When reproducibility was not assured they resorted to two rays traversing the model simultaneously in nearly normal incidence at the same point. A preliminary test was needed for E_o. They obtained the same results as Bohler and Schumann.

These direct photometric methods are accompanied by complementary measurements for stress separation. For known and constant principal directions Schwieger and Reimann as well as Kuske and Robertson have suggested measurements in oblique incidence in the principal planes. Oheix and Lagarde |10| |11| have also applied oblique incidence using two circular polariscopes with different wavelengths and with an arrangement assuring compensation for depolarisation on entry to and exit from the model. A newly given formulation does not demand prior knowledge of the principal directions. We shall not present these solutions since oblique incidence is unsuitable for study of rapid dynamic events, where it introduces an "integrating effect" in space and time.

Bohler and Schumann have given a solution for complementary information opposite to dynamic measurement. It employs, for the first time to our knowledge, the power and coherence of a fine laser beam to obtain interference of the rays reflected from the model surfaces without recourse to a reflective layer. The circularly polarised beam impinges in normal incidence on the model. The whole of the reflected rays, especially from multiple reflections on the surfaces, is separated from the incident ray by a half mirror and directed onto a photodetector, which receives an intensity of the form :

$$E'' = \frac{1}{2} E''_o \left(1 - \cos \Phi \cos (\psi + \psi_o)\right)$$

$$\text{with} \quad \psi = \frac{2\pi e}{\lambda} (C''_1 + C''_2)(\sigma_1 + \sigma_2) \quad ; \quad \psi_o = \frac{4\pi n_o e}{\lambda}$$

$$C''_1 = C_1 - \frac{\nu}{E} n_o \qquad C''_2 = C_2 - \frac{\nu}{E} n_o$$

n_o being the refractive index of the unloaded medium.

Determination of E_o is rather delicate. The authors note that ψ varies at a higher rate than Φ so that envelopes of E'' are given by curves $E_o\tilde{E}$ and $E_o(1 - \tilde{E})$, where $\tilde{E}$ is the normed signal derived from the circular polariscope signal. Thus they trace two families of affine curves $\partial\tilde{E}$ and $a(1 - \tilde{E})$

with a the parameter defining their relationships. The value of E_o corresponds to that value of a for which one curve in each family forms one of
the two envelopes of E".

   With the function cos Φ previously determined, the value of $\cos(\psi+\psi_o)$
is then found. As the cosine function repeats itself, the authors can find
the values of σ_1 and σ_2 only in so far as they can deduce a physical interpretation of the phenomenon or peruse spatial continuity from a point
where the magnitudes are known.

   Before briefly presenting our methods, we recall the definition of
the principal direction Ox_1 as that inclined to Ox at an angle α such that
$|\alpha| \leqslant \pi/4$. This convention, which clearly defines the principal axes, may
lead to a discontinuity in α of $\pm \pi/2$ when α attains the value $\pm \pi/4$, together with a discontinuity in σ_1 and σ_2 and in their difference but not
in their sum. Such discontinuities arising from notation have clearly no
physical significance.

3.   The method of Oheix-Lagarde

   This method uses four rays of two wavelengths λ_1, λ_2 with initial right-handed polarisation and with analyser orientations Ox and Ox' that
differ by $\pi/4$, the transmitted intensities have the form :

$$E_1 = \frac{1}{2} E_{01} (1 + \sin 2\alpha \sin \Phi_1) \; ; E'_1 = \frac{1}{2} E'_{01} (1 + \cos 2\alpha \sin \Phi_1)$$

$$E_2 = \frac{1}{2} E_{02} (1 + \sin 2\alpha \sin \Phi_2) \; ; E'_2 = \frac{1}{2} E'_{02} (1 + \cos 2\alpha \sin \Phi_2)$$

where the subscripts 1, 2 correspond to the wavelengths λ_1, λ_2. The angle
α is unambiguously obtained from the relations :

$$\tan 2\alpha = \frac{2E_1 - E_{01}}{2E'_1 - E'_{01}} = \frac{2E_2 - E_{02}}{2E'_2 - E'_{02}}$$

The wavelength yielding the better precision may be chosen.

   Knowledge of α permits us to calculate sin Φ_1 and sin Φ_2 and hence
to deduce the values of

$$\psi_1 = \text{arc sin} (\sin \Phi_1) \qquad |\psi_1| < \pi/2$$
$$\psi_2 = \text{arc sin} (\sin \Phi_2) \qquad |\psi| < \pi/2$$

From the values of ψ_1 and ψ_2 the relative retardation δ can be determined analytically $|12|$ $|13|$ as long as

$$|\delta| < \frac{1}{2} \frac{\lambda_1 \lambda_2}{|\lambda_1 - \lambda_2|}$$

The last condition may be circumvented by a graphical method $|14|$ by drawing a nomogram with ψ_1 abscissa and ψ_2 as ordinate. The locus of the representative points, as a function of δ, is a broken line of slope $\pm \lambda_1/\lambda_2$, which can be graduated in relative retardations (fig. 1). Given $\psi_1 = a$, $\psi_2 = b$, the value of δ can be read. At points where branches of the broken line intersect, two values are possible. This ambiguity is resolved by recourse to the continuity of measurements in time. Knowing the sign of d sin Φ_1/dt and of d sin Φ_2/dt one may obtain the sign of cotg $\Phi_2 \cdot \dfrac{d\phi_2}{d\phi_1} = \dfrac{d\psi_2}{d\psi_1}$ and see if the point representing the birefringence is located on the positive or negative slope of the graph.

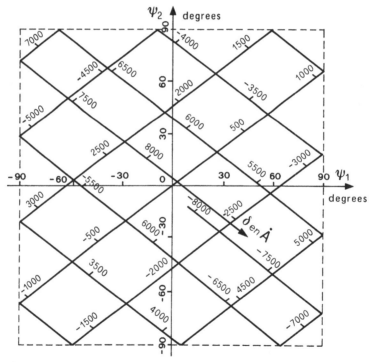

Fig. 1 Nomogram for birefringence measurements

Let us regard the problem of measuring the level E_o, in the case of reproducible phenomena. The signals E and E' for the same wavelength are received by the same photodetector. As they result from beam passing through the same elements they are implicitly calibrated, which means

that their maximal amplitudes are the same ($E'_o = E_o$) and are recorded on
the same negative film. The points of intersection of the traces $E(t)$ and
$E'(t)$ are readily seen and correspond to $\sin \phi = 0$. As their ordinate
equals $E_o/2$ the required quantity is thus obtained very easily and without
preliminary test. If the value $\Phi = \pm \pi$ is not reached, it suffices to measure, for the unloaded model, the light intensities E and E'' for two analyser directions at $\pi/2$ to each other. Then $E + E'' = E_o$. Note that owing to
residual birefringence, for example in Araldite, E_o is not double the light
intensity transmitted through the unloaded model.

Results obtained are satisfactory : the angle α is found within less
than two degrees and the uncertainty in δ is about 0,02 of the mean wavelength.

In practice, the range for measurement of δ is limited by nomogram
line density which increases with $|\delta|$ and may cause interpretation errors,
but it remains at several wavelengths.

The method permits determination of α and δ for any given time t,
practically without any need to follow the development of these quantities.

4. The method of Grellier-Lagarde |15-16-17|

The new method proposed by Grellier and Lagarde |15| |16| on the con-
trary, can determine the above quantities only by taking account of their
evolution in time. However it uses the previous method in an easier form,
with only a single wavelength. It resorts to Bohler and Schumann's complementary measurement but simplifies evaluation by its choice of set-up.

Devised specially for the study of transients, it takes into consideration the instants τ_o and τ_p which mark the start and end of the wave passage over the point considered and at which the material is just unloaded.
The quantities Φ and ψ are thus zero outside the time interval (τ_o, τ_p).

In accordance with the foregoing, the method is based on measurement
of the three signals.

$$E = \frac{1}{2} E_o (1 + \sin 2\alpha \sin \phi) \qquad (1)$$

$$E' = \frac{1}{2} E_o (1 + \cos 2\alpha \sin \phi) \qquad (2)$$

$$E'' = \frac{1}{2} E_o (1 - \cos \Phi \cos (\psi + \psi_o)) \qquad (3)$$

with $\phi = 2\pi\delta/\lambda$ $\delta = Ce\,(\sigma_2 - \sigma_1)$

The level E_o is obtained as previously indicated. Fig. 2 reproduces recorded traces of E and E'.

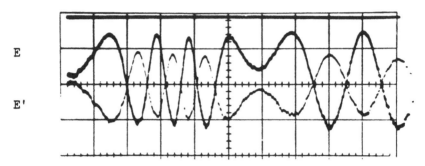

E

E'

Fig. 2 Time-dependence of the intensity E and E' during a transient, at
a point

4.1 Determination of principal directions

The angle α may be obtained using the relation

$$tg\ 2\alpha = \frac{2E - E_o}{2E' - E_o}$$

Let us note that when sin Φ = 0, the calculation is not possible ;
usually, this is not disturbing because the phase difference ϕ has a fas-
ter variation than the orientation of principal directions. Except the
points where $|\alpha| = \pi/4$ and tan 2α changes its sign, the variation of α
with time is continuous. It is enough to make the calculation at points
where the accuracy is considered acceptable.

4.2 Determination of the difference of principal stresses

Let us emphasize first that the description is longer than the process itself ; in order to simplify it, we shall neglect for the moment,
the residual birefringence which means :

$$\phi\,(\tau_o) = 0$$

At the begining, we separate the values ϕ and α appearing in (1) and
(2). In order to do so, we calculate at every moment the value $\tilde{E}$ defined
as follows :

$$\tilde{E} = \frac{E_o}{2} \left(1 + \sin \phi\right)$$

using :

$$\tilde{E} = \frac{E_o}{2} + \frac{1}{\sin 2\alpha} (E - E_o/2) \text{ and } \tilde{E} = \frac{E_o}{2} + \frac{1}{\cos 2\alpha} (E' - E_o/2)$$

The calculation of $\tilde{E}$ is simplified because E (or E') and E_o are experimentally determined and the angle α was previously calculated. Selection of one of the above relations depends on the value of α in order to obtain a better accuracy.

The next line of argument allows the determination of ϕ by knowing $\tilde{E}$. If the variation of α with time is small, it is simpler to operate directly on E or E' without calculating $\tilde{E}$.

First, one has to locate, using the graph, all the points where the angular birefringence has the value $k \frac{\pi}{2}$, by observing where $\tilde{E}$ has a maximal value E_o, a minimal value zero, or a value $E_o/2$.

The next step is to determine the time τ_m when the angular birefringence ϕ is extremal, which means $\frac{\partial \phi}{\partial t} (\tau_m) = 0$.

For $\phi (\tau_m) \neq 2k\pi \pm \frac{\pi}{2}$ the experimental value of $\tilde{E}$ is clearly different from values corresponding to $\sin \phi = \pm 1$ because it does not appear at the same level.

If $\phi (\tau_m) = 2k\pi \pm \frac{\pi}{2}$, $\tilde{E}$ being zero or E_o, it is difficult to distinguish the extremum $\tilde{E} (\tau_m)$ from the extremal values for $\sin \phi = \pm 1$, because they appear at the levels zero and E_o.

We propose several techniques to solve this problem :

- when the birefringence has an important time-gradient the diagram of $\tilde{E}$ presents a succession of maximal and minimal values similar to a "sinusoïdal curve", whose period is lower as $d\phi/dt$ is greater. If $d\phi/dt$, for example, decrease and pass through zero, the peaks of the "sinusoïdal curve" keep spreading. It is easy to locate such a zone on the graph of E. Moreover, the passage through zero of $d\phi/dt$ is accompanied by an obviously flattened extremum.

For low values of birefringence, when the phenomenon is repeated, we may add a small birefringence (a fraction of wavelength) using a Bravais compensator oriented as needed. These simple variation, brings $\phi (\tau_m)$ to a value distinct from $2k\pi \pm \dfrac{\pi}{2}$ and the extremum is then easy to observe.

For low values of birefringence, if the phenomenon is not repeated, one may :

(i) calculate at several points, near by the extremum, the value :

$$\bar{\phi} = arc\ sin\ (\dfrac{2\tilde{E}}{E_o} - 1)$$

(ii) calculate the value : $\left|\dfrac{d\bar{\phi}}{dt}\right|$ which is equal to $\left|\dfrac{d\phi}{dt}\right|$. If this deriva-
tive passes through zero, an extremum of ϕ is found.

Locating, this way, all the moments τ_m where ϕ is passing through an
extremum and through $k\pi/2$, it is important now to determine the value of
k, considering the fact that between two successive moments τ_m, ϕ (and
therefore k) have a monotone variation.

Previously, one has to know the direction of the ϕ variation at the
start of the phenomenon. As already mentioned, the residual birefringence
is small enough to be neglected. In other words, we have at the moment τ_o,
$\phi = 0$ and $\tilde{E} = \dfrac{E_o}{2}$.

Let us put $\dfrac{d\phi}{dt} = \varepsilon\ \left|\dfrac{d\phi}{dt}\right|$ with $\varepsilon = \pm\ 1$

and study the direction of the $\tilde{E}$ variation at the very start of the phenomenon.

- If $\tilde{E}$ increases from $E_o/2$ to E_o, then sin ϕ has a positive value. Taking into consideration the fact that the function sine is uneven we may
infer that ϕ will be positive and increasing between τ_o and τ_1 (τ_1 being
the moment when ϕ is passing for the first time through an extremum) hence
$\varepsilon = +\ 1$.

- If $\tilde{E}$ decreases between $E_o/2$ and zero then ϕ will be negative and decreasing between τ_o and τ_1 ($\varepsilon = -\ 1$).

Hence, the sense of variation and the sign of ϕ between τ_o and τ_1, can
be easily read out from the evolution of $\tilde{E}$.

Let $t_{1,k}$ with $k = 1,\ldots$ k_1 be the instants between τ_o and τ_1 where we
have $\phi = k.\dfrac{\pi}{2}$. We obtain $\phi\ (t_{1,k})$ by means of the following relations :

$$\phi\ (t_{1,1}) = \varepsilon\ \dfrac{\pi}{2}\ ;\ \phi\ (t_{1,k}) = \phi\ (t_{1,k-1}) + \varepsilon\ \dfrac{\pi}{2}$$

Let $t_{2,k}$ with $k = 1,\ldots k_2$ be the instants comprised between τ_1, and τ_2 where $\phi = k\frac{\pi}{2}$; $\frac{d\phi}{dt}$ having changed its sign ; we have :

$$\phi\ (t_{2,1}) = \phi\ (t_{1,k1})\ ;\ \phi\ (t_{2,k}) = \phi\ (t_{2,k-1}) - \epsilon\ \frac{\pi}{2}$$

Generalizing for $t_{m,k}$ comprised between τ_{m-1} and τ_m :

$$\phi\ (t_{m,1}) = \phi\ (t_{m-1},\ k_{m-1})$$

$$\phi\ (t_{m,k}) = \phi\ (t_{m,k-1}) \pm \epsilon\ \frac{\pi}{2}$$

with the sign + if m is an odd value and the sign - if m is an even value.

Fig. 3 is a typical illustration of exploration .in the case where α is constant (we can then make deductions from E and E' without recourse to the calculation of $\bar{E}$). Two extremes $\phi\ (\tau_1)$ and $\phi\ (\tau_2)$ are visible. Between τ_0 and τ_1, there are six points ($k_1 = 6$) where ϕ goes through a value which is a multiple of $\frac{\pi}{2}$.

There are nine between τ_1 and τ_2. Proceeding in the described manner, we have determined at these points the value of ϕ.

After having singled-out all the instants t_k where $\phi\ (t_k) = k\frac{\pi}{2}$ and determined the value of k at these instants, it remains to calculate ϕ at any instant t knowing that at this instant, ϕ has a value included between $k\frac{\pi}{2}$ and $(k + \epsilon)\ \frac{\pi}{2}$, $\epsilon = \pm 1$ according to whether ϕ is increasing or decreasing. The calculation can be carried out with the help of the following relations :

for k = 4p (p integer) :
$$\phi = k\ \frac{\pi}{2} + \text{arc sin}\ (\frac{2\bar{E} - E_o}{E_o})$$

for k = 4p+1
$$\phi = (k + 2\epsilon - 1)\ \frac{\pi}{2} - \text{arc sin}\ (\frac{2\bar{E} - E_o}{E_o})$$

for k = 4p+2
$$\phi = (k + 2\epsilon - 2)\ \frac{\pi}{2} - \text{arc sin}\ (\frac{2\bar{E} - E_o}{E_o})$$

for k = 4p+3
$$\phi = (k + 4\epsilon - 3)\ \frac{\pi}{2} - \text{arc sin}\ (\frac{2\bar{E} - E_o}{E_o})$$

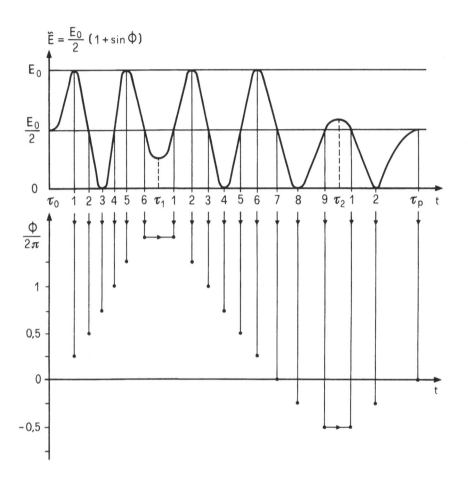

Fig. 3 Determination of the discretised values of the phase difference
        φ from a typical Ẽ diagram.

Knowledge of ϕ at random instants is not very useful when the instants t_k are numerous and close together, the tracing of the curve $\phi(t)$ being carried out without any difficulty. However it proves necessary :

- when the birefringence evolves very little in time
- to determine with precision the values $\phi(\tau_i)$ at the extremes.

Knowledge of $\phi(t)$ enables us to determine the evolution of the difference of the principal stresses at the measurement point during the phenomenon (by taking a law of elastic behaviour)

$$\sigma_2 - \sigma_1 = \frac{\lambda}{2\pi Ce}\, \phi$$

where C is the photoelastic constant of the material.

We have seen previously that the selection operated in the definition of the principal axes can lead to discontinuities on the orientation α. It is the same for the birefringence. Let τ_d be the instant where the first discontinuity occurs. The method enables us to know the value $\phi(\tau_d^-)$ and afterwards the value of $\phi(\tau_d^+)$ which is its opposite. In practice, we note that the slopes of ϕ at τ_d^- and τ_d^+ are also in opposite directions. The sense of variation of ϕ has thus changed. The calculation of ϕ is then carried out as previously indicated.

If a residual birefringence does exist it superposes upon the state of stress induced by the wave. The two tensors having generally different directions, it is possible |15| to determine for these two tensors the birefringence and the direction of the axes. In practical cases, it is not necessary to execute this operation. The residual birefringences are low with respect to the maximum birefringence induced by the wave. This may signify big variations of α at the start and at the end of the phenomenon |16| and indicates the advantage of the method.

4.3 Determination of the sum of principal stresses

We shall use the recording of signal E'' which contains some information : (such as cosine) about the quantity ψ, an affine function of $\sigma_1 + \sigma_2$

$$E'' = \frac{E''_o}{2}\left(1 - \cos\phi \cos(\psi + \psi_o)\right)$$

with

$$\psi = \frac{2\pi}{\lambda} \, e \, (C'_1 + C'_2) \, (\sigma_1 + \sigma_2)$$

The method of data processing from the recording is based on the fact that the time-gradient of ψ (in absolute value) is much higher than the one of ϕ. Please note that this condition was obtained when the selected material was PSM1. Then the cosine of $\psi + \psi_o$ is modulated by the cosine of ϕ, E" goes through a series of maxima and minima corresponding to the values of $\psi + \psi_o = k\pi$.

The extremes are situated on two envelopes given by :

$$E_a = \frac{E''_o}{2} (1 - \cos \phi) \qquad \text{for } \psi + \psi_o = 2k\pi$$

$$E_b = \frac{E''_o}{2} (1 + \cos \phi) \qquad \text{for } \psi + \psi_o = (2k+1) \, \pi$$

The curve E" presents a succession of nodes and anti-nodes corresponding to the values $\phi = (2k+1) \frac{\pi}{2}$, $\phi = k\pi$ respectively.

The instants τ_m where $\psi + \psi_o$ goes through an extreme (where $\frac{\partial \psi}{\partial t} (\tau_m)=0$) are localized by observing the zones where the "period" of E" increases significantly. If $\psi(\tau_m) + \psi_o \neq k\pi$ we have a characteristic "swelling" in the recording of E" which does not reach the envelope (point C of fig. 4). If $\psi(\tau_m) + \psi_o = k\pi$ this swelling reaches the envelope for a certain period of time (point B of fig. 4). Between two successive extremes, $\psi + \psi_o$ has a uniform variation ; the different values of the integer k are then easily determined. We are not going to describe the process any further, because it differs but little from the one previously described for the determination of ϕ.

In all the experimental cases which we have dealt with we generally limited ourselves to reading out the instants where $\psi + \psi_o = k\pi$ (k being an integer) and to determining k, which was sufficient, considering the important variations of ψ.

From the expression of E" only, it is not possible, considering the evenness of the cosine function, to determine the sense of variation of $\psi + \psi_o$, that is to say we do not know at the instant τ_o the sign of $\frac{\partial \psi}{\partial t}$ or again the sign of $\varepsilon = \pm 1$ with $\varepsilon \frac{d\psi}{\partial t} > 0$.

At the instant τ_o we have $\phi = \psi = 0$ and afterwards we determine $|\psi_o|$.
The sign of ψ_o in function of ε is easily obtained by observing the variation of E". Indeed $\varepsilon \, \psi_o$ is positive or negative according to whether E"
increases or decreases at the start.

In order to determine ε we can proceed by a physical interpretation
in the simple cases or, by a spatial continuity starting from a point (free
boundary for example) where this sign can be known.

We propose, for the study of reproducible phenomena and when the angle α does not vary too much, a technique which resorts to complementary
measurements. Let us give the following lemma, easy to demonstrate. Let
two real magnitudes be a and b of which $|a|$, $|b|$ and a-b are known.

- if $|a| > |b|$ a+b and a-b have the same sign
- if $|a| < |b|$ a+b and a-b have opposite sign

We use a polarised incident light successively orientated along the
two principal directions (known beforehand). We record the two intensities

$$E''_1 = \frac{E''_{o1}}{2} (1 - \cos \phi_1)$$

$$E''_2 = \frac{E''_{o2}}{2} (1 - \cos \phi_2)$$

We can easily obtain $|\phi_1|$ and $|\phi_2|$. Knowing $\phi_1 - \phi_2 = 2\phi$ we deduce
from the lemma the sign of $\psi_1 + \psi_2 = 2(\psi + \psi_o)$. The value of ε can be derived from it immediately, hence that of ψ_o and then that of ψ.

We then have the sum of the principal stresses :

$$\sigma_1 + \sigma_2 = \frac{\lambda}{2\pi e \ (C'_1 + C'_2)} \, \psi$$

Note : to determine the sign of ψ it is necessary to know $|\phi_1|$ and $|\phi_2|$
over a time interval $|\tau_o, \, \tau_o + \Delta\tau|$. Experience has shown to us that this
determination can be carried out without any risk of errors if the angle
α does not vary more than about ten degrees during this interval of
time. It is always possible to restrain $\Delta\tau$ to be sure this condition is
satisfied.

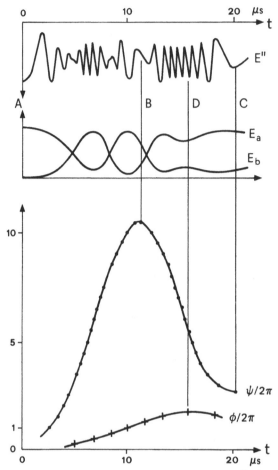

Fig. 4   Determination of ψ from a recording of E'' (E_a and E_b are the
enveloping curves of E'')

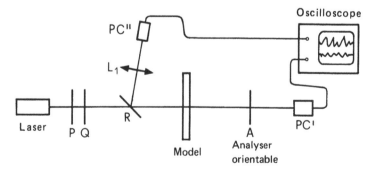

Fig. 5 : Optical set up for a consecutive recording of E and E'.

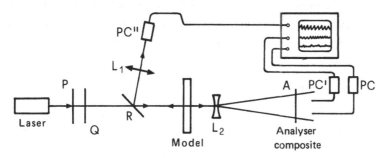

Fig. 6 : Optical set up for a simultaneous recording of E and E'

P+Q circular polarisers

A   linear analyser (suitably orientated)

R   half mirror - L_1 and L_2 lenses

PC PC' PC" photocells

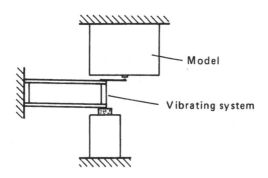

Fig. 7 Mechanical set-up

4.4 Application of the method |15|

4.4.1 Optical set up

Two alternative ways are proposed for the recording of the signals E and E' : using or not for the point analysis, repetition of the phenomenon.

A laser source S provides a light beam of 4880 Å wavelength which is plane polarised. A suitably orientated quarter-wave plate then gives a right handed circular light vibration (Fig. 5). The beam reflected from the model surfaces is separated from the incident beam by a half-mirror. It is collected by the photocell PC" placed in the image plane of the model formed by a lens L_1. The lens L_1 plays an important role in ensuring convergence of the ordinary and extraordinary rays which issue from the same point of the model and owing to the gradient in refractive index emerge with orientations slightly different from that of the ray reflected from the entry surface. For transmitted rays, J.T. Pindera has already pointed out this situation. The light intensity E" received by the tube is transformed into a proportional current which is recorded on the first channel of an oscilloscope TEKTRONIC 555.

- If the phenomenon is reproduced, the transmitted beam is rectilinearly analysed before it is collected by a photomultiplier. A catch-system on which the rectilinear analyser A is fixed enables the latter to be oriented along two directions at an angle of 45° one from the other. This device is itself dependent on a set-up allowing the rotation of the whole system around the optical axis. The two light intensities E and E' are recorded one after the other on the second channel of the oscilloscope.

The determination of E_o is the same as the one in the previous method ; knowledge of E_o is necessary for the calculation of the orientation α and of $\bar{E}$.

- For the study of a non-reproducible phenomenon, it is necessary to obtain simultaneously the two signals E and E'. For this, a composite analyser consisting of two juxtaposed polarisers, at 45° relative orientation, is used (see fig. 6). The transmitted light proceeding from each of these analysers is directed to two adjustable photocell (PC and PC') through fiber optics conductors. It is essential to adjust them so that the same maximum intensity will be obtained for E and E'. In the case, only the birefringence and the principal directions are needed, it is enough to eliminate the half mirror R.

In order to study the influence of a defective quarter-wave, let us call ε the error of the retardation and η the angle of its fast axis with the analyser. It may be demonstrated $|14|$ that for $\eta = 0$ and $\eta = \frac{\pi}{2}$ the absolute value of the error is increased to $2|\varepsilon|$. Moreover, this error is eliminated by carrying out two consecutive measurements with $\eta = 0$ and $\eta = \frac{\pi}{2}$. It is interesting to observe that the measurement of $E_o/2$ is not influenced by the defect of the quarter-wave if $\eta = 0$ and if $\eta = \frac{\pi}{2}$. Hence, in the case of repetitive phenomena, it is not necessary to use a high quality quarter wave but one which may be obtained using on adjusted Bravais-compensator.

4.4.2 Mechanical set-up

The generator of brief repetitive shocks consists of a vibratory system with two degrees of freedom with a mass attached through a flexible
blade to a system of elastic plates. The whole system is excited sinusoïdally at one of its resonant frequencies (40 Hz). The two plates and the exci-
ting coil are attached to a heavy and rigid frame (In order to porperly
isolate the vibratory system). Fig. 7.

During each cycle, the mass hammers on a small anvil cemented to the
model to avoid progressive damage. A reproducible impact thus occurs 40
times per second.

A very simple electric circuit, connecting the anvil and the little
mass permits the setting of the time basis for the oscilloscope in synchronization with the onset of the shock, that is to say with the instant
where the mass comes into contact with the anvil.

The advantage of such a system lies in the fact that a phenomenon of
perfect reproducibility is obtained, which allows a great facility for the
recording of the three light intensities, necessary for the application of
our method. Its repetitive character and the simplicity of the manipulations to be performed allow measurements to be carried out at a great number of points in the model.

5. Application (example) |18|

To illustrate the method let us show the possibility to study the evolution of the state of stress in the matrix of a composite material, with
parallel filters during propagation of a wave of stress due to a mechanical
shock.

The structure is a plate loaded in its plane, and the fibers are
straight cylinders perpendicular to the plane and uniformly distributed.

To characterize such a material one has to define :

- the mechanical characteristics of the matrix and of the fibers (coefficients of Poisson ν_m and ν_f, Young's modulus E_m, and E_f, the characteristic
mass ρ_m and ρ_f)
- the density of fibers V_f (ratio of the sectional area of fibers versus the total area)
- the fiber distribution

5.1 Similitude

Let us present briefly some basic principles of similitude, in order to justify the use of a model to anticipate the stress distribution in a prototype.

For a dynamic problem of elastostatic it is well known, that one has to obtain the equality of four non-dimensional ratios :

$$\pi_i = {}_M\pi_i$$

where π_i is related to the prototype and ${}_M\pi_i$ to the model. The geometrical similitude is expressed by $\pi_1 = \dfrac{x_i}{L^*}$, the ratio between an ordinary length x_i and the characteristic length L^*. The equality $\pi_1 = {}_M\pi_1$ is expressing actually a similitude between the shape of the prototype and the shape of the model. The ratio $\gamma = {}_ML^*/L^*$ may be considered as an enlargement in our case. We take for the characteristic length the fiber diameter.

The elasticity similitude makes use of :

$$\pi_2 = \frac{E}{\sigma} \cdot \frac{u^*}{L^*} \quad , \qquad \pi_3 = \nu \qquad \gamma = \frac{{}_ML^*}{L^*} = {}_ML^*/L^*$$

The value u^* is a characteristic displacement. The expression π_2 allows to determine the stress state σ at a point x_i of the prototype, by knowing the state of stress ${}_M\sigma$ in a corresponding point ${}_Mx_i$:

$$\sigma(x_i) = {}_M\sigma \, ({}_Mx_i) \, \frac{E}{{}_ME} \cdot \frac{u^*}{{}_Mu^*} \cdot \gamma$$

The expression π_3 shows that the coefficient of Poisson has to be the same for the model and the prototype. The consequence is the equality of ratios of transversal and longitudinal velocities (C_T and C_L) :

$$\frac{{}_MC_T}{C_T} = \frac{{}_MC_L}{C_L} = \frac{{}_MC}{C}$$

with $C = (\frac{E}{\rho})^{1/2}$

The dynamic similitude introduces a fourth relation :

$$\pi_4 = C \, \frac{t^*}{L^*}$$

t^* being a characteristic time (for example, it can be the time-length of a shock or the impact time).

In the case of a composite material with parallel fibers, the geometrical similitude involves two conditions :

- the same bulk density of fibers
- the same disposition of the fibers in the matrix. As a characteristic length L^*, one may select the diameter d of the fiber.

If these two conditions are fulfilled, one may simulate (model) a composite material having fibers of about ten μm diameter (graphite for example) with a material having fibers of one mm diameter. In this case, the enlargement, geometric scale, γ will be then of 10^2.

Generally, the Young moduli of the fibers and of the matrix are different. One may, however, define an average modulus $\bar{E}$ using the law :

$$\bar{E} = E_f V_f + E_m (1 - V_f)$$

and an average characteristic mass :

$$\bar{\rho} = \rho_f V_f + \rho_M (1 - V_f)$$

which allows to define a velocity of propagation :

$$\bar{C} = (\bar{E}/\bar{\rho})^{1/2}$$

The equality of ratios π may be then written :

$$\frac{\bar{C}}{{}_M\bar{C}} \cdot \frac{t^*}{{}_M t^*} \cdot \gamma = 1$$

This expression relates the three parameters which have to govern the selection in the process of modeling.

- the velocity of the wave in the material
- the type of shock (impact-time)
- the enlargement (geometric scale)

5.2 Building and testing the model

The model is made of a rectangular plate of PSM1 of 6,25 mm thickness fixed on one side on a stiff support, the impact being applied in the middle of the opposite side.

The vibrating loading system allows to apply an impact of reduced amplitude and short time (about 50 to 80 µs). The size of the model is sufficient to avoid too fast reflections on the sides. The geometric scale allows impact-times which are realistic for the prototype.

The fibers are small steel cylinder glued in holes, cut into the plate. The fiber diameter of one mm allows a good adhesion. The distances between fibers are 3 mm which represents materials with law bulk density of reinforcement.

As an example let us indicate some results obtained for a grid oriented at $\pi/4$ versus the sides of the model and for measurement points located on the loading axis. One may check that this direction is a principal direction. Figure 9 shows the evolution of principal stresses for a point at 3 mm from the loaded side. The stress σ_1, is always negative and much higher in absolute value than σ_2. This shows that the main phenomenon is a wave of longitudinal compression.

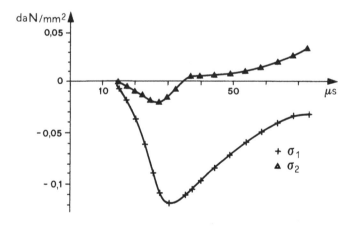

Fig. 9 Principal stresses for a point at 3 mm from the loading side

References

1. Becker, H. "Simplified Equipment for Photoelastic studies of Propagating Stress Waves", Proc. of S.E.S.A., 18, 1961.

2. Sapaly, J. "Contribution à l'étude de la photoextensométrie statique et dynamique", Thèse Paris 1961.

3. Robert, A., Doré, A., "Méthode de photoélasticimétrie dynamique", R.F.M., 1974, n° 50-51.

4. Royer, J. "Contribution à l'ellipsométrie automatique appliquée en photoélasticimétrie", (Thèse d'Etat - Université de Nantes - 1974).

5. Lagarde, A. "Etude du comportement mécanique et photoélastique d'un barreau biréfringent soumis à des vibrations", Thèse d'Etat, Paris 1962 Publication Scientifique et Technique du Ministère de l'Air, N.T. n° 127 (1963).

6. Schwieger, H., Reimann, V. "Der Bregestoss aufeine elastishe Rechteckplatte", Forsch. Ing. Wes. 1964, 90, n° 5, p. 140-145.

7. Lagarde, A. Oheix, P. "Determination of neutral axis and algebric value of birefringence in a plate by a static and dynamic point method I.U.T.A.M., Symposium of the photoelastic effect and its applications", Bruxelles, 1973.

8. Boehler, P., Schumann, W. "On the determination of dynamics states of stress", Experimental Mechanics, 1968, 8-3, p. 115-121.

9. Kuske, Robertson "Photoelastic stress analysis", John Wiley Sons, 1974, p. 473-475.

10. Lagarde, A. "Sur la photoélasticimétrie dynamique", Symposium International Waterloo-Ontario, Canada, 1972.

11. Lagarde, A. Oheix, P. "Construction et mise au point d'un photoélasticimètre permettant la détermination ponctuelle des contraintes pour des problèmes plans statiques et dynamiques et un matériau élastique biréfringent", Contrat D.G.R.S.T., n° 69 01853, 1975.

12. Lagarde, A., Oheix, P. "Méthode ponctuelle statique et dynamique de détermination des directions et de la différence des contraintes principales pour un matériau biréfringent", C.R.Ac.Sc., Paris, t. 270, Série A, p. 441-444, Septembre 1973.

13. Oheix, P. "Contribution à l'étude des méthodes de mesure en photoélas-
    ticimétrie bidimensionnelle", Thèse de Docteur Ingénieur, Poitiers,
    26 Juin 1975.

14. Lagarde, A., Oheix, P. "Méthodes ponctuelles statiques et dynamiques de
    photoélasticimétrie pour des problèmes bidimensionnels", Proc. 5th Int.
    Conf. Exp. Stress Analysis, Udine, 1974, Ed. by G. Bertolozzi (Tech-
    noprint-pitagora, Bologna) p. 121-128.

15. Grellier, J.P. "Méthode de photoélasticimétrie et d'interférométrie
    ponctuelle et à champ complet. Application à la détermination des con-
    traintes planes en régime dynamique transitoire", Thèse Docteur
    Ingénieur, Poitiers, 1982.

16. Grellier, J.P., Lagarde, A. "Point measurement method for the determina-
    tion of the orientation and algebraic values of principal stresses in
    a plane model submitted to transient loading" to be published "Solide
    Mechanics Archives Martinus Nijhoff Publishers.

17. Grellier, J.P., Lagarde, A. "Punctual method for determination of stres-
    ses in a photoelastic plane model subjected to an impact", Mec. Res.
    Com., Vol. 9, n° 2, 71-76, 1982.

18. Grellier, J.P., Lagarde A. "Propagation d'ondes de contraintes provo-
    quées par un choc dans un matériau hétérogène". Contrat D.R.E.T.
    n° 81046.

DYNAMIC PHOTOELASTICITY AND ITS APPLICATION TO STRESS WAVE PROPAGATION, FRACTURE MECHANICS AND FRACTURE CONTROL

James W. Dally
Mechanical Engineering Department
University of Maryland

CHAPTER 1

RECORDING SYSTEMS FOR DYNAMIC PHOTOMECHANICS

1.1 Introduction

Since research in the field of dynamic photoelasticity was initiated by Tuzi[1] in 1928, there has been a continuous development of new and improved high-speed photographic systems. With the development of new higher speed films, more intense light sources, ingenious camera designs, and reliable electronic circuitry, continuous improvement has been made in the quality of the photographs of dynamic events with objects or images propagating at high velocity. Three different photographic systems have been specially adapted for application involving dynamic photomechanics which provide whole field representation of the data. All three of these systems may be considered adequate for photographing high-density fringe patterns propagating at velocities as high as 100,000 in/sec (2540 m/sec), but each system exhibits advantages and disadvantages.

The first system utilizing a high speed framing camera such as the Beckmann and Whitley was introduced by Feder et al.[2] in 1956. The application of this camera to dynamic photoelasticity has been markedly improved by Flynn et al.[3-5] and used in a number of interesting problems. The optical system for this camera is illustrated in Fig. 1.1 and a view of the back of the assembly is given in Fig. 1.2. The light from the object is collected by a relatively long focal length lens and focused on a rotating mirror. This mirror is driven at high speed by a light gas turbine and the image of the object is swept about the drum. Framing is accomplished by using a series of slits and relay lenses which receive the image from the rotating triangular mirror. The slit width is adjusted so that an image is transmitted to the film plane for about 1/3

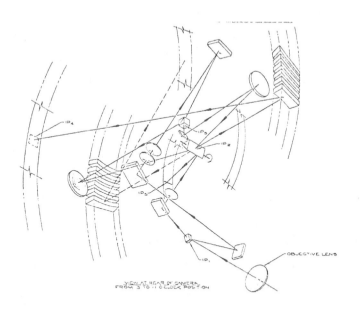

Fig. 1.1   Schematic of Optical System of a Framing Camera

Fig. 1.2   View of Back Framing Camera Showing Major Components

of the interframe interval. The relay lenses serve to focus the image of
the object from the mirror to the film plane. The film usually 35 mm in
width is positioned around the drum-like case for the camera.

Framing rates are controlled by adjusting the speed of the turbine.
Framing rates R from about 100,000 to 2,000,000 frames/sec are common.
Exposure times, t_e, are related to framing rates by

$$t_e = \frac{1}{3R} \tag{1.1}$$

Since exposure times of less than 1 µsec are usually desirable in
recording stress wave and crack propagation, the camera should be
operated at rates exceeding 333,000 frames/sec. The number of frames
which can be recorded vary with the individual camera manufacture and the
model; however, 30 to 80 frames is common. These cameras are expensive
with costs usually in the range of $100,000 to $200,000 for a complete
system which includes lighting and synchronization circuits.

The second high-speed photographic system entails the use of a
multiple spark camera developed originally by Cranz and Schardin[6] in
1929. This camera has been used extensively in Europe over the past
forty-five years in fluid dynamics, ballistics, and fracture; however,
its application in dynamic photoelasticity has been more limited.
Christie[7] in England described the first application of the
Cranz-Schardin camera to photo-elasticity in 1955, and Wells and
Post[8] introduced the camera in the United States in 1957. More
recently, Dally and associates[9-14] have employed this system in a
number of applications of dynamic photoelasticity related to problems in
geophysics and fracture. The multiple spark camera is moderate in cost,
easy to operate and well suited to dynamic photo-elasticity and caustic
studies of fracture or stress wave propagation. The multiple spark
camera will be described in considerable detail later in this chapter.

The third method of recording involves the use of a pulsed ruby
laser. The optics of this system are presented in Fig 1.3 where the
details of the laser cavity are presented. In dynamic recording, the
laser cavity is Q switched to produce a very short burst of coherent
monochromatic light. The exposure time is so short that it is difficult

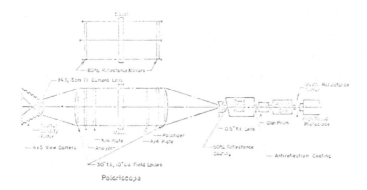

Fig. 1.3   Schematic of Optics of a Pulsed Ruby Laser

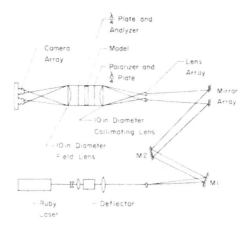

Fig. 1.4   Laser Photograph of a Dynamic Fringe Pattern

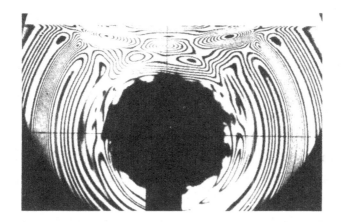

Fig. 1.5   Hybrid Laser-Schardin Recording System

to measure – less than 0.1 μsec. As is evident in Fig. 1.4, the quality of the photographs recorded is outstanding.

The disadvantage of most laser recording systems is the fact that only single frames can be recorded and it is usually necessary to repeat the experiment with carefully controlled time shifts to record the entire dynamic event. Taylor and Rowlands[15,16] have coupled a sequentially pulsed ruby laser with a high speed framing camera; however, synchronization problems are very difficult, costs become prohibitive and resolution is sacrificed. Taylor et al[17]. have also developed a hybrid system using an acousto-optic deflector to divert the beam of a sequentially pulsed laser and multiple lenses to produce up to five images as illustrated in Fig. 1.5. This system does offer promise but it is complex as both the electronics and optics are elaborate.

More recently Dally and Sanford[18,19] constructed a multiple ruby laser system for high speed photography. The system is basically a modification of the Cranz-Schardin arrangement where the traditional spark-gaps have been replaced by Q-switched ruby lasers. The light output from the lasers is transmitted to the optical system with fiber optic light guides as illustrated in Fig. 1.6. A microprocessor is incorporated into the control circuits to dynamically program in real time the firing of the flash lamps and the Q-switching of the lasers to obtain extremely precise sychronization with the dynamic event.

The advantages of this system include:

(1)   very short exposure times 35 ns

(2)   extremely high framing rates 10×10^6 fps

(3)   significant improvement of available light

(4)   highly monochromatic light which is free of coherent noise

(5)   improved control of the synchronization with event

(6)   adaptable to varied optical uses with fiber optics

The disadvantages are costs estimated in excess of $40,000 per frame and availability since the system is not produced commercially.

This chapter reviews the requirements of an optimum high-speed photo-

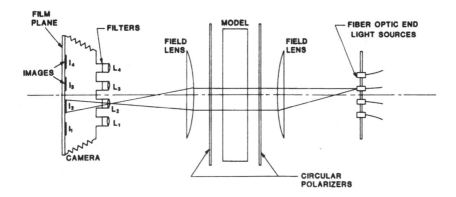

Fig. 1.6   Cranz-Schardin System with Fiber Optic Modification

graphic system for dynamic photomechanics and compares the characteristics of the Cranz-Schardin camera with these requirements.   Design
principles associated with the construction of a typical spark type
camera are given in detail.   Finally, a number of photoelastic fringe
patterns are presented to illustrate the dynamic results which can be
obtained by using this recording system.

1.2   Requirements of an Optimum Photographic System in Dynamic Photo-
mechanics

The characteristics which determine the adaptability of a high-speed
recording system to dynamic photoelasticity include: framing rate, resolution, exposure, synchronization, size of field and recorded image, and
total cost of the system.   The objective of a dynamic photomechanics
study is to obtain a prescribed time-sequenced series of negatives exhibiting sharply defined fringe or caustic patterns.   The patterns represent the propagation of several types of stress waves or the propagation
of a crack tip which occurs during a dynamic event.   The velocities of
the fringes being recorded may vary considerably in a high-modulus photoelastic model.   For instance, typical stress wave velocities for the

dilatational (P) wave, shear (S) wave, and Rayleigh (R) wave in CR-39 are
75,000, 43,000, and 40,000 in/sec (1900, 1100, and 1000 m/sec) respectively. In order to adequately record these high-velocity fringes, a
camera is required which operates in the range between 20,000 and 500,000
frames/sec. The lower framing rates are used to study crack tip propagation over reasonably large model distances. The higher rates are used in
examining dilatational waves interacting with model boundaries where
changes in the fringe pattern occur very rapidly and fringe velocities
exceed those of the incident wave.

Resolution of a dynamic fringe pattern is a function of the exposure
time t_e, the velocity of the fringe packet c_f, and the fringe gradient G.
The fringe movement s during the exposure period which is approximated by
$s = c_f t_e$ must be small relative to 1/G to maintain contrast and to prevent wash-out or blurring on the negative. To minimize s, exposure times
approaching zero are required, which represents a condition impossible to
achieve. In practice, values of t_e are held as short as possible, with
0.5 to 1.0 μs considered acceptable, providing fringe gradients do not
exceed 10 to 20 fringes/in (250 to 500 fringes/mm).

Proper exposure of film in dynamic photoelasticity is usually a more
difficult problem than in other high-speed photographic applications
because of the large light losses in the optical system. A typical dynamic polariscope utilizing two sheets of circular polaroid (HNCP-38) will
absorb about 84 percent of the incident light. Also, a significant
amount of light is absorbed by the notched optical filter which must be
employed to give a close approximation to monochromatic light. In most
cases, the intensity of light transmitted to the film is less than 1 percent of the intensity of the light source. For this reason, it is common
practice to use exceptionally bright sources in conjunction with pre-
fogged high-speed film to obtain adequate exposure for sharp negatives.

In a dynamic event, the time interval of interest ranges from about
50 μsec, as a lower limit for high-velocity waves in small-sized models,
to about 1000 μsec, as an upper limit for low-velocity cracks propagating
in relatively large models. The dynamic event is often initiated by a
small explosive charge which loads the model. The initiation of the

camera is usually delayed to permit the fringe pattern to propagate from
the load source to the region of interest in the model. The pre-selected
delay times usually range from 10 to 100 µs. The method of introducing
delay into the recording system must be variable and the control such
that the actual delay period be within ± 1 or 2 µs of the preset delay
period.

The required field size of a dynamic polariscope is usually large in
order to investigate a wide range of problems which arises in geophysical
and engineering applications. Field diameters from 10 to 15 in.
(250-375 mm) are usually sufficiently large to adequately cover the
region of interest even if models of much larger size are employed. The
size of the image on the film must also be large, since negatives on
high-speed film exhibit grain-size effects and do not produce fine
quality prints when the amount of enlargement is high. Usually an image
to field magnification ratio of about 1/10 produces images 1 to 2 in. in
diameter which can be subsequently enlarged by a factor of 5 to 10.

Finally, the total cost of the high-speed photographic system must be
kept within reasonable bounds. High-speed photography is inherently
expensive since high-performance light sources, complex mirror and lens
systems, electronic synchronization circuits, and other specialized components entail large costs. Usually, the cost of a high-speed spark
system adapted for dynamic photoelastic applications ranges between
$50,000 to $100,000. The cost of a laser system would be higher by
nearly a factor of 10.

1.3 Design of a Spark-type Cranz-Schardin Recording System

Multiple spark cameras can be designed with a single trigger which
initiates the sequential firing of the array of spark gaps or with
multiple triggers to fire each gap independently. This description per-
tains to the design of a system with a single trigger which the author
has found to be simple to operate and easy to maintain. Several different systems have been constructed by the author, and the system
described here represents a composite of design features which have
operated satisfactorily.

The Cranz-Schardin camera is comprised of three basic subsystems

which include: the spark assembly, the optical bench, and the control
circuits for synchronization. Each of these subsystems is described in
detail in the following subsections.

1.3.1 The Spark Assembly

An array of spark gaps provides a time sequenced series of high-
intensity light flashes used both to illuminate the object and to shutter
the camera. Physically each spark gap consists of two co-axial electrodes spaced with a gap of 7 to 9 mm. One electrode is hollow to accommodate the placement of a fiber optic light guide in close proximity to
the spark. In the ready mode of operation, both electrodes of each gap
are at the same potential; hence, the stability of the system is
excellent.

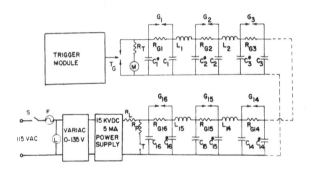

Fig. 1.7   Series Connected L-C Line Energizing an Array of Spark Gaps

The array of spark gaps is energized by a series of connected L-C
lines illustrated in Fig. 1.7. The capacitors (0.05 µF – 20 kV dc) C_1,
C_2, C_3 ... and C_1^*, C_2^*, C_3^* ... are charged to 15 kV dc with a 5 mA power
supply. The firing sequence of the spark gaps G_1, G_2, G_3 ... is initiated
by applying a 20-kV pulse to the trigger gap. When the trigger gap fires,
the capacitor C_1 discharges through the gap first to and subsequently below
ground potential. As the voltage on C_1^* becomes negative, a large potential difference develops across spark gap G_1 (about 25 kV), causing the air
in the gap to ionize and the energy stored in capacitor C_1^* to discharge
through the spark gap G_1. This discharge produces a short (0.5 µsec)

intense flash of light (10 joules).

The time between firing the first and second spark gaps is a function of the inductance L_2 in the $C_1^* L_2 C_2$ loop. When the spark gap G_1 fires, the voltage on capacitor C_2^* begins to decay sinusoidally with time. When the voltage on capacitor C_2^* decays to a negative value nearly equal in magnitude to the voltage on capacitor C_2^*, spark gap G_2 fires and activates the next loop in the line $(C_2 L_3 C_3^*)$. This process repeats until all of the spark gaps fire in a time controlled sequence.

The framing rate R is approximately equal to the reciprocal of the time required for the maximum negative voltage to develop in the sinusoidal oscillation of a given CLC loop. Thus, with a fixed capacitor bank, the framing rate can be controlled by varying the inductance as indicated below:

$$R = [1/LC]^{-0.5}/\pi \qquad\qquad (1.2)$$

where C is the total capacitance in μF

and L is the inductance in μH.

Framing rates varying in discrete steps from 20,000 to 500,000 frames/sec require inductance of 10,000 μH to 16 μH respectively. Inductors employed by the authors are formed with single windings on laminated plastic cores. The inductors are tapped at twelve locations and wired into a patch board. Jumper wires are used to insert the required inductance in the circuit to adjust the framing rate.

1.3.2 The Optical Bench

The optical system associated with the Cranz-Schardin camera which is illustrated in Fig. 1.6 performs three distinct functions, including polarization, image separation, and magnification. A dynamic light field polariscope is incorporated on the bench by mounting two circular polarizers (HNCP-38) between the field lenses. No provision is made for rotating either polarizer since the dynamic event is so short as to preclude the possibility of rotation.

The images due to the spark gaps are separated by the geometric positioning of both the fiber-optic ends and the camera lenses relative to

the optical axis of the bench. As indicated in Fig. 1.6, the light from
fiber optic E_2 is collected by the field lens and focused on camera
lens ℓ_2. In a similar manner the light from the fiber optics E_1, E_3,
E_4 ... is focused on camera lenses L_2, L_3, L_4, ... With this optical
arrangement, it is possible to establish distinct images on the film
plane, with each image formed by the light from one and only one fiber
optic. With this image separation method, a time-sequenced series of
n photographs is obtained on a fixed sheet of film in a commercial view
camera where n is the number of sparks and lenses incorporated in the
system.

To minimize the angle of incident of the light on both the model and
the camera lenses, it is important to reduce the distance of the fiber
optics from the optical axis of the system. A 18 mm center to center
distance was the smallest convenient spacing which could be achieved in
this design. This spacing produced a maximum angle of incidence for the
corner spark gaps of 3.7 degrees with the first field lenses fixed at its
focal length of 700 mm.

A wide variety of optical arrangements can be designed with either
single or double field lens arrangements. Camera lenses are selected to
provide images 25 to 50 mm in diameter on a single sheet of 200 x 250 mm
film.

The film utilized in recording the image is of the blue sensitive
type. This film has a relatively slow emulsion with a very fine grain
structure. As this film is sensitive to light between 3500 and 5100 Å
and the output of the spark peaks in this same region, the film speed is
much higher than the speed rating indicates. Several different filters
(Kodak Wratten type 4, 8, 34A, and 47B) have been employed in conjunction
with this film to obtain a transmitted band sufficiently narrow to
approximate monochromatic light.

1.3.3 Synchronized Circuits

In dynamic photoelastic applications involving propagating stress
waves event times normally range from about 50 to 300 μsec. In this
brief interval, the model must be loaded, the camera delayed and then
started, and the time of the individual frames recorded. All of these

functions must be initiated on a microsecond time scale if the dynamic event is to be satisfactorily recorded. The synchronization function is performed by using a number of sequenced electronic circuits shown in the block diagram in Fig. 1.8.

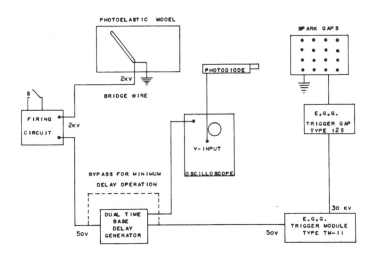

Fig. 1.8   Synchronization Circuits to Control the System Timing

The dynamic event, which is for example the detonation of an explosive charge, is initiated by activating the switch S in the firing circuit. The firing circuit then simultaneously emits three pulses: a 2-kV pulse to the bridge wire detonator, a 10-V pulse to trigger an electronic counter, and a 10-V pulse to initiate a delay generator. The 2-kV pulse passes through a small diameter constantan bridge wire which explodes and initiates detonation of a lead azide explosive charge with less than 2-μ jitter time. The detonating lead azide charge explosively loads the photoelastic model.

Simultaneously, the two 10-V pulses activate the electronic counter and the delay generator. After a predetermined delay time, the delay generator emits a pulse which is used to turn off the electronic counter, trigger an oscilliscope, and initiate the camera. To initiate the camera, it is necessary to amplify the pulse out of the delay generator from 10-V to about 20-kV. This pulse amplification is accomplished with pulse transformers. The output from the pulse transformer activates a trigger gap and initiates the firing sequence of the camera.

The time of delay is read out directly on the electronic counter, and the time of the individual frames is determined by sensing the intensity of the light output from the fiber optic array with a high frequency photodiode. The voltage out of the photodiode is recorded on the oscilliscope to provide an intensity time trace. The peaks on this trace are used to establish the time associated with each of the individual photographs.

1.4 Characteristics of the Cranz-Schardin Camera

The camera assemblies constructed by the author have been capable of either $3 \times 3 = 9$, $4 \times 4 = 16$ or $4 \times 5 = 20$ frames. Framing rates from 20,000 to 800,000 frames/sec have been achieved although neither the upper or lower limits of the camera have been established. A typical optical bench provides 300 mm field with a corresponding image diameter of 45 mm and a magnification ratio of -0.15.

The static resolution of the optical system exceeds 80 lines/in. on the model and about 530 lines/in. on the film plane. The dynamic resolution is determined by the exposure time which is 0.4 μs[1] as illustrated by the intensity time trace presented in Fig. 1.9. The overall duration of the light from the spark gap is approximately 2 μs. The dynamic resolution was established experimentally by generating a close-packed array of fringes which increase their relative spacing as they propagate (i.e., center of dilatation in an entire plane). By photographing this array at a high framing rate, it is possible to establish the first frame where the pattern becomes distinct. Typical results characterizing the dynamic resolution of the Cranz-Schardin camera are shown in Fig. 1.10. Analysis of the fringe pattern permits the determination of the fringe order as a function of position, and the maximum slope of this curve

---

[1]Where time is measured at the 1/3 peak intensity level.

gives the upper limit on dynamic resolution. In this case, the limit on dynamic resolution was 0.9 lines/mm for a P wave propagating at 1900 m/s which can be expressed more concisely as 1.7×10^6 lines/s.

With typical synchronization circuits, the time of the individual photographs can be determined to ± 1 µsec for framing rates in excess of 100,000 fps and to ± 2 µs for framing rates less than 100,000 fps.

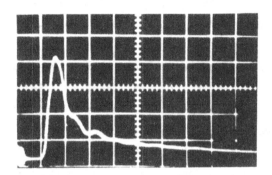

(.2 µs / cm.)

Fig. 1.9   Intensity-time Trace of Light Pulse From an Open Air Spark Gap

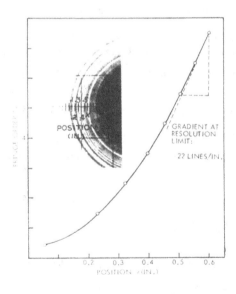

Fig. 1.10   Dynamic Resolution Determination

References

1. Tuzi, Z. "Photographic and Kinematographic Study of Photoelasticity", Jour. Soc. Mech. Eng., Vol. 31, No. 136, 1928, p. 334-339.

2. Feder, J.C., Gibbons, R.A., Gilberg, J.T., and Offenbacher, E.L., "The Study of the Propagation of Stress Waves by Photoelasticity", Proc. Soc. Expl. Stress Analysis, Vol. 14, No. 1, 1956, p. 109-122.

3. Flynn, P.D., Gilberg, J.T., and Roll, A.A., "Some Recent Developments in Dynamic Photoelasticity", Jour. SPIE, Vol. 2, No. 4, 1964, p. 128-131.

4. Flynn, P.D., "Photoelastic Studies of Dynamic Stresses in High Modulus Materials", Journ. SMPTE, Vol. 75, 1966, p. 729-734.

5. Flynn, P.D., "Dynamic Photoelastic Stress Patterns from a Simplified Model of the Head", Head Injury Conference Proceedings, J.B. Lippincott Co., Philadelphia, 1966, p. 344-349.

6. Cranz, C. and Schardin, H., "Kinematographic auf ru hendem Film und mit extrem hoher Bildfrequenz", Zeits, f. Phys., Vol. 56, 1929, p. 147.

7. Christie, D.G., "A Multiple Spark Camera for Dynamic Stress Analysis", Journ. Phot. Sci., Vol. 3, 1955, p. 153-159.

8. Wells, A.A. and Post, D., "The Dynamic Stress Distribution Surrounding a Running Crack – A Photoelastic Analysis", Proc. Soc. Expl. Stress Analysis, Vol. 16, No. 1, 1957, p. 69-92.

9. Dally, J.W. and Riley, W.F., "Stress Wave Propagation in a Half-Plane Due to a Transient Point Load", Developments in Theoretical and Applied Mechanics, Vol. 3, Pergamon Press, New York, 1967, p. 357-377.

10. Riley, W.F. and Dally, J.W., "A Photoelastic Analysis of Stress Wave Propagation in a Layered Model", Geophysics, Vol. 31, No. 5, 1966, p. 881-889.

11. Dally, J.W. and Thau, S.A., "Observations of Stress Wave Propagation in a Half-Plane with Boundary Loading", International Journal of Solids and Structures, Vol. 3, 1967, p. 293-308.

12. Dally, J.W. and Lewis, D., "A Photoelastic Analysis of Propagation of Rayleigh Waves Past a Step Change in Elevation", Bulletin of the

Seismological Society of America, Vol. 58, No. 2, 1968, p. 539-563.

13. Reinhardt, H.W. and Dally, J.W., "Dynamic Photoelastic Investigation of Stress Wave Interaction with a Bench Face", Transactions of AIME, No. 250, 1971, pp. 35-42.

14. Henzi, A.N. and Dally, J.W., "A Photoelastic Study of Stress Wave Propagation in a Quarter-Plane", Geophysics, Vol. 36, No. 2, 1970, pp. 296-310.

15. Rowlands, R.E., Taylor, C.E., and Daniel, I.M., "Ultra-High Speed Framing Photography Employing a Multiply Pulsed Ruby Laser and a Smear Type Camera; Application to Dynamic Photoelasticity", Presented at 8th International Congress on High-Speed Photography, Stockholm 1968.

16. Rowlands, R.E., Taylor, C.E., and Daniel, I.M., "Multiple-Pulsed Ruby Laser System for Dyamic Photomechanics; Applications to Transmitted - and Scattered - Light Photoelasticity", Experimental Mechanics, Vol. 9, No. 9, 1969, pp. 385-393.

17. Hendley, D.R., Turner, J.L., and Taylor, C.E., "A Hybrid System for Dynamic Photoelasticity", Experimental Mechanics, Vol. 15, No. 8, 1975, p. 289-294.

18. Dally, J.W. and Sanford, R.J., "A New High Speed Photographic System for Experimental Mechanics", Mechanics Research Communications, Vol. 9, No. 5, 1982, p. 337-342.

19. Dally, J.W. and Sanford, R.J., "Multiple Ruby Laser System for High Speed Photography", Optical Engineering, Vol. 21, No. 4, 1982, pp. 704-708.

CHAPTER 2

FIDELITY OF DYNAMIC RECORDING METHODS

2.1 Introduction

Regardless of the photographic system employed in dynamic photoelasticity, the fidelity of the fringe pattern recorded is an important
consideration. Parameters such as resolution, contrast, distortion of
amplitude and time shift must be accounted for in judging the adequacy of
a given system. These fidelity parameters are dependent upon the characteristics of the stress field (its velocity and stress gradient with
respect to position) and the intensity-time profile of the light source
in the recording system. Particular importance is attached to the exposure time, and to other related quantities such as film speed, film
contrast and the transmission of the dynamic polariscope.

A general theory is presented which gives the recorded exposure of a
dynamic fringe pattern associated with a propagating stress wave with
both ramp and triangular stress distributions. The solutions obtained
for three recording systems are given in terms of dimensionless parameters. Fidelity parameters are defined and computed. Finally, the method
for establishing fidelity parameters, including resolution, contrast,
distortion and time-shift, associated with a specified spark-gap camera,
is illustrated.

2.2 General Theory for Dynamic Exposure

The theory for the dynamic exposure recorded in a light-field circular polariscope with a photoelastic model may be established by considering a plane-stress wave traveling with velocity c in the x
direction. Let the difference in the principal stresses in this plane-
stress wave be given by $\tau(x,t)$ and note that the instantaneous intensity

emerging from the polariscope, which is defined here as the transmission coefficient $T(x,t)$, is given by

$$T(x-ct) = \frac{T_o}{2} \left\{ 1 + \cos \left[\frac{2\pi h}{f_\sigma} \tau(x - ct) \right] \right\} \tag{2.1}$$

where

   t = time

   h = model thickness

   f_σ = dynamic material-fringe value

   T_o = transmission coefficient of the polariscope and model for $\tau = 0$.

   The record of the dynamic fringe pattern characterized by the transmission coefficient is strongly dependent on the light source-camera combination which controls the recording intensity, $I(t)$, since the exposure of the film, $E(x)$, is expressed by the integral with respect to time of the product of the recording intensity and transmission coefficient. Thus:

$$E(x) = \int_0^\infty I(t) \; T(x - ct) dt \tag{2.2}$$

Although the representation of the exposure $E(x)$, given in eq. (2.2), is adequate for calculations involving a specific stress distribution, $\tau(x)$, and a specific intensity function, $I(t)$, it is more convenient to employ the normalized integral expression given below when comparative calculations are made.

$$\frac{E(x^*)}{I_o T_o t_e} = \frac{1}{2} \left\{ 1 + \int_0^\infty i(t^*) \quad \cos \left[2\pi \tau^*(x^* - t^*) dt^* \right\} \tag{2.3}$$

the normalized parameters are defined as:

   $t^* = t/t_e$

   $x^* = x/ct$

$$i(t^*) = I(t^*)/I_o \tag{2.4}$$

$$I_o = \int_0^\infty I(t^*)dt^*$$

$$\tau^* = h\tau/f_\sigma$$

The characteristic exposure time of the light source, t_e, is a constant which is defined for a given operating condition with a particular light source-camera combination.

2.3 Characterization of Recording System

Three different recording systems have been developed in the past decades which give adequate results in dynamic photoelastic investigations employing high-modulus models. These systems include the high-speed framing camera, the multiple-spark assembly and the Q-spoiled laser system. The intensity-time traces in normalized coordinates for each of these three systems are given in Fig. 2.1 with equivalent exposure time t_e^*, equal to unity.

Analytically, the normalized intensity-time functions are approximately represented by:

$$i(t^*) = \begin{cases} 0 & t^* \leqslant 0 \\ 1 & 0 \leqslant t^* \leqslant 1 \\ 0 & 1 \leqslant t^* \end{cases}$$

for the high-speed framing camera   (2.5)

$$i(t^*) = \begin{cases} 0 & t^* \leqslant 0 \\ \\ \ln 3 e^{-t^* \ln 3} & 0 \leqslant t^* \end{cases}$$

for the spark-gap source   (2.6)

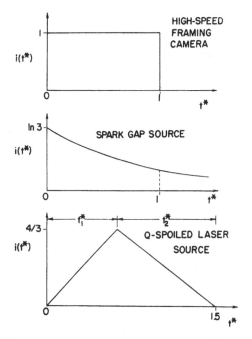

Fig. 2.1   Representative Intensity-Time Functions

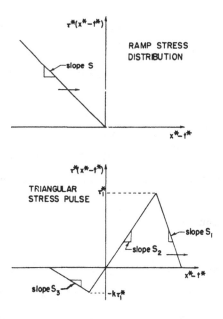

Fig. 2.2   Stress Distributions

$$i(t^*) = \begin{cases} 0 & t^* \leqslant 0 \\[2mm] \dfrac{2}{t_1^* + t_2^*} \dfrac{t^*}{t_1^*} & 0 \leqslant t^* \leqslant t_1^* \\[4mm] \dfrac{2}{t_1^* + t_2^*} \left(1 - \dfrac{t^* - t_1^*}{t_2^*} \right) & t_1^* \leqslant t^* \leqslant t_2^* \\[4mm] 0 & t_1^* + t_2^* \leqslant t^* \end{cases} \qquad (2.7)$$

for the Q-spoiled laser

Where t_1^* and t_2^* are defined in Fig. 2.1.

The high-speed framing camera operates with a constant-intensity source with an exposure time t_e which varies with framing rate (the is approximately 1/3 of the interframe interval). The spark light source provides an exponentially decaying intensity independent of framing rate. The exposure time t_e, established at the 1/3 peak-intensity level points, is typically 0.5 μs. Similarly, the Q-spoiled laser also provides a triangular pulse of light independent of framing rate; however, the value of t_e established at the 1/3 peak-intensity level is here typically 0.05 μs.

The ability of each of these photographic systems to accurately record a dynamic photoelastic-fringe pattern is a function of the stress-wave velocity and the stress distribution. To illustrate this point, two specific categories of stress distributions are considered here. These categories, which include a ramp stress distribution and a triangular stress pulse, are illustrated in Fig. 2.2, and are analytically described below:

Ramp Stress Distribution

$$\tau^*(x^* - t^*) = \begin{cases} -S(x^* - t^*) & \text{for } x^* - t^* \leqslant 0 \\[2mm] 0 & x^* - t^* \leqslant 0 \end{cases} \qquad (2.8)$$

where S is the slope of the ramp function, or the dimensionless stress

gradient.

Triangle Stress Pulse

$$\tau^*(x^* - t^*) = \begin{cases} 0 \\ -S_1(x^* - t^*) + \tau_1^*\left(1 + \dfrac{S_1}{S_2}\right) \\ +S_2(x^* - t^*) \\ -S_3(x^* - t^*) - k\tau_1^*\left(1 + \dfrac{S_3}{S_2}\right) \\ 0 \end{cases} \qquad (2.9)$$

$$\text{for} \begin{cases} \tau_1^*\left(\dfrac{1}{S_2} + \dfrac{1}{S_1}\right) \leqslant x^* - t^* \\[2mm] \dfrac{\tau_1^*}{S_2} \leqslant x^* - t^* \leqslant \tau_1^*\left(\dfrac{1}{S_1} + \dfrac{1}{S_2}\right) \\[2mm] -\dfrac{k\tau_1^*}{S_2} \leqslant x^* - t^* \leqslant \dfrac{\tau_1^*}{S_2} \\[2mm] -k\tau_1^*\left(\dfrac{1}{S_3} + \dfrac{1}{S_2}\right) \leqslant x^* - t^* \leqslant -\dfrac{k\tau_1^*}{S_2} \\[2mm] x^* - t^* \leqslant -k\tau_1^*\left(\dfrac{1}{S_2} + \dfrac{1}{S_3}\right) \end{cases}$$

where S_1, S_2, S_3 are the slopes, and τ^* and $-k\tau^*$ are the magnitudes of
the triangular stress pulse given in Fig. 2.2.

Substitution of eqs. (2.8) or (2.9) together with one of eqs. (2.5),
(2.6), or (2.7) into eq. (2.3) gives the solution for the exposure
$E^*(x^*)$. Of the six possible solutions which may be obtained in this
manner, five are listed in reference 1.

2.4 Typical Results for Normalized Exposure E*(x*)

To illustrate the application of the relationships listed in
Reference 1, consider a specific case of a triangular stress distribution
with the following parameters: $\tau^* = 5$, $S_1 = 1.0$, $S_2 = 0.5$, $S_3 = 0.2$,
$k = 1$ recorded with a spark-gap system. Substituting the stress-wave
parameters into the suitable equation for the solution and performing the
numerical computations, the results for the normalized exposure as a
function of normalized position are given as shown in Fig. 2.3a. From
this figure, it is clear that the normalized exposure varies between
maximum and minimum values with a range of variation inversely proportional to the values of the slopes S_1, S_2, and S_3. The positions of the
maximum and minimum values of E* correspond to the positions of the light
and dark fringes on the photographic record and, hence, permit the determination of the recorded fringe-order positions as illustrated in Fig.
2.3b. The true distribution of the triangular pulse is also shown in
Fig. 2.3b. Qualitative comparison between the recorded pulse and the
true pulse shows that the maximum fringe order is accurately recorded,
pulse distortion is quite negligible, and time or position shift is
small.

To obtain a more quantitative evaluation of the fidelity of a dynamic
photoelastic recording system, three measures of the accuracy of representation of the record must be defined. These criteria which include
time shift, distortion and contrast are defined as indicated in subsequent paragraphs.

First, consider time shift and note that, at some time t_s after the
initiation of the exposure and during the exposure interval, the leading
0.5 order fringes[1] associated with the true and recorded stress will
occur at the same position. The time t_s is then defined as the time
shift which can be represented as

---

[1]The definition is based on the location of the 0.5 order fringe rather
than the front of the wave because of the difficulty encountered in
establishing the leading zero-order fringe from a dynamic fringe pattern.

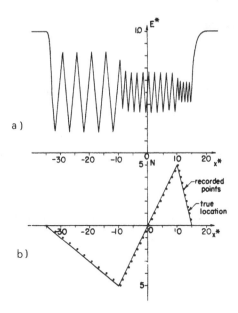

Fig. 2.3a   Normalized Exposure as a Function of Normalized Position

Corresponding to a Triangular Stress Distribution

Fig. 2.3b   True and Recorded Fringe Order as a Function of Normalized

Position

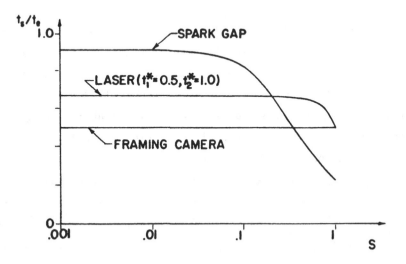

Fig. 2.4   Time Shift t_s/t_e as a function of Stress Gradient S

$$t_s = [x_R^* - x_T^*]_{N=0.5} t_e \qquad (2.10)$$

where

x_R^* = normalized position of the recorded fringe

x_T^* = normalized position of the true fringe

Since $t_s > 0$, the recorded fringe pattern will be ahead of the true pattern [see Fig. 2.3b] because the time of the record is defined at the initiation of the exposure.

Next, consider distortion of the wave shape as the difference in position of the maximum recorded fringe order and the true maximum at time t_s during the exposure interval. This distortion D is given by:

$$D = [x_R^* - x_t^*]_{N_{max}} ct_e \qquad (2.11)$$

Finally, consider contrast C as the difference in the normalized exposure between adjacent maxima and minima; thus:

$$C = |E^*(x^*_{N/2}) - E^*(x^*_{[N+1]/2})| \qquad (2.12)$$

The contrast is extremely important in the resolution of a fringe pattern. If the contrast is sufficiently high that each fringe order can be clearly established, then the value of the maximum fringe order can be accurately determined and no amplitude distortion will result.

The time shift was computed for the three different recording systems by considering a ramp-stress-distribution function and using suitable equations from reference 1. The results obtained are presented in Fig 2.4 where t_s/t_e is given as a function of the slope of the ramp function (i.e., the stress gradient) for each system. With the high-speed framing cameras, the time shift is independent of S and equal to $t_e/2$. For the Q-spoiled laser source ($t_1^* = 0.5$ and $t_2^* = 1$), the time shift is essentially $2t_e/3$ for $S < 0.2$ and decreases to $0.5\ t_e$ as S approaches 1. The time shift occurring with a spark system is $0.91\ t_e$ for small stress gradients and decreases monotonically to $0.23\ t_e$ as S goes to one.

The contrast as defined in eq. (2.12) was established for a ramp
stress distribution considering the three different recording systems.
For each system, the contrast is constant with respect to x* and the
value of the constant is a function only of the slope of the ramp.
Results showing the dependence of C on the slope S are presented in Fig.
2.5.

For small slopes, S < 0.02, or low stress gradients, all three
recording systems give a maximum contrast of 1.0. For $0.02 < S < 0.9$ the
spark system exhibits lower contrast than the other two systems. With
S > 0.9, the contrast of the spark system decreases monotonically with S.
Also, there is zero contrast with the high-speed framing camera for
integer values of S and zero contrast with the Q-spoiled laser for even
integer values of S.

The distortion D was established only for the spark system with the
triangular stress pulse. The results presented in Fig. 2.6 indicate that
the distortion increases as the ratio of S_1/S_2 increases, indicating that
rapid changes in slope are of primary importance in producing distortion.
It is also noted that the distortion decreases as S_1 increases.

2.5 Analysis of the Fidelity of the Spark Camera

The results presented in the previous section indicate general
characteristics of the three recording systems in terms of normalized
parameters. To effectively employ these results, the specific parameters
of a particular recording system must be determined. To illustrate the
process of evaluating the fidelity of a dynamic recording system, consider the multiple-spark camera used by the author.

Intensity-time traces for the sparks in this system indicate that the
exposure time, t_e, was 0.5 μsec. The limit of resolution of this system
was established by photographing a fringe pulse which exhibited a
decreasing fringe gradient as it propagated into the model. It was
established that a maximum fringe gradient of 22 fringes/in. (0.88
fringes/mm) could be resolved with an associated fringe velocity of
72,000 in/sec (1830 m/sec). The film employed was Kodak Gravure Positive
developed to maximum contrast (γ = 1.8).

The fringe gradient is related to the slope S of the dimensionless

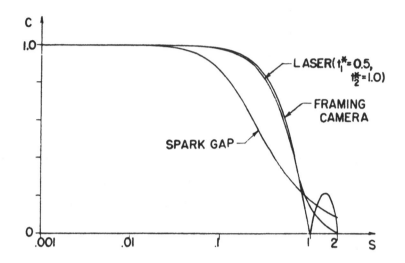

Fig. 2.5   Contrast C as a Function of Stress Gradient S

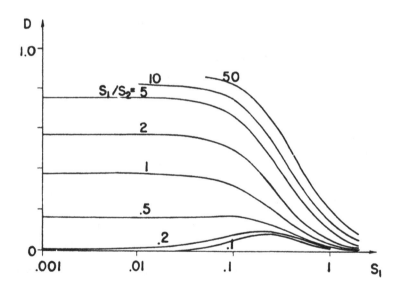

Fig. 2.6   Distortion D as a Function of S_1, with S_1/S_2 as a Parameter

ramp stress distribution, since at some fixed time:

$$S = \frac{d\tau^*}{dx^*} = \left(\frac{ct_e h}{f_\sigma}\right)\left(\frac{d\tau}{dx}\right) = ct_e \frac{dN}{dx} \qquad (2.13)$$

From eq. (2.11) and the parameters c, t_e and dN/dx established for the recording system, it is evident that the value of S which can be resolved is about 0.8. Referring now to Fig. 2.5, the corresponding minimum value of contrast C is about 0.21. It should be noted here that the minimum value of C associated with the resolution limit is dependent upon the film employed. For a film with a higher contrast, $\gamma > 1.8$, the minimum value of C would be lower.

The maximum value of the fringe gradient which can be resolved is a function of the velocity of the stress wave. Thus, eq. (2.13) can be employed to establish the maximum fringe gradient which can be resolved for any type of stress wave in any material

$$\left(\frac{dN}{dx}\right)_{max} = \frac{S_{max}}{ct_e} = \frac{1.6 \times 10^6}{c} \qquad (2.14)$$

where the wave velocity c is expressed as inches per second. Resolution limits for the spark camera and two widely used material, CR-39 and epoxy, are given in Table 2.1.

Computation of time shift and distortion for the spark system is illustrated here by considering the fringes associated with a Rayleigh wave propagating along the surface of a half-plane. The fringe pattern and a graph showing fringe order as a function of position are presented in Fig. 2.7. The wave speed in this instance is about 45,000 in/sec (1140 m/sec) and the fringe gradient at the leading edge is about 22-fringes/in (0.87 fringes/mm). From eq. (2.13) the value of $S_1 = 0.5$ is established and, from Fig. 2.4, the time-shift parameter $t_s/t_e = 0.41$ is determined. With $t_e = 0.5$ μs, the time shift is approximately 0.2 μs.

To estimate distortion, the Rayleigh wave is approximated with a

triangular stress distribution where S_1 = 0.5, as established previously.
The slope S_2 = 0.54 which corresponds to a fringe gradient of

TABLE 2.1

Resolution Limits on Fringe and Stress Gradients
with the Spark Camera

Material	Type of Wave	Wave Velocity in/sec	m/sec	Fringe Gradient fringe/in	fringe/mm	Stress Gradient† psi/in	MPa/mm
CR-39[*]	P	72000	1830	22	0.87	9700	2.63
CR-39	R	35000	890	45	1.77	19800	5.38
Epoxy[**]	P	80000	2030	20	0.79	4800	1.30
Epoxy	R	45000	1140	35	1.38	8400	2.28

† For model thickness of h = 0.25 in (6.35 mm)

* f_σ = 110 psi-in/fringe (19.3 MPa-mm/fringe)

** f_σ = 60 psi-in/fringe (10.5 MPa-mm/fringe)

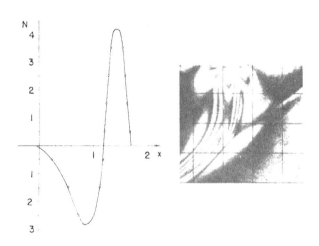

Fig. 2.7 Photoelastic Representation of Rayleigh Wave Propagation

24-fringes/in (0.94 fringes/mm), represents the stress gradient on the
declining portion of the Rayleigh pulse. The distortion parameter

D/ct_e = 0.09 is obtained from Fig. 2.6. Thus, the distortion, as expressd by a shift of position of the maximum fringe order, is 0.002 in. (0.05 mm) and is considered to be negligible in this case.

2.6 Summary

A general theory has been developed to evaluate the effect of exposure time and the intensity-time profile on the quality of a dynamic photoelastic fringe pattern. The theory has been developed in normalized coordinates for three different recording systems which are currently employed in dynamic photoelasticity. All three systems can adequately be applied to a wide range of elasto-dynamic problems; however, a detailed comparison of the three systems was not possible except in terms of normalized parameters. This detailed comparison requires the experimental determination of the maximum fringe gradient which can be resolved for a stress wave propagating at a known velocity. This resolution limit is a characteristic of each recording system and will depend on the intensity of the light source, speed and contrast to the film, exposure time, and transmission characteristics of the optical system. However, with a few relatively simple measurements, an individual recording system can be characterized and fidelity parameters, such as contrast ratio, time shift and distortion, can be calculated as illustrated here for the spark system.

References

1. Dally, J.W., Henzi, A. and Lewis, D., "On the Fidelity of High-speed Photographic Systems for Dynamic Photoelasticity", Experimental Mechanics, Vol. 9, No. 9. 1969, pp. 394-399.

CHAPTER 3

DATA ANALYSIS IN DYNAMIC PHOTOELASTICITY

3.1 Introduction

Methods to increase the information which can be extracted from a
dynamic isochromatic fringe pattern by separating the principal stresses
have been developed by a number of investigators. Riley and
Durelli[1] have utilized moiré patterns together with isochromatic patterns to separate the stresses. Flynn and Frocht[2] have shown that the
oblique-incidence method is applicable in dynamic photoelasticity.
Kuske[3] has suggested a separation technique based on the stress
equations of motion where particle accelerations are determined using
moiré data. Post[4] has demonstrated that a series interferometer can be
used to obtain the sum of the principal stresses in the dynamic problem.
Taylor et al[5] have obtained both isopachics and isochromatic patterns
simultaneously using combined holography and photoelasticity.

This lecture treats a method of analysis of dynamic fringe patterns
(both isochromatic and isoclinic) to yield individual values of the principal stresses σ_1 and σ_2 as well as the dynamic displacements. The
approach followed is quite simple in concept; namely, to find a second
equation to employ together with the stress-optic law which permits
separation of the principal stresses. This second equation can be
obtained in two ways. First, geometric properties of the model, such as
a free or rigid boundary or symmetry conditions, were used to give an
independent displacement-strain equation. Second, properties of the
dilatational (P) wave (no rotation) and the shear (S) wave (no dilatation) were used to give the independent equation. The method described
here is not general since it is restricted to those lines or regions in

the model where the second equation is valid. Nevertheless, a significant amount of information can be obtained for a number of important wave-propagation problems in a simple and direct manner by using these procedures.

3.2 Data Analysis Employing Model Characteristics

Case (I) Free Boundary

Consider any type of wave or combination of wave types propagating along a free boundary as illustrated in Fig. 3.1. It is evident that the x axis defines the free boundary and σ_{xx} and σ_{yy} are principal stresses with $\tau_{xy} = \sigma_{yy} = 0$. The stress-optic law:

$$\sigma_1 - \sigma_2 = \frac{(f_\sigma)_d}{h} N \tag{3.1}$$

where

$(f_\sigma)_d$ = the dynamic material fringe value

N = the fringe order

h = the model thickness

reduces to:

$$\sigma_{xx} = \pm \frac{(f_\sigma)_d}{h} N \tag{3.2}$$

and the separation of stresses is immediate.

The displacement along the boundary u_x can also be obtained from the isochromatic fringe pattern. From the stress-strain relations and eq. (3.2) the strain ε_{xx} can be related to the fringe order N by:

$$\varepsilon_{xx} = \frac{\sigma_{xx}}{E} = \frac{(f_\sigma)_d}{Eh} N \quad \text{and} \quad \varepsilon_{yy} = -\nu\varepsilon_{xx} \tag{3.3}$$

where E is the modulus of elasticity and ν is Poisson's ratio. Since the strains are by definition (linear theory)

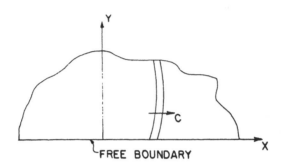

Fig. 3.1   Free Boundary Model

$$\varepsilon_{xx} = \frac{\partial u_x}{\partial x} , \ \varepsilon_{yy} = \frac{\partial u_y}{\partial y} , \ \gamma_{xy} = \frac{\partial u_x}{\partial y} + \frac{\partial u_y}{\partial x} \tag{3.4}$$

combining eqs. (3.3) and (3.4) and integrating along x with y = 0 gives:

$$u_x = \pm \frac{(f_\sigma)_d}{Eh} \int_B N \partial x \tag{3.5}$$

This integration can be performed quite accurately from a curve showing N as a function of position along the boundary. The displacement is equal to the area under the curve from the front of the wave to the position x along the boundary.

Case (II) Rigid Boundary

Consider a dilatational wave propagating in a photoelastic model bonded to a rigid half plane as illustrated in Fig. 3.2. Along the interface defined by the x axis, the rigidity condition imposes the following constraints on the displacement field:

$$\begin{aligned} u_x &= 0 \\ u_y &= 0 \quad \text{along x} \\ \omega_z &= 0 \end{aligned} \tag{3.6}$$

where

$$\omega_z = \frac{1}{2} \left(\frac{\partial u_y}{\partial x} - \frac{\partial u_x}{\partial y} \right) \ldots \text{rigid-body rotation}$$

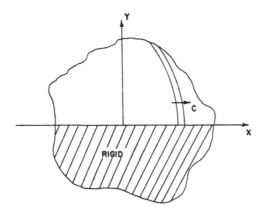

Fig. 3.2   Rigid Boundary Model

From eq. (3.6), it is evident that

$$\frac{\partial u_x}{\partial x} = 0, \ \frac{\partial u_y}{\partial x} = 0, \ \frac{\partial u_y}{\partial x} = \frac{\partial u_x}{\partial y} = 0$$

and hence the strain-displacement relations become:

$$\varepsilon_{xx} = 0, \ \varepsilon_{yy} = \frac{\partial u_y}{\partial y}, \ \gamma_{xy} = 0 \tag{3.7}$$

In this instance, σ_{xx} and σ_{yy} are the principal stresses and the stress-optic law becomes

$$\sigma_{yy} - \sigma_{xx} = \frac{(f_\sigma)_d}{h} N \tag{3.8}$$

The second equation for the separation process is obtained from eq. (3.7) and the stress-strain relations. Noting that $\varepsilon_{xx} = 0$ implies

$$\sigma_{xx} = \nu\sigma_{yy} \tag{3.9}$$

Combining eqs. (3.8) and (3.9) gives:

$$\sigma_{xx} = \frac{\nu}{1-\nu} \frac{(f_\sigma)d}{h} N$$

$$\tag{3.10}$$

$$\sigma_{yy} = \frac{1}{1-\nu} \frac{(f_\sigma)d}{h} N$$

Case (III) Axisymmetric-radial-wave Propagation

Consider a point source in an entire plane as shown in Fig. 3.3 which produces a displacement field:

$$u_r = f(r) \quad \text{and} \quad u_\theta = 0 \tag{3.11}$$

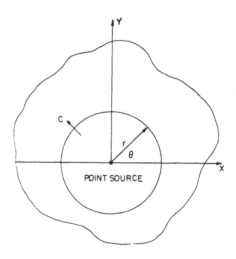

Fig 3.3   Radially Expanding Dilatational Wave

In this instance, all radial lines from the point source are axes of symmetry and the rigid body rotations $\omega_z = 0$. Thus, the stress wave propagates as a dilatational wave. The strains in polar coordinates are:

$$\varepsilon_{rr} = \frac{du_r}{dr} \quad \varepsilon_{\theta\theta} = \frac{u_r}{r} \quad \gamma_{r\theta} = 0 \tag{3.12}$$

The difference in the principal stresses is related to the difference in the principal strains by the stress-strain relations as:

$$\sigma_{\theta\theta} - \sigma_{rr} = \frac{E}{1 + \nu} [\varepsilon_{\theta\theta} - \varepsilon_{rr}] \tag{3.13}$$

Combining eqs. (3.1), (3.12) and (3.13) gives

$$r \frac{d}{dr} \left(\frac{u_r}{r} \right) = - \frac{1 + \nu}{E} \frac{(f_\sigma)_d}{h} N$$

Which upon integration yields

$$\varepsilon_{\theta\theta} = \frac{u_r}{r} = - A \int_r^{} \frac{N}{r} dr \tag{3.14}$$

where

$$A = \frac{1 + \nu}{E} \frac{(f_\sigma)_d}{h}$$

Integration of eq. (3.14) provides the radial displacement and the tangential strain $\varepsilon_{\theta\theta}$. The expression for the radial strain ε_{rr} is obtained from eqs. (3.1) and (3.13) as:

$$\varepsilon_{rr} = \varepsilon_{\theta\theta} - AN \tag{3.15}$$

and the individual values of the principal stresses are obtained from the stress-strain relations:

$$\sigma_{rr} = \frac{E}{1 - \nu^2} [\varepsilon_{rr} + \nu\varepsilon_{\theta\theta}]$$

$$\tag{3.16}$$

$$\sigma_{\theta\theta} = \frac{E}{1 - \nu^2} [\varepsilon_{\theta\theta} + \nu\varepsilon_{rr}]$$

3.3 Data Analysis Employing Stress-Wave Characteristics - Two-dimensional

Case (IV) Plane-dilatational Wave

Consider a plane-dilatation wave propagating in, say, the x direction. This wave may be represented by a displacement field

$$u_x = f(x) \quad \text{and} \quad u_y = g(x) \tag{3.17}$$

which is independent of y. The dilatational wave propagates without rotations; hence, the expression for ω_z leads to

$$\frac{\partial u_x}{\partial y} = \frac{\partial u_y}{\partial x} = 0 \tag{3.18}$$

The irrotational conditions of eq. (3.18) imply that $u_y = 0$ and, thus, the strain-displacement relations reduce to:

$$\varepsilon_{xx} = \frac{du_x}{dx}, \; \varepsilon_{yy} = \gamma_{xy} = 0 \tag{3.19}$$

Note, the x-y directions are principal. Since $\varepsilon_{yy} = 0$, a condition of plane strain exists; hence, the stress relations yield

$$\sigma_{yy} = \nu\sigma_{xx} \tag{3.20}$$

Combining eqs. (3.1) and (3.20) gives the expressions for the individual values of the principal stresses σ_{xx} and σ_{yy} as:

$$\sigma_{xx} = \frac{1}{1 - \nu} \frac{(f_\sigma)d}{h} N$$

$$\tag{3.21}$$

$$\sigma_{yy} = \frac{\nu}{1 - \nu} \frac{(f_\sigma)d}{h} N$$

The displacement u_x is obtained from the stress-strain relations and eqs. (3.1) and (3.19) which can be manipulated to give

$$\sigma_{xx} = \nu \; \sigma_{yy} = \frac{E}{1 + \nu} \; \frac{du_x}{dx} = \frac{(f_\sigma)_d}{h} \; N \qquad\qquad (3.22)$$

Equation (3.22) is integrated along x to give

$$u_x = A \int_x Ndx \qquad\qquad u_y = 0 \qquad\qquad (3.23)$$

The principal stresses and displacements can be obtained from eqs. (3.21) and (3.23) in a simple and direct fashion. These equations are valid when a plane-dilatational wave occurs alone. They are not valid in regions of the model where two different types of waves are superimposed.

Case (V) Plane-Shear Wave

Consider a plane-shear wave propagating along, say, the x-axis with an associated displacement field given by:

$$u_x = f(x) \quad \text{and} \quad u_y = g(x) \qquad\qquad (3.24)$$

Since the shear wave propagates with no volume change, the first invariants of stress I_1 and strain J_1 are

$$I_1 = J_1 = 0 \qquad\qquad (3.25)$$

and the separation of the principal stresses can be accomplished with no difficulty. From eqs. (3.1) and (3.25) it is evident that

$$\sigma_1 = - \; \sigma_2 = \frac{(f_\sigma)_d}{2h} \; N \qquad\qquad (3.26)$$

The directions of the principal stresses can be established from Mohr's circle. To construct the circle it is necessary to establish σ_{xx} and σ_{yy} from the displacement and volume conditions. From eqs. (3.24), (3.25), and the stress-strain relations, it can be shown that:

$$\varepsilon_{xx} = \varepsilon_{yy} = \sigma_{xx} = \sigma_{yy} = u_x = 0 \qquad (3.27)$$

Mohr's circle construction illustrated in Fig. 3.4 shows that the directions of σ_1 and σ_2 make a 45-deg. angle with the x axis and that

$$\tau_{xy} = \tau_{max} = \frac{\sigma_1 + \sigma_2}{2} = \frac{(f_\sigma)d}{2h} N \qquad (3.28)$$

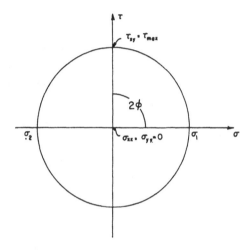

Fig. 3.4   Mohr's Circle for a Plane Shear Wave

The remaining displacement u_y is obtained from the relation between
the shear strain γ_{xy} and the displacements. From eqs. (3.27) and (3.28)
it is clear that

$$\frac{du_y}{dx} = AN$$

which upon integration along x yields the displacement u_y as:

$$u_y = A \int_x Ndx \qquad (3.29)$$

Case (VI) General Two-dimensional Dilatational Wave

Consider a general two-dimensional dilatational wave propagating in a plate with a displacement field represented by:

$$u_x = f(x,y), \quad u_y = g(x,y) \tag{3.30}$$

Since $\omega_z = 0$ for dilatational-wave propagation

$$\frac{\partial u_x}{\partial y} = \frac{\partial u_y}{\partial x} \tag{3.31}$$

and

$$\gamma_{xy} = 2 \frac{\partial u_x}{\partial y} = 2 \frac{\partial u_y}{\partial x}$$

From the stress equations of transformation and the stress-strain relations of together with eq. (3.31) it is evident that

$$\tau_{xy} = \frac{\sigma_2 - \sigma_1}{2} \sin 2\phi = -\frac{(f_\sigma)d}{2h} N \sin 2\phi = \frac{E}{1 + \nu} \frac{\partial u_x}{\partial y} = \frac{E}{1 + \nu} \frac{\partial u_y}{\partial x}$$

$$\tag{3.32}$$

where ϕ is the angle between the σ_1 direction and the x axis obtain from isoclinic data. Integrating eq. (3.32) yields

$$u_x = \frac{A}{2} \int_y N \sin 2\phi \, \partial y$$

$$\tag{3.33}$$

$$u_y = \frac{A}{2} \int_x N \sin 2\phi \, \partial x$$

Stress separation is accomplished by using the first stress invariant, the stress-strain and the strain-displacement relations to show:

$$\sigma_1 + \sigma_2 = \sigma_{xx} + \sigma_{yy} = \frac{E}{1 - \nu} \left[\frac{\partial u_x}{\partial x} + \frac{\partial u_y}{\partial y} \right] \tag{3.34}$$

Since the displacement field is established by eq. (3.33), the sum of the principal stresses is obtained from eq. (3.34). Separation of the individual values of σ_1 and σ_2 is completed by combining the results of eqs. (3.1) and (3.34).

Case (VII) General Two-dimensional Shear Wave

Consider a general two-dimensional shear wave propagating in a plate with a displacement field represented by:

$$u_x = f(x,y), \quad u_y = g(x,y), \quad u_z = 0 \tag{3.35}$$

Since $I_1 = J_1 = 0$, for the shear wave:

$$\varepsilon_{xx} = -\varepsilon_{yy}$$
$$\tag{3.36}$$
$$\sigma_{xx} = -\sigma_{yy}$$

Next, the stress equations of transformation, the stress-strain relations, and the stress-optic law can be used with eq. (3.36) to yield:

$$\frac{\partial u_x}{\partial x} = \frac{A}{2} N \cos 2\phi$$

and upon integrating along x,

$$u_x = \frac{A}{2} \int_x N \cos 2\phi \, \partial x \tag{3.37}$$

In a similar manner the displacement, u_y is obtained as:

$$u_y = -\frac{A}{2} \int_y N \cos 2\phi \, \partial y \tag{3.38}$$

Thus, the displacement field is established by integrating products of isochromatic and isoclinic data along lines parallel to the x and y axes.

The separation of the principal stresses is performed in the same manner as in Case V by using eq. (3.26).

3.4 <u>Discussion</u>

The geometric characteristics of either the model or the stress wave or both have been utilized to provide additional information for dynamic photo-elastic analyses. This information permits the separation of principal stresses and the determination of dynamic displacements. The method is not general and cannot be used at any point in a model where two or more stress waves are superimposed. Nevertheless, many problems arise where the stress waves separate and the method can be utilized. Ten different cases have been illustrated here and the possibilities of application have not be exhausted.

The integration process often involved the multiplication of an area integral by a constant $A = [(1 + \nu)/E][(f_\sigma)_d/h]$. In establishing the material properties for this term, the dilatational and shear-wave velocities measured on the model under investigation were employed to determine rate averaged values of E and ν. This procedure appears to be satisfactory for the high-modulus photoelastic materials such as CR-39 or epoxy which can be considered essentially elastic over the predominant range of frequencies encountered in a transient photoelastic analysis. The material fringe value $(f_\sigma)_d$ used was also a rate averaged value as determined by Clark[6,7].

References

1. Riley, W.F. and Durelli, A.J., "Application of Moiré Methods to the Determination of Transient Stress and Strain Distributions", <u>Journal of Applied Mechanics</u>, 26 (1), (1962), 23-29.

2. Flynn, P.D. and Frocht, M.M., "On the Photoelastic Separation of Principal Stresses Under Dynamic Conditions by Oblique Incidence", <u>Journal of Applied Mechanics</u>, 38 (1), (1961), 144-415.

3. Kuske, A., "Photoelastic Research on Dynamic Stresses", Experimental Mechanics, 6 (2), (1966), 105-112.

4. Wells, A. and Post, D., "The Dynamic Stress Distribution Surrounding a Running Crack - A Photoelastis Analysis", <u>Proceedings of the Society for Experimental Stress Analysis</u>, 16 (1), (1958), 69-92.

5. Holloway, D.C., Ranson, W.F. and Taylor, C.E., "A Neotaric

Interferometer for Use in Holographic Photoelasticity", Experimental Mechanics 12 (10) (1972), 461-465.

6. Clark, A.B.J., "Static and Dynamic Calibration of Photoelastic Model Material, CR-39", Proc. SESA, 14 (1) (1956), 195-204.

7. Clark, A.B.J. and Sanford, R.J., "A Comparison of Static and Dynamic Properties of Photoelastic Materials", Proc. SESA, 20 (1), (1963), 148-151.

CHAPTER 4

LABORATORY METHODS FOR STRESS-WAVE AND FRACTURE EXPERIMENTS

4.1 Introduction

The dynamic experiment performed to study either stress waves or
fracture must be planned and executed with extreme care. Unlike a static
experiment the observer cannot interact during the observation period to
adjust the model, loading fixture, instruments, etc. or to make any
corrections. Since the observation period in a dynamic experiment is
measured in tens or hundreds of μs the observer simply cannot interact
with the experiment except to initiate the dynamic event by depressing a
start switch. Moreover, loading the model and operation of the high
speed recording system must be carefully sychronized to ensure that the
photographic records are obtained which cover the entire duration of the
event. As the model is usually destroyed in these experiments it is
essential that the experiment be conducted with great precision and with
success on the first attempt.

Some of the laboratory techniques involved in conducting an experiment with precision are discussed in the following sections.

4.2 Model Preparation

Models for stress wave experiments are usually fabricated from relatively large sheets of CR-39 or Homalite 100. These materials are
available in large sizes with excellent surface finish and they are
crystal clear. Epoxy sheets are also available commercially but the surface finish is not comparable to CR-39 or Homalite 100. Also, the light
transmissibility of many of the epoxies is reduced due to its color and
light absorption and as a result proper film exposure is often difficult
to achieve.

The model size should be large in stress wave experiments to permit the various waves to separate and to individually reflect from prescribed surfaces. The velocity[1] of the commonly observed stress waves are

$$c_L^2 = E/\rho(1-\nu^2) \tag{4.1}$$

$$c_2^2 = E/2\rho(1+\nu) \tag{4.2}$$

$$c_o^2 = E/\rho \tag{4.3}$$

$$c_R^2 = f(\nu)c_2^2 \tag{4.4}$$

where c_L is the velocity of the dilatational wave in a plate

c_2 is the velocity of the shear wave

c_o is the velocity of the rod wave

c_R is the velocity of Rayleigh or edge wave

E   is the dynamic modulus of elasticity

ν is the dynamic Poisson's ratio

ρ is the mass density.

The wave velocities differ to some degree from one material to another and from one lot to another lot and should be measured for each experiment. Typical values for the velocities are given in Table 4.1.

TABLE 4.1

Typical Wave Velocities for Common Photoelastic Materials (m/s)

Material	c_L	c_2	c_o	c_R
CR-39	1780	1040	1690	960
Homalite 100	2150	1230	2040	1130
Epoxy	1970	1130	1870	1040

As the dilatational wave speed c_L is the highest, it controls the timing of the experiment and/or the model size. The time required for the dilatation wave to propagate 100 mm is 56, 47 and 51 µs in CR-39, Homalite 100 and Epoxy respectively. In large models, say 600 mm in size with

symmetrical loading, observation periods of about 150 µs can be anticipated before reflections from the model boundaries occur.

4.3 Model Loading

Dynamic loading is usually accomplished with explosives, projectile
impact, shock tube, stress bar or electrical impulse. Each of these
loading methods requires special equipment and/or skills and most laboratories tend to concentrate on only one or two of the loading methods.

Small explosive charges offer a relatively easy way to apply point
loads as indicated in Fig. 4.1. Lead azide PbN_6 can be produced in small
quantities by mixing aqueous solutions of sodium azide NaN_6 and lead
nitrate $PbNO_3$. Charge weights of 20 to 100 mg are sufficient to produce
high level stress waves and localized fracturing. The velocity of detonation varies with packing density of the lead azide and is about 3000
m/s. A cylindrical charge 4 mm in diameter with a detonator positioned
at its centerline will detonate in 0.67 µs. The decay time is much
longer and depends entirely on the containment of the charge.

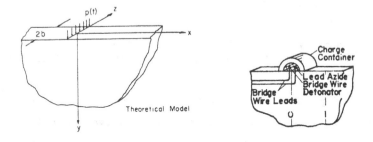

Fig. 4.1 Modeling a Load on a Half-Plane with an Explosive Charge

Precise timing of the detonation is essential to properly synchronize
the high speed recording system with the dynamic event. Ordinary hot
wire detonators are not adequate because their response time is too slow
and too variable. Bridge wire detonators are necessary where very fine

diameter wire (say 50 gage) about 2 mm long with a resistance of 0.1 to 1 Ω is vaporized to detonate the charge. The vaporization of the bridge wire can be accomplished in 1 or 2 μs within ± 1 μ jitter time by switching it across a 5000 V potential from a 25 joule source. The control which can be achieved with bridge wire detonation of lead azide charges is demonstrated in Fig. 4.2 where a photoelastic fringe pattern is shown for a half-plane loaded with two charges which were detonated simultaneously. The symmetry of the fringe pattern shows that both the magnitude and the time of initiation of the detonation was the same.

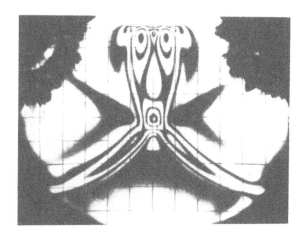

Fig. 4.2   Symmetry of the Isochromatic Fringe Pattern Indicates the Control of both the Magnitude of the Load Pulse and the Simultaneity of the Dual Initiation.

Projectile impact is another commonly employed method of loading where either a rifle or an air gun is used to drive the projectile. When a rifle is used with a soft lead bullet, the bullet will act like a fluid and the impact stress and its variation with time can be determined. For normal impact the duration t_d of the pulse is given by

$$t_d = L/V \qquad\qquad (4.5)$$

where L is the length of the projectile
       V is the velocity
and the impact force P is

$$P = \rho V^2 A \tag{4.6}$$

where ρ is the density of the projectile material

A is the contact area

If the shape of the projectile is adjusted, the impact area A in eq. (4.6) is controlled to give specific force-time profiles. Some examples of the dynamic load characteristics for 0.22 in. caliber rifle bullets are given by R. M. Davies[1] as

TABLE 4.2

Dynamic Load Characteristics from 22 Caliber Rifle Bullets

Type of Bullet	ICI Short	Palma Long	ICI Hornet
Length L (mm)	9.4	11.5	13.9
Velocity V (m/s)	274	329	777
Duration t_d (μs)	34.3	34.9	17.9
Maximum Force P(KN)	21.0	30.2	168.5
Mass (g)	1.95	2.56	2.87
Momentum (kg-m/s)	5.35	8.43	22.3

Air guns are also used to propel rigid projectiles into striker plates fabricated from lead or a soft polymer. A typical air gun is fabricated with a fast-acting solenoid valve for actuation, an accumulator, the air gun barrel and projectiles. Proximity switches are often mounted at the end of the barrel to measure exit velocity. Air guns permit a wider range of velocity control than rifles and are safer to use since the exit velocities are usually much lower.

Synchronization can be accomplished with projectile impact by placing two metal foils separated by a thin sheet isolator on the surface of the model at the point of impact. As the projectile contacts the surface it cuts through the isolator sheet, closes the switch formed with the two metal foils and initiates the high-speed recording system.

Shock tubes may also be used to apply pressure pulses to select points on a model. The velocity V_s of the shock pulse[2] is given by

$$V_s = \sqrt{kp/\rho} \tag{4.7}$$

where k is the isentropic exponent of the gas

   p is the pressure

   ρ is the density

The magnitudes of the pressure pulse delivered by the tube is controlled
by adjusting the pressure of the gas in the driver section. The duration
of the pulse is a function of the velocity V_s and the length of the
driver section of the tube.

   Another method of loading is to use a Hopkinson[3] pressure bar to
apply the imput pulse. The bar can be instrumented with strain gages
which can be monitored to give the profile of the applied pulse. An
example of this method of loading is illustrated in Figs. 4.3 and 4.4.

4.4 Synchronization

   Timing is extremely critical in conducting a dynamic experiment where
the entire duration of event is in the range from 50 to 500 µsec.
Synchronization is accomplished with electrical circuits which initiate
the high speed recording system at some pre-selected delay time after the
application of a loading pulse. The multiple spark recording system is
well suited to synchronization since it can be initiated with a low
voltage trigger pulse with an inherent delay of only about 5 µs.

   A block diagram for a typical synchronization circuit for explosively
loaded models is illustrated in Fig. 1.8. The experiment is initiated by
closing the switch on the firing circuit which vaporizes the bridge wire
and detonates the lead azide in 1 to 2 µs. A part of this pulse triggers
an oscilliscope and a delay generator. The delay generator is used if
the high speed recording system is not initiated until later in the
event. The output from the delay generator is amplified to produce a
20-30 KV pulse which is used to initiate the camera. The light flashes
are monitored by a photodiode. The output from the photodiode is
recorded on the oscilliscope. The oscilliscope trace provides the precise instant of time when each frame was recorded during the dynamic
event.

   Synchronization for a fracture experiment is similar as indicated by
the block diagram illustrated in Fig. 4.5. In this experiment the frac-

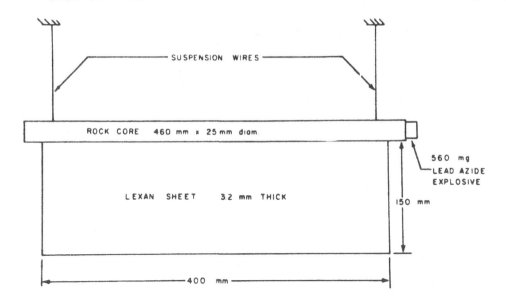

Fig. 4.3   Hopkinson Pressure Bar and Attached Photoelastic Model

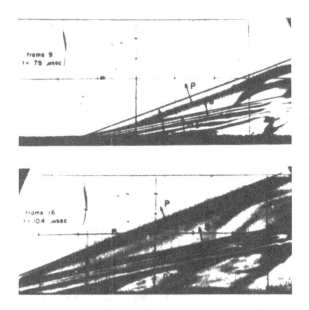

Fig. 4.4   Stress Waves Produced in the Photoelastic Model by Dynamic

Loading of the Pressure Bar

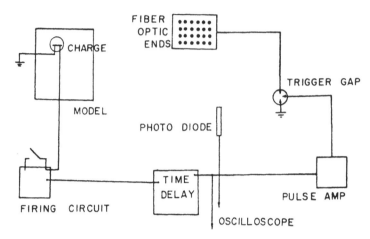

Fig. 4.5 Synchronization Circuit for a Dynamic Loading Produced with an
          Explosive Charge

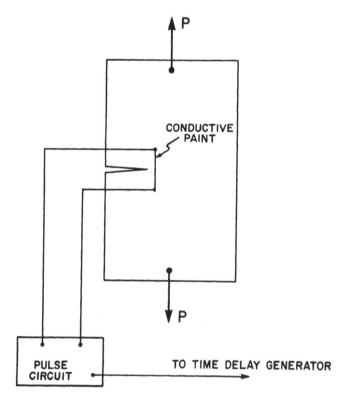

Fig. 4.6 Synchronization Circuit for a Fracture Specimen with a Starter
          Crack

ture model is subjected to an increasing load P until the starter crack
is initiated. The propagating crack cuts through a conductor which is
painted on the surface of the specimen. Interrupting this conductor produces a pulse from the circuit described in Fig. 4.6 and initiates a time
delay generator. The remainder of the sequence to start the camera is
identical to that described previously.

4.5 Dynamic Calibration

Dynamic calibration of photoelastic materials[4] to determine
$(f_\sigma)_d$ is usually accomplished by impacting the end of a long rod with a
square or rectangular cross-section by a projectile fired from an air
gun. Strain gages (two element rosettes) are mounted in the center of
the rod on the top and bottom surfaces and a beam of polarized light is
transmitted through the vertical surfaces of the rod at this same location. The intensity of light from the polariscope is monitored with a
photodiode to obtain $N(t)$ and the strain gages provide $\varepsilon_a(t)$ and $\varepsilon_t(t)$.
The signals are recorded simultaneously on an oscilloscope. Strain
gages are also mounted at a second station so that the velocity of propagation of the bar wave can be measured. The optical and mechanical properties of the material are obtained from this data by using the
following equations:

$$\nu = -\varepsilon_t(t_1)/\varepsilon_a(t_1) \tag{4.8}$$

$$c_o = s/t_* \tag{4.9}$$

where t_* is the transit time for the bar wave to travel the distance s
between the two strain gage stations, and c_o is the bar wave velocity.
And, from eq. (4.3) and the stress and strain optic laws, one obtains

$$E = c_o^2 \rho \tag{4.10}$$

$$(f_\varepsilon)_d = \frac{h(1 + \nu)}{N(t)} \varepsilon_a(t) \tag{4.11}$$

$$(f_\sigma)_d = \frac{E}{1 + \nu} (f_\varepsilon)_d \tag{4.12}$$

The value of $(f_\sigma)_d$ is usually slightly higher than the static determination of the material fringe value. Under static loading, usually
several hundred seconds are required for calibrating the photoelastic
polymers. During this loading period the polymers undergo viscoelastic
deformation which produces a retardation and in effect reduce f_σ. Under
dynamic loading where load durations are of the order of 10^{-5} to 10^{-4}
seconds viscoelastic deformation is much smaller and as a consequence
$(f_\sigma)_d > (f_\sigma)_s$.

The wave speeds in plates which can be measured photoelastically can
also be used to determine the dynamic values of the elastic constants.
Refer to eqs. (4.1) and (4.2) and note that:

$$(c_2/c_L)^2 = (1-\nu^2)/2(1+\nu) = (1-\nu)/2$$

and then

$$\nu = 1 - 2(c_2/c_L)^2 \tag{4.13}$$

and since ν is known

$$E = \rho c_L^2 (1-\nu^2) \tag{4.14}$$

or

$$E = 2\rho(1+\nu)c_2^2 \tag{4.15}$$

It is possible to measure the inclination angle θ of the von-Schmidt
wave with the boundary of a half plane model as shown in Fig. 4.7. This
angle θ is related to the ratio c_2/c_L of the wave velocities by:

$$\sin\theta = (c_2/c_L) \tag{4.16}$$

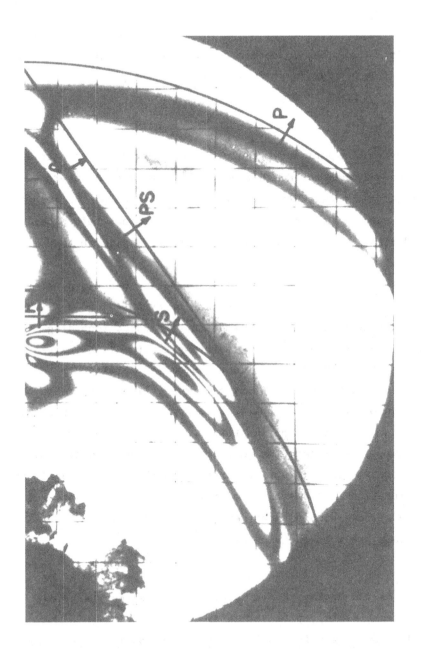

Fig. 4.7   Photoelastic Patterns in a Half Plane Showing the P, S and PS

(von-Schmidt) waves

Substituting eq. (4.16) with eq. (4.13) yields

$$\nu = 1-2 \sin^2\theta = \cos 2\theta \qquad\qquad (4.17)$$

and shows that it is possible to determine Poisson's ratio from a single photograph of a dynamic isochromatic pattern in a half plane which clearly shows the P, S and PS (von-Schmidt) waves.

4.6 Summary

Successful dynamic experiments require careful attention to detail as the models are usually destroyed and there is no time for operator interaction. The well planned experiment entails a model of sufficient size to permit the basic wave types P, S and R to separate before reflection from boundaries creates additional waves which complicate the data analysis.

Synchronization is essential with initial time t = 0 defined as of the instant of crack initiation or the instant of dynamic load application. The operator interaction is limited and high speed recording must be initiated automatically by incorporating electronic circuits which incorporate pre-selected delay times and pulse amplifiers.

Loading is accomplished with explosives, projectiles shock tubes or Hopkinson pressure bars. All methods require some additional equipment and considerable skill by the experimentalist.

It is necessary to calibrate the model material under dynamic conditions which closely parallel that encountered in the actual experiment. The dynamic calibration includes E, ν and f_σ. All of these quantities can be determined from a simple experiment with a rod where optical response and strain are measured simultaneously.

References

1. Batchelor, G.K. and Davies, R.M., *Surveys in Mechanics*, Cambridge University Press, (1956), p. 67.

2. Hunsaker, J.C. and Rightmire, B.G., *Engineering Applications of Fluid Mechanics*, McGraw-Hill, (1947), p. 166.

3. Hopkinson, B. "A Method of Measuring the Pressure Produced in the Detonation of High Explosives or by the Impact of Bullets", Roy. Soc. Phil. Trans. A, Vol. 213, (1914), p. 437.

4. Clark, A.B.J. and Sanford, R.J., "A Comparison of Static and Dynamic Properties of Photoelastic Materials", Proc. SESA, 20, No. 1, (1963), pp. 148-151.

CHAPTER 5

APPLICATION OF DYNAMIC PHOTOELASTICITY TO STRESS WAVE

PROPAGATION

5.1 Introduction

Dynamic photoelasticity is an extremely effective method to employ in
the investigation of a wide variety of dynamic problems in many fields.
In an effort to show the versatility of the method, four different applications will be briefly introduced and results summarized. The areas of
application include geophysics, mechanics, non-destructive testing and
mining.

5.2 Geophysics Application

The first application deals with the geophysics problem of stress
wave propagation in a two-layered model[1]. The experiment consisted of
the fabrication of a model with a layer photoelastic material CR-39
bonded to a sheet of aluminum as defined in Fig. 5.1. The impedance
mismatch between the two layers was 6 to 1. The model was loaded with a
charge of lead azide and photographs were taken with the multiple spark
camera. The experiment was repeated and photographs of the fringe patterns in the region near the explosive charge and also in the region well
removed from the charge were recorded. Typical results of the dynamic
isochromatic fringe patterns recorded are presented in Fig. 5.2.

Wave propagation in the two layered model is extremely complex and
nine different waves are predicted to occur in the top layer. These
include the incident dilatational wave P_1, the reflected dilatational
waves P_1P_1 at both the free boundary and the interface, the reflected
shear wave P_1S_1 at the boundary and the interface, and the four head
waves generated by the refracted waves propagating in the lower layer

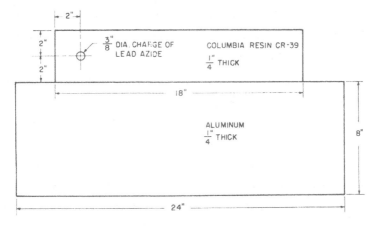

Fig. 5.1   Dimensions of Two Layer Model

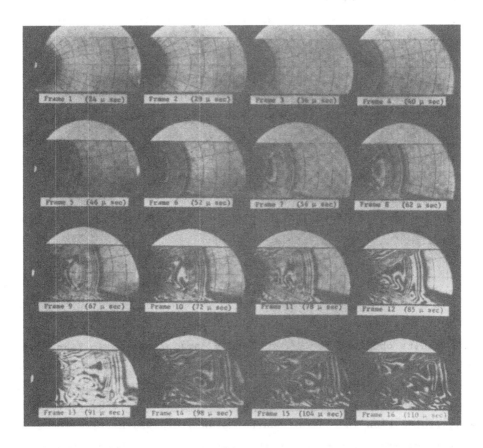

Fig. 5.2   Stress Wave Propagation in a Layered Model

$P_1P_2P_1, P_1P_2S_1, P_1S_2P_1, P_1S_2S_1$. The fronts of these various waves can be easily determined by the elementary laws governing reflection and refraction of stress waves as indicated in Fig. 5.3.

Six of these nine admissable wave fronts are identified in the fringe patterns shown in Figs. 5.4, 5.5 and 5.6. Early in the dynamic event, the incident P_1 wave and the reflected P_1S_1 waves exhibited the largest amplitude. However, late in the event, the three head waves $P_1P_2S_1$, $P_1S_2P_1$ and $P_1S_2S_1$ were dominant waves as they completely eroded the cylindrical wave front associated with the incident P_1 wave. The $P_1P_2P_1$ and P_1P_2 waves were not of sufficient magnitude to produce a photoelastic response and as such they are of minor importance for this case where the impedance mismatch was 6 to 1.

The radial displacement u_r associated with the incident P wave was determined from:

$$\frac{u_r}{r} = - \left(\frac{1+\nu}{E} \right) \frac{(f_\sigma)}{h} \int_r \frac{N}{r} \, dr \tag{5.1}$$

and the principal stresses were determined from:

$$\sigma_2 = \sigma_r = \frac{E}{1-\nu^2} [\varepsilon_r + \nu\varepsilon_\theta] \tag{5.2}$$

where $\varepsilon_\theta = u_r/r$ and $\varepsilon_r = \varepsilon_\theta - \frac{1+\nu}{E} \left(\frac{N}{h} \right) (f_\sigma)_d \tag{5.3}$

The results obtained for u_r, σ_r, and σ_θ are presented in Figs. 5.7 and 5.8 respectively.

5.3 Mechanics Application – Stress Wave Propagation in a Quarter Plane

Wave propagation in elastic quarter-planes and wedges has been a topic of considerable interest in the last decade. Although the first rigorous solutions by Lamb[2] of half-plane problems date back to the

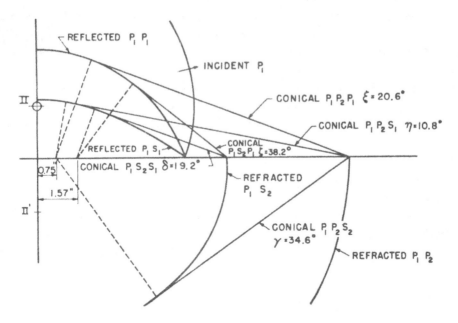

Fig. 5.3 Wave Propagation in the Period t > 33 μsec

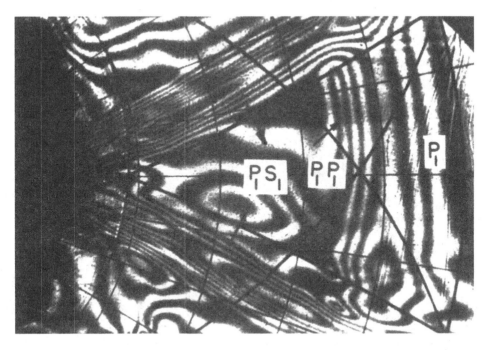

Fig. 5.4 Comparison of Position and Shape of the Reflected P_1P_1 and
P_1S_1 Waves with Photoelastic Fringe Patterns

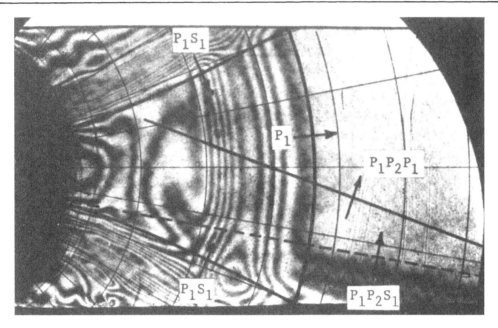

Fig. 5.5   Comparison of the Relative Positions of the $P_1P_2S_1$ and
$P_1P_2P_1$ Headwaves and the P_1S_1 Reflected Waves

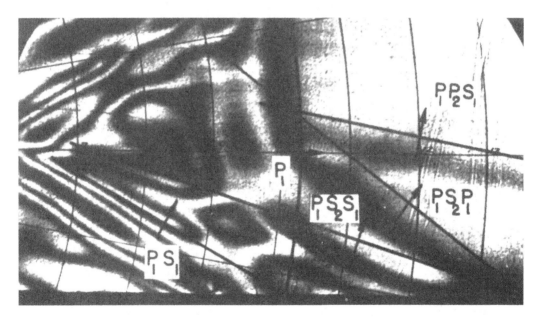

Fig. 5.6   Comparison of the Position and Shape of the $P_1P_2S_1$, $P_1S_2P_1$ and
$P_1S_2S_1$ Headwaves with the Photoelastic Results

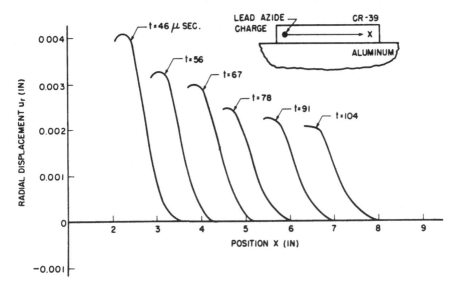

Fig. 5.7   Radial Displacement u_r Associated with the Incident P_1 Wave

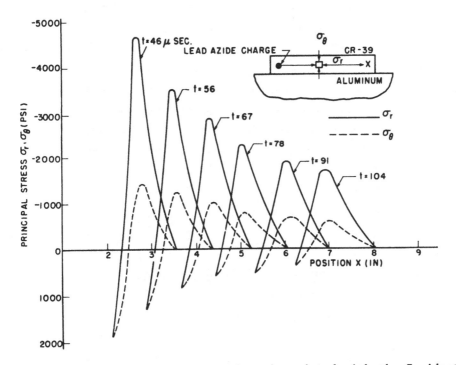

Fig. 5.8   Principal Stresses σ_r and σ_θ Associated with the Incident P_1 Wave

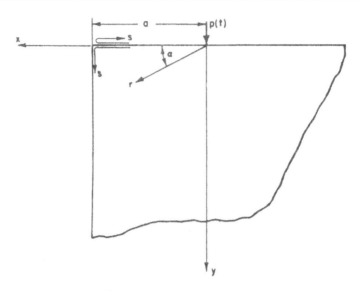

Fig. 5.9   Elastodynamic Quarter-Plane

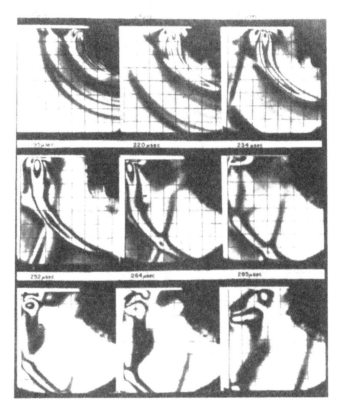

Fig. 5.10   Selected Isochromatic Fringe Patterns Representing the
Dynamic Event

early part of this century and although many other half-plane solutions
have been obtained, the quarter-plane and wedge geometries have proved
intractable to exact analytical treatment. Yet, the quarter-plane (see
Fig. 5.9) or wedge geometry is but the first step in complexity from the
half-plane geometry.

This study of a quarter-plane problem utilizes dynamic photoelasticity to obtain full field measurements of the stresses at various times
after the application of an explosively generated pulse to a point on one
of the boundaries of a quarter-plane.

Typical fringe patterns showing the reflection from the corner of the
quarter-plane are shown in Fig. 5.10. Enlargements of selected fringe
patterns presented in Figs. 5.11 and 5.12 show the reflection process.
The early response to the dynamic load consists of a leading dilatational
wave (P), a following equivoluminal wave (S), an equivoluminal "von
Schmidt" wave (PS), and a Rayleigh wave (R).

Dilatational and equivoluminal wave speeds determined from these
fringe patterns are

$$c_L = 71,500 \text{ inches/sec (1820 m/sec)}$$
$$c_2 = 41,500 \text{ inches/sec (1050 m/sec)}$$

indicating the dynamic value of Poisson's ratio

$$\nu_d = 1 - 2(c_2/c_L)^2 = 0.33.$$

The Rayleigh wave speed is

$$c_R = 38,200 \text{ inches/sec (970 m/sec)}.$$

Interactions of the waves with the corner and the unloaded edge,
indicated in Fig. 5.11 may be grouped into three separate sets of
interactions: P and PS-wave interactions, S-wave interactions, and R-
wave interactions. In the P and PS-wave interaction set, the response
waves are dilatational (PP) and equivoluminal (PS*) reflections of the P
wave from the unloaded edge, an equivoluminal von Schmidt wave (PPS)
generated by the PP wave as it grazes along the loaded edge, an equivoluminal reflection (PSS) of PS wave from the unloaded edge, and equivoluminal wave (S) generated at the corner in response to P wave arrival. The
fronts of these waves are indicated in Fig. 5.12A.

The major response waves arising in the S-wave interactions are the

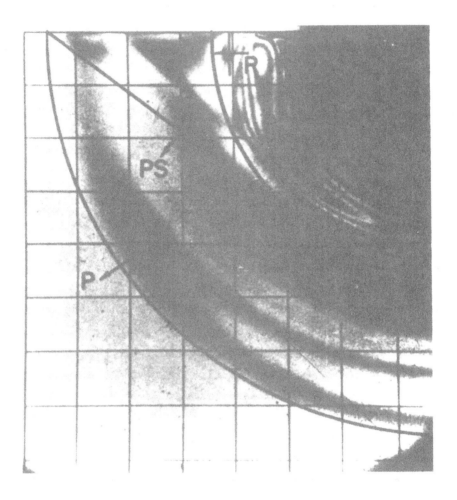

Fig. 5.11   Isochromatic Fringe Patterns Showing Early Response of the
            Quarter-Plane

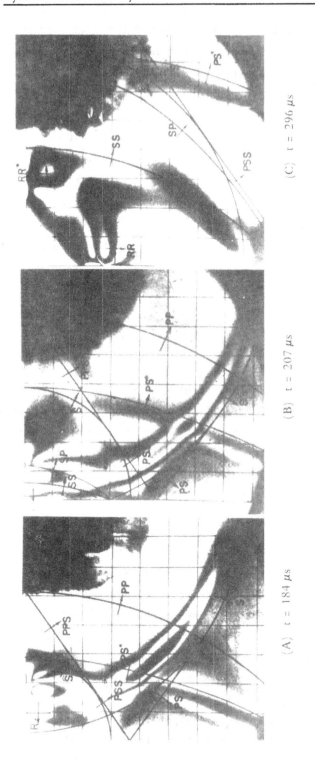

(A) t = 184 μs

(B) t = 207 μs

(C) t = 296 μs

Fig. 5.12 Isochromatic Fringe Patterns Showing Reflection Process

dilatational (SP) and equivoluminal (SS) reflections of the S wave from
the unloaded edge. The photoelastic response of these waves as well as
the locations of their fronts are indicated in Fig. 5.12B

Finally, the interaction of the Rayleigh wave (R), with the corner
gives rise to a Rayleigh wave transmitted around the corner (RR) and a
Rayleigh wave reflected back from the corner (RR*), as shown in Fig.
5.12C.

Data from the absolute retardation experiments allowed quantitative
measurements of the reflection of the P wave from the unloaded edge, in
which the PP and PS* wves are generated, and also for the transmission
and reflection of the R wave at the corner, which produces the RR and RR*
waves.

5.3.1 Results

P-wave reflection

These results show that the principal waves arising in the P-wave
reflection from the corner of the quarter-plane are the PP and PS* waves.
The relative magnitudes and spatial variations are in good qualitative
agreement with predictions from the theory of reflection of a plane dilatational wave from a free edge. Energy reflected at almost normal incidence goes mainly into the dilatational reflection (PP), whereas at
smaller angles of incidence, an increasing proportion of the P-wave
energy goes into the equivoluminal reflection (PS*).

Rayleigh wave response

Surface energy transmission and reflection coefficients from this
study and several previous investigations are listed in Table 5.1.
Coefficients from this study are based upon energies computed from
tangential surface stresses, whereas those of de Bremaecker[3] and
Knopoff and Gangi[4] are based upon energies computed from normal surface
displacements. The coefficients listed for Pilant, Knopoff and
Schwab[5] and for Mal and Knopoff[6] are based upon the squares of the
ratios of normal surface displacement amplitudes.

The energy transmission ratio $E_{R,RR}$ from this study agrees well with
previous experimental values obtained by Knopoff et al (1960,1964) but is

much greater than that found in the analytical study for $\nu = 0.25$.

The reflection coefficient $E_{R,RR*}$ found in our study is much greater than that reported in the earlier studies. Since the values differ by a factor ranging from 2 to 5, the difference cannot be attributed to experimental error in either the photoelastic or the transducer studies. It is believed that the reflection coefficient differences are attributable to the significant differences in the frequency spectrum between the incident Rayleigh wave (R) obtained from the explosive charge utilized in this study and that attained with the $BaTiO_3$ transducers employed previously.

TABLE 5.1

Surface Energy Transmission and Reflection Coefficients

	Type of Measurement	ν	Coefficients	
			Transmission	Reflection
Dally and Henzi[7]	Photoelastic	0.33	$E_{R,RR} = 0.44$	$E_{R,RR*} = 0.31$
de Bremaecker	$BaTiO_3$ Trans.	0.20	$E_{R,RR} = 0.40$	$E_{R,RR*} = 0.15$
Knopoff and Gangi	$BaTiO_3$ Trans.	0.36	$E_{R,RR} = 0.55$	$E_{R,RR*} = 0.06$
Pilant et al	$BaTiO_3$ Trans.	0.33	$A_{R,RR}^2 = 0.45$	$A_{R,RE*}^2 = 0.06$
Mal and Knopoff	Analytical	0.25	$A_{R,RR}^2 = 0.23$	$A_{R,RR*}^2 = 0.18$

The transmitted wave was found to exhibit a transition layer, as does the incident wave, whereas the reflected wave exhibits no such layer. Experimental results indicate that in the Rayleigh wave interaction with the corner, there is a significant transfer of energy from the surface region to positions well below the surface. This energy appears in the transmitted wave rather than in the reflected wave. Energy associated with the reflected wave is localized much more closely to the boundary.

In the deep subsurface regions, the transmitted wave is more highly dilatational than the incident wave and is tensile, whereas the incident wave is compressive. Moreover, the reflected wave exhibits a more highly dilatational subsurface character than the incident or transmitted waves and is tensile.

5.4 Application to Mining Problems

The next example shows the application of dynamic photoelasticity to a mining problem[8]. In this study, the interaction of stress waves generated by a distributed line charge, at the free surface of a quarry bench with a fixed bottom was investigated. The groove in the model employed, shown in Fig 5.13 was loaded with lead azide (2.5 mg per mm. to length) to simulate a charge in a borehole. Four identical models were examined where the ignition procedure was varied to include: firing from the top, firing from the bottom, firing from the center and firing, simultaneously, from both ends.

A typical fringe pattern representing the stress wave produced by a line charge is shown in Fig. 5.14. This frame was recorded during Test 1 at a time 66 μs after the line charge was ignited at the top. Inspection shows that the entire line charge was detonated and that two distinct incident waves have been generated. The first wave is a dilatational or P wave which exhibits a nearly plane portion inclined 40° to the line charge and the second is a shear or S wave which is inclined 22°.

The velocity of detonation c_d of the line charge was found to be 110,000 in/sec (2800 m/sec) by using the relation:

$$c_d = c_L/\sin \alpha \tag{5.4}$$

where c_L is the dilatational wave velocity given in Table 2 and α is the angle of inclination of the wave front.

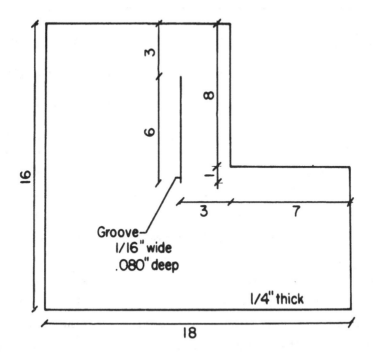

Fig. 5.13 Model of Bench Face

TABLE 5.2
Dynamic Characteristics of CR-39

Specific Weight	1.30
Modulus of Elasticity, E_d	535,000 psi (3690 MPa)
Poisson's Ratio, ν_d	0.32
Tensile Strength	5000-7000 psi (35-48 MPa)
Wave Velocities	
Dilatational, c_L	70,000 ips (1780 m/sec)
Shear, c_s	41,000 ips (1040 m/sec)
Material fringe value*, f_σ	110 psi-in. (19.3 MPa-mm)

* In blue light with λ = 5000Å.

Representative fringe patterns from each of the four tests are shown in Figs. 5.15 to 5.18 to illustrate the different characteristics of the stress waves produced by varying the ignition location. The fringe patterns for Test 1 with the charge ignited at the top is presented in Fig. 5.15. The first frame taken at 40 μs after ignition shows that undistorted shape of the incident waves prior to interaction with the boundary. The undistorted wave exhibits maximum fringe orders of 7.5 for the P and 8 for the S wave. The next frame at 60 μs shows the undistorted pattern to the left and the beginning of the interaction process at the free boundary where the incident P wave has reflected, producing PP and PS (reflected dilatational and shear) waves. The P wave at the face exhibits a compression of about two orders which is converted to a tension of two orders in the region behind the reflected waves.

The next frame recorded at 82 μs shows that the dominant wave at the vertical face is the incident S wave striking the surface and producing a tensile peak of about N = 5. The incident P wave and the reflected PS wave are in the process of turning the lower corner. On the left side it is evident that the incident waves have attenuated with the P wave exhibiting N_{max} = 5.5 and the S wave N_{max} = 6.5.

In the final frame of Fig. 5.15 recorded at 115 μs the major part of the interaction process has been completed. The incident S wave shows a maximum of 4.5 orders on the vertical face and the incident P wave is now

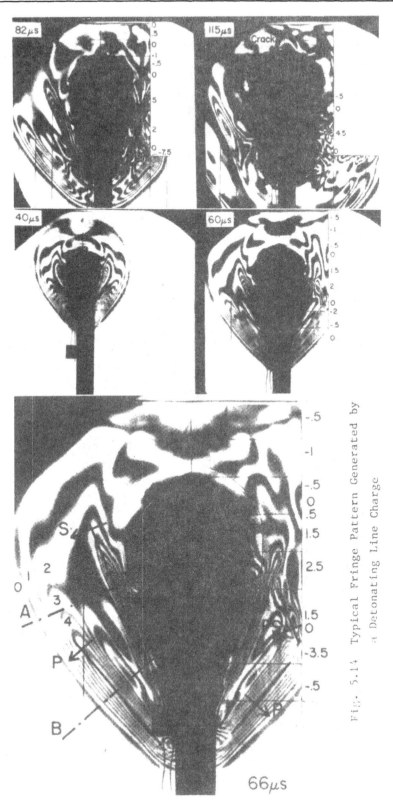

Fig. 5.15 Fringe Patterns for Top Ignition

Fig. 5.14 Typical Fringe Pattern Generated by a Detonating Line Charge

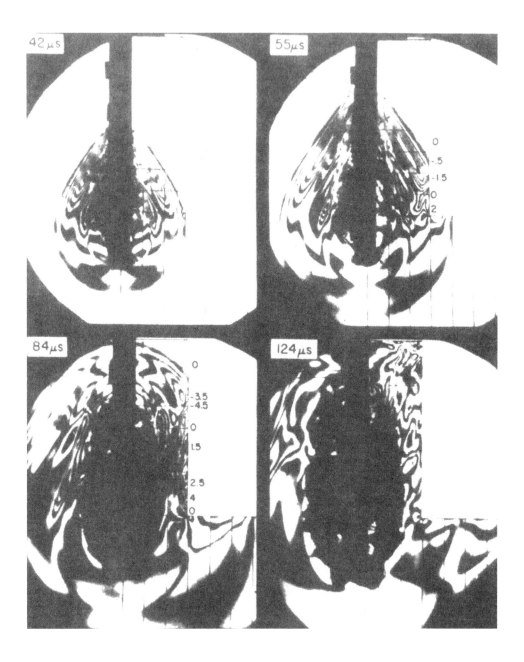

Fig. 5.16 Fringe Patterns for Bottom Ignition

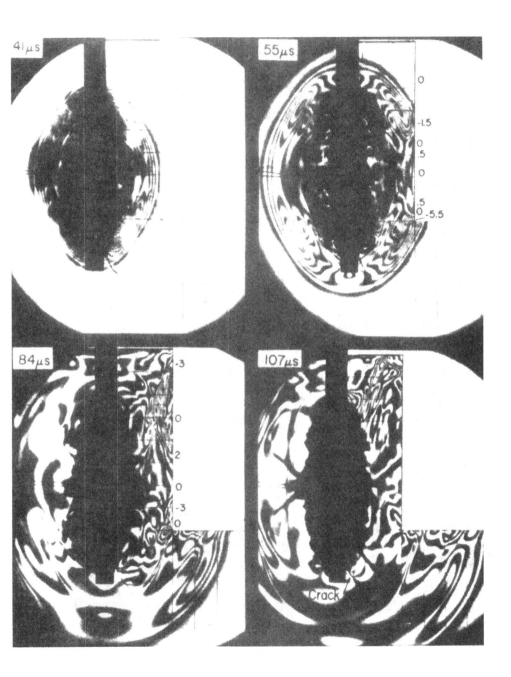

Fig. 5.17   Fringe Patterns for Center Ignition

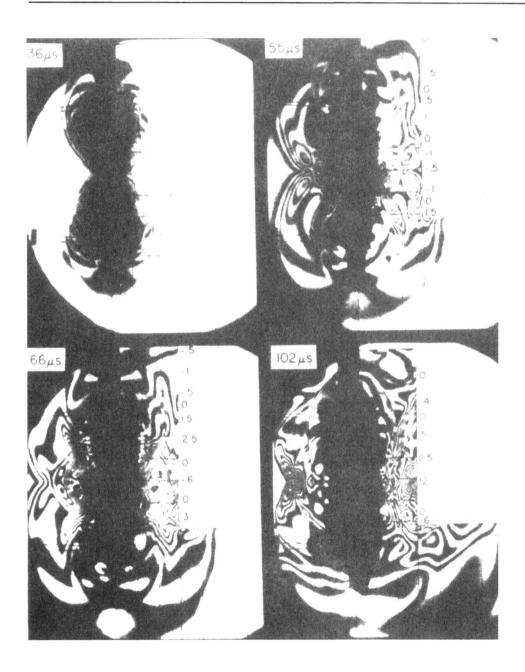

Fig. 5.18   Fringe Patterns for Simultaneous End Ignition

propagating along the horizontal bottom surface of the bench. The incident
P wave has attenuated to 3.5 orders and the S wave to 5 orders. A crack
tip can be identified emerging from smoke at the top end of the line
charge.

The results for Test 4 illustrated in Fig. 5.18 show a very complex
and interesting wave interaction when the line charge is detonated
simultaneouly at both ends. The initial wave pattern shown in the first
frame indicates that the burning is complete in less than 36 µs. The
next frame at 55 µs corresponds to the initial P wave interaction with
the boundary and the interaction of the two P waves at interior points.
In the next record at 6´. µs, a maximum compressive fringe order of 6 is
recorded when the P waves reinforce each at the boundary. The final
frame at 102 µs indicates that a great reinforcement occurs when the two
shear waves are simultaneously interacting and reflecting from the vertical face to produce a tensile peak of 12 fringes. On the left side of
the line charge, the reinforcement of the two S waves is also evident as
a peak of 9.5 fringes is obtained. The PS wave near the top of the face
exhibits a compressive fringe order of 4.

The fringe patterns described previously give a visual comparison of
the four different experiments. However, a more complete and accurate
comparison can be made if the fringe order and the stresses are shown as
a function position along the vertical face at discrete times during the
dynamic event. The fringe order–position functions are presented for
Test 4 in Fig. 5.19, and a comparison of the maximum stresses along the
boundary for the four tests are presented in Fig. 5.20.

Dynamic photoelastic tests have provided comprehensive field data,
giving in effect a visual display of stress waves propagating from a line
charge and interacting with the free vertical face of a fixed bottom
bench. The photoelastic fringe patterns clearly show the reflection process, the generation of PS waves, and the reinforcing which occurs when
two waves overlap. The surface stresses along the vertical face were
examined and it was noted that the incident S wave consistently produced
tensile stresses of significant magnitude. These tensile stresses are
large enough to produce failure in a low tensile strength material like

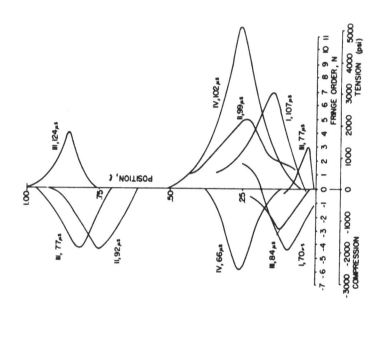

Fig. 5.20   Maximum Stresses as a Function
of Position (all tests)

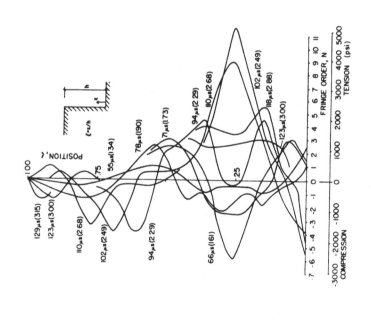

Fig. 5.19   Tangential Stresses as a Function
of Position and Time   (Test 4)

rock and should be important in improving fragmentation.

The P wave produced compressive stresses on the boundary and only will be effective in comminution by inducing scabbing-type failures. Even in this case the subsurface tensile stresses required for scabbing probably will occur in the PS wave which trails the P wave.

Comparison of the four methods of igniting the line charge clearly indicates that simultaneous ignition at both ends is the most effective technique and that ignition in the center was clearly the least effective technique.

The results indicate a new technique for investigating ignition patterns and charge distributions. Further photoelastic experiments will be required with other geometries and charge patterns. Scaling laws must be considered and verification of the results by testing in rock will represent interesting developments in the future.

5.5 Application to Non-destructive Testing

The final example[9] relates to non-destructive testing where Rayleigh waves are interacted with surface flaws as a means to detect the presence of these flaws. In this study dynamic photoelastic analyses were performed on half-plane models containing slits of different lengths as defined in Fig. 5.21. Rayleigh waves generated by small explosive charges propagated along the surface of the model and interacted with the slit. Features of the interaction process with shallow flaws are illustrated by the dynamic isochromatic fringe patterns given in Fig. 5.22. For these shallow flaws it is evident that a significant portion of the R wave is transmitted in all three instances. There is a marked change in the reflected and scattered waves as the ratio of the flaw depth to wave length d/λ is increased. For $d/\lambda = 1/32$ there is no visible reflected Rayleigh wave RR and the scattered RS is just discernable. As d/λ increases to 1/16 and 1/8 the RR wave becomes clearly evident and significant amounts of energy are scattered by means of RP (Rayleigh induced dilatational wave), RS (Rayleigh generated shear wave) and RPS (shear-wave induced by the RP waves).

Transmission and reflection coefficients associated with the

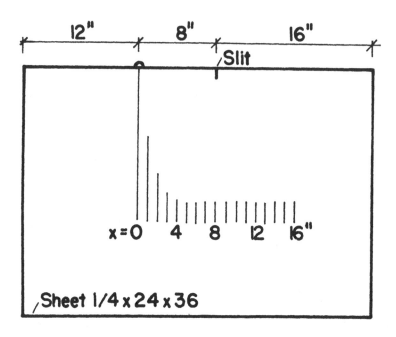

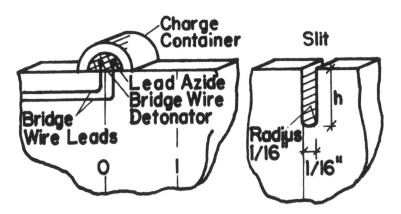

Fig. 5.21   Definition of Model and Details of Slit and Charge

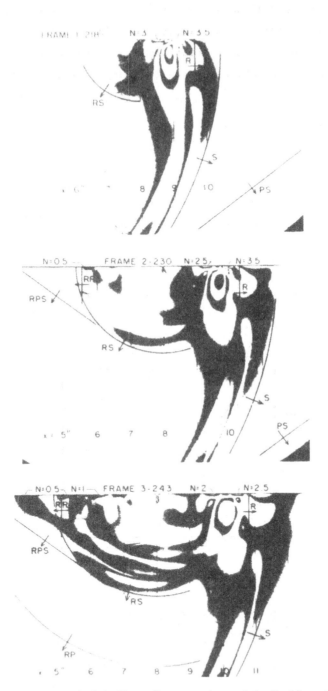

Fig. 5.22   Rayleigh Wave Interaction with Shallow Flaws

transmitted and reflected components of the Rayleigh wave were
established from the fringe patterns as shown in Fig. 5.23. It is evident that the transmission coefficient A_t approaches zero as d/λ
increases; however, the reflection coefficient A_r will not approach unity
because of the energy lost due to the generation of body waves such as
the RP, PS, and PRS waves.

In this case dynamic isochromatic fringe patterns permit a visual
examination of the complex interaction which occurs when a Rayleigh wave
encounters a slit type of surface flaw. The results are useful in
interpreting results in crack detection tests which use ultrasonic
reflection and transmission techniques to predict the existence and size
of surface flaws.

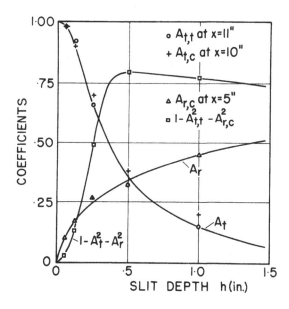

Fig. 5.23 Energy Loss, Reflection Coefficient and Transmission
Coefficient as a Function of Slit Depth

References

1. Riley, W.F. and Dally, J.W., "A Photoelastic Analysis of Stress Wave Propagation in a Layered Model", Geophysics, Vol. 31, No. 5, (1966), pp 881-899.

2. Lamb, M., "On the Propagation of Tremors Over the Surface of an Elastic Solid", Phil. Trans. Roy. Soc. London, Series A, Vol. 203, (1904), p. 1.

3. de Bremaecker, J. Cl, "Transmission and Reflection of Rayleigh Waves at Corners", Geophysics, Vol. 23, (1958), p. 253-266.

4. Knopoff, L. and Gangi, A.F., "Transmission and Reflection of Rayleigh Waves by Wedges", Geophysics, Vol. 25, (1960), p. 1203-1214.

5. Pilant, W.L., Knopoff, L. and Schwab, F., "Transmission and Reflection of Surface Waves at a Corner, 3" Rayleigh Waves (experimental)", J. Geophys. Res., Vol. 69, (1964), p. 291.

6. Mal, A.K. and Knopoff, L., "Transmission of Rayleigh Waves at a Corner", Bull. Seis. Soc. Am, Vol. 56, (1966), p. 455.

7. Henzi, A.N. and Dally, J.W., "A Photoelastic Study of Stress Wave Propagation in a Quarter-Plane", Geophysics, Vol. 36, No. 2, (1970), pp. 296-310.

8. Reinhardt, H.W. and Dally, J.W., "Dynamic Photoelastic Investigation of Stress Wave Interaction with a Bench Face", Transactions Society of Mining Engineers, AIME, Vol. 250, (1971), p. 35-42.

9. Reinhardt, R.W. and Dally, J.W., "Some Characteristics of Rayleigh Wave Interaction with Surface Flaws", Materials Evaluation, Vol. 28, No. 10, (1970), p. 213.

CHAPTER 6

FRACTURE MECHANICS

6.1 Introduction

Fracture mechanics is a relatively new technical approach which has
been developed to analyze bodies containing cracks. Pre-existing cracks
or newly formed cracks are of critical importance because these cracks
markedly reduce the strength and/or load capacity of a structure.
Fracture mechanics is an important approach with applicable concepts
which can be used in materials science, applied mechanics, materials
engineering and engineering design. The illustration presented in Fig.
6.1 shows the broad range of fracture mechanics and relates the applications to a feature size ranging from 10^{-10} to 10^{2} m.

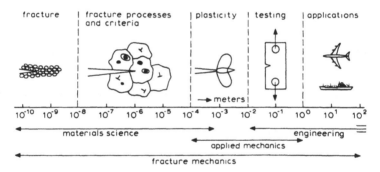

Fig. 6.1 Application of Fracture Mechanics to a Wide Range of Disciplines

From an engineering viewpoint, fracture mechanics is used as a basis
for predicting

1. Critical crack size.

2. Strength of a structure as a function of crack size (length).

3. Crack growth rates as a function of cycles of load or time under
    load.

4. Inspection requirements pertaining to size of an admissible
    crack and the period of time between inspection.

In mechanics the concern is with the stress and/or strain distribution near the crack tip. This problem is complex because the stresses at the crack tip approach infinity as the radius of the crack tip approaches zero. This stress singularity at the crack tip impeded the development of fracture mechanics until 1921 when Griffith[1] introduced a concept for crack instability. Griffith using an energy approach indicated that a crack would become unstable when the decrease in potential energy due to an increase in the crack surface area is greater than the increase in the surface energy due to an increase in the crack surface area.

The Griffith criterion did not accommodate ductile metals where a significant amount of energy is dissipated at the crack tip due to localized plastic deformation. Irwin[2,3] modified the Griffith theory and added terms to accommodate external work and internal plastic energy dissipation and defined a new measure to predict the stability of cracks in structures. Irwin defined a crack extension force as:

$$\mathcal{G} = \frac{dU_T}{dA} \qquad \text{(system isolated)} \qquad (6.1)$$

where U_T is the total stress field energy. The "system isolated" requirement implies that no energy is transmitted to or from the body at the load-points during crack extension dA.

When

$$\mathcal{G} > \mathcal{G}_c \qquad (6.2)$$

where $\mathcal{G}_c$ is a critical crack extension force, the crack becomes unstable and propagates at high speed causing failure. The critical crack extension force $\mathcal{G}_c$ is a function of material, material properties, temperatures, environment and specimen thickness.

Later, the stress intensity factor K which will be discussed in Section 6.2 replaced $\mathcal{G}$ as a measure of the criticality of the stresses at the crack tip. However, K is related to $\mathcal{G}$ by

$$K^2 = \mathcal{G} E \qquad \text{for plane stress} \qquad (6.3)$$

where E is the modulus of elasticity.

For this reason the stability criterion

$$K > K_c \qquad\qquad (6.4)$$

is equivalent to that given in eq. (6.2).

6.2 Westergaard Stress Function for Two-Dimensional Problems

Consider the crack of length 2a in an infinite plate as presented in Fig. 6.2 as the basic problem to illustrate the important features of fracture mechanics. Of particular interest is the relationship between the stresses, the stress intensity factor K and the position co-ordinates defined in Fig. 6.2.

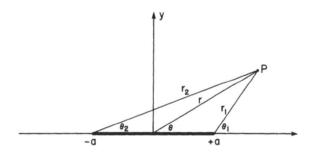

Fig. 6.2 Co-ordinates for a Crack of Length 2a in an Infinite Plane.

The problem is approached in a classical manner with the Airy's stress function ϕ where the stresss

$$\begin{aligned}
\sigma_{xx} &= \partial^2 \phi / \partial y^2 \\
\sigma_{yy} &= \partial^2 \phi / \partial x^2 \\
\tau_{xy} &= -\partial^2 \phi / \partial x \partial y
\end{aligned} \qquad\qquad (6.5)$$

where the Airy's stress function ϕ is selected to satisfy the biharmonic equation:

$$\nabla^4 \phi = 0 \qquad\qquad (6.6)$$

where

$$\nabla^2 = \frac{\partial^2}{\partial x^2} + \frac{\partial^2}{\partial y^2}$$

Solutions of eq. (6.6) can be written in the form

$$\phi = f_1 + xf_2 + yf_3 + (x^2 + y^2)f_4 \qquad (6.7)$$

where the functions f_1, f_2, f_3 and f_4 satisfy the harmonic equation

$$\nabla^2 f = 0 \qquad (6.8)$$

Westergaard[4] considered a restricted form of eq. (6.7)

$$\phi = f_1 + yf_3 \qquad (6.9)$$

and selected f_1 and f_3 so that $\tau_{xy} = 0$ when $y = 0$.
    To follow the Westergaard approach note in the following equations that a complex function $g(z)$ where

$$z = x + iy \qquad (6.10)$$

may be written as

$$g(z) = \text{Re } g + i \text{ Im } g \qquad (6.11)$$

where Re g is the real part of g
       Im g is the imaginary part of g.
Note

$$\frac{\partial g}{\partial x} = \frac{\partial}{\partial x} \text{Re } g + i \frac{\partial}{\partial x} \text{Im } g \qquad (6.12)$$

$$g' = \text{Re } g' + i \text{ Im } g' \qquad (6.13)$$

$$\frac{\partial g}{\partial y} = \frac{\partial}{\partial y} \text{Re } g + i \frac{\partial}{\partial y} \text{Im } g \qquad (6.14)$$

$$i \ g' = i \ \text{Re} \ g' - \text{Im} \ g' \tag{6.15}$$

Matching the real and imaginary parts of eqs. (6.12) and (6.13) gives

$$\frac{\partial}{\partial x} \ \text{Re} \ g = \text{Re} \ g' \quad \text{and} \quad \frac{\partial}{\partial x} \ \text{Im} \ g = \text{Im} \ g' \tag{6.16}$$

and similarly from eqns. (6.14) and (6.15) one obtains

$$\frac{\partial}{\partial y} \ \text{Re} \ g = - \ \text{Im} \ g' \quad \text{and} \quad \frac{\partial}{\partial y} \ \text{Im} \ g = \text{Re} \ g' \tag{6.17}$$

Equations (6.16) and (6.17) are the Cauchy-Riemann relations which provide the rules for the derivatives of $g(z)$ regardless of the selection of $g(z)$

Returning to the Westergaard solution one notes that f_1 and f_3 in eq. (6.9) were selected as

$$f_1 = \text{Re} \ \overline{\overline{Z}} \tag{6.18}$$

$$f_3 = \text{Im} \ \overline{\overline{Z}} \tag{6.19}$$

where Z', Z, $\overline{Z}$ and $\overline{\overline{Z}}$ are functions of the complex variable z with the following definitions of the notation

$$Z' = \frac{d}{dz} \ Z$$

$$Z = \frac{d}{dz} \ \overline{Z} \tag{6.20}$$

$$\overline{Z} = \frac{d}{dz} \ \overline{\overline{Z}}$$

Combining eqs. (6.5), (6.9), (6.18), (6.16), (6.17), (6.18), (6.19) and (6.20) one can write

$$\sigma_{xx} = Re\ Z - y\ Im\ Z' \tag{6.21}$$

$$\sigma_{yy} = Re\ Z + y\ Im\ Z' \tag{6.22}$$

$$\tau_{xy} = -\ y\ Re\ Z' \tag{6.23}$$

It is evident from eq. (6.23) that $\tau_{xy} = 0$ when $y = 0$ as required by boundary conditions on the crack surface regardless of the choice of Z provided Z' is finite when $y = 0$. Thus, Z is selected to satisfy other boundary conditions consistent with the geometry of the two dimensional body and the loading conditions.

6.3 Solutions to Crack Problems

The solution to a crack problem of the type illustrated in Fig. 6.2 can be obtained by writing the Westergaard stress function Z as

$$Z = \sigma z/(z^2 - a^2)^{1/2} \tag{6.24}$$

and noting that

$$Z' = -\ \sigma a^2/(z^2 - a^2)^{3/2} \tag{6.25}$$

From Fig. 6.2 it is clear that

$$z = re^{i\theta}$$
$$z - a = r_1 e^{i\theta_1} \tag{6.26}$$
$$z + a = r_2 e^{i\theta_2}$$

Substituting eq. (6.26) into (6.24) and (6.25) gives

$$Z = \frac{\sigma\ r}{\sqrt{r_1 r_2}}\ \left[\cos\left(\theta - \frac{\theta_1 + \theta_2}{2} \right) + i\ \sin\left(\theta - \frac{\theta_1 + \theta_2}{2} \right) \right] \tag{6.27}$$

and

$$Z' = -\ \frac{\sigma a^2}{(r_1 r_2)^{3/2}}\ \left[\cos\frac{3}{2}(\theta_1 + \theta_2) - i\ \sin\frac{3}{2}(\theta_1 + \theta_2) \right] \tag{6.28}$$

Substituting eqs. (6.27) and (6.28) into (6.21), (6.22) and (6.23) gives

$$\sigma_{xx} = \frac{\sigma}{\sqrt{r_1 r_2}} \left[r \cos \left\{ \theta - \frac{\theta_1 + \theta_2}{2} \right\} - \frac{a^2 y}{r_1 r_2} \sin \frac{3}{2}(\theta_1 + \theta_2) \right] \tag{6.29}$$

$$\sigma_{yy} = \frac{\sigma}{\sqrt{r_1 r_2}} \left[r \cos \left\{ \theta - \frac{\theta_1 + \theta_2}{2} \right\} + \frac{a^2 y}{r_1 r_2} \sin \frac{3}{2}(\theta_1 + \theta_2) \right] \tag{6.30}$$

$$\tau_{xy} = \frac{\sigma a^2 y}{(r_1 r_2)^{3/2}} \cos \frac{3}{2}(\theta_1 + \theta_2) \tag{6.31}$$

Note that angles θ, θ_1 and θ_2 are restricted to the range of $-\pi$ to π and the required symmetry is obtained in the solution above. Also, it is evident from eqs. (6.27) and (6.28) that

Re $Z \rightarrow \sigma$

y Re $Z' \rightarrow 0$ $\tag{6.32}$

y Im $Z' \rightarrow 0$

with $r \rightarrow \infty$ for all values of θ. This implies that the applied load at infinity is $\sigma_{xx} = \sigma_{yy} = \sigma$.

Consider next the crack with $y = 0$ and x ranging between $-a$ and $+a$.

To examine the crack line let $y = 0$, $\theta = 0$ or π, $\theta_1 = \pi$ and $\theta_2 = 0$. Note that $\tau_{xy} = 0$ because $y = 0$ and $\sigma_{yy} = 0$ because the $\cos (\theta - \frac{\theta_1 + \theta_2}{2}) = 0$ along the crack line.

6.4 Near Field Approximation

The stresses expressed in eqs. (6.29) - (6.31) can be simplified by considering a small region very close to the right leading edge of the crack. Let

$$z = a + z_1 \tag{6.33}$$

where the small region near the crack tip is depicted by z_1 as indicated in Fig. 6.3.

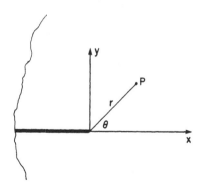

Fig. 6.3   Near Field Region - Co-ordinates

If z_1 is very small compared to a then

$z = a + z_1 \approx a$

$z - a = z_1$ (6.34)

$z + a = 2a + z_1 \approx 2a$

Substituting eq. (6.34) into (6.24) yields

$$Z \approx \sigma a/[2az_1]^{1/2} = \frac{\sigma\sqrt{\pi a}}{\sqrt{2\pi z_1}} = \frac{K}{\sqrt{2\pi z_1}} \qquad (6.35)$$

where the term $K = \sigma\sqrt{\pi a}$ is known as the stress intensity factor.

The stress field near the crack tip can be derived directly from

$$Z = K/\sqrt{2\pi z} \qquad (6.35\text{bis})$$

where the subscript has been dropped from z_1 by shifting the co-ordinates to the crack tip. Note that,

$$Z' = - \frac{K}{2r\sqrt{2\pi r}} \left[\cos \frac{3\theta}{2} - i \sin \frac{3\theta}{2} \right]$$

$$y \operatorname{Im} Z' = \frac{K}{\sqrt{2\pi r}} \left[\sin \frac{\theta}{2} \cos \frac{\theta}{2} \sin \frac{3\theta}{2} \right] \tag{6.36}$$

$$\operatorname{Re} Z = \frac{K}{\sqrt{2\pi r}} \cos \frac{\theta}{2}$$

Substituting eq. (6.27) into eqs. (6.21) - (6.23) gives

$$\sigma_{xx} = \frac{K}{\sqrt{2\pi r}} \cos \frac{\theta}{2} \left[1 - \sin \frac{\theta}{2} \sin \frac{3\theta}{2} \right] \tag{6.37}$$

$$\sigma_{yy} = \frac{K}{\sqrt{2\pi r}} \cos \frac{\theta}{2} \left[1 + \sin \frac{\theta}{2} \sin \frac{3\theta}{2} \right] \tag{6.38}$$

$$\tau_{xy} = \frac{K}{\sqrt{2\pi r}} \sin \frac{\theta}{2} \cos \frac{\theta}{2} \cos \frac{3\theta}{2} \tag{6.39}$$

The form of eqs. (6.37) - (6.39) is simplified and for this reason these equations are usually employed instead of eqs. (6.29) - (6.31).

It is interesting to continue the analysis and write the principal stresses

$$\sigma_1 = \frac{\sigma_{xx} + \sigma_{yy}}{2} + \sqrt{\left(\frac{\sigma_{yy} - \sigma_{xx}}{2}\right)^2 + \tau_{xy}^2}$$

$$= \frac{K}{\sqrt{2\pi r}} \cos \frac{\theta}{2} \left(1 + \sin \frac{\theta}{2}\right) \qquad\qquad (6.40)$$

$$\sigma_2 = \frac{\sigma_{xx} + \sigma_{yy}}{2} - \sqrt{\left(\frac{\sigma_{yy} - \sigma_{xx}}{2}\right)^2 + \tau_{xy}^2}$$

$$= \frac{K}{\sqrt{2\pi r}} \cos \frac{\theta}{2} \left(1 - \sin \frac{\theta}{2}\right) \qquad\qquad (6.41)$$

and the maximum shear stress is:

$$\tau_m = \left(\frac{\sigma_1 - \sigma_2}{2}\right) = \frac{K}{\sqrt{2\pi r}} \sin \frac{\theta}{2} \qquad\qquad (6.42)$$

Equation (6.42) indicates that photoelastic isochromatics will be in the form of closed ovaloids as indicated in Fig. 6.4.

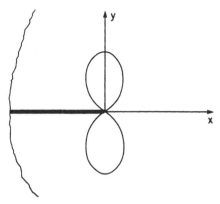

Fig. 6.4   Isochromatics Representing τ_m at a Crack Tip

6.5   The Generalized Westergaard Equations

Frequently the stress state near the crack tip cannot be adequately described by the Westergaard equations. Sanford[5] recognized the short coming of the Westergaard approach and has derived a generalized approach for more accurately characterizing the stress field near a crack tip. Sanford used the Goursat-Kolosov[6] complex representation of the plane problem and the potential functions ϕ and ψ defined in terms of stresses as

$$\sigma_{xx} + \sigma_{yy} = 4 \text{ Re } \phi' \tag{6.43}$$

$$\sigma_{yy} - \sigma_{xx} + 2i \, \tau_{xy} = 2 \left[\, \bar{z}\phi'' + \psi' \, \right] \tag{6.44}$$

Separating eq. (6.44) into its real and imaginary parts gives

$$\tau_{xy} = x \text{ Im } \phi'' - y \text{ Re } \phi'' + \text{ Im } \psi' \tag{6.45}$$

Impose symmetry conditions that $\tau_{xy} = 0$ or $y = 0$ gives from eq. (6.45)

$$x \text{ IM } \phi'' + \text{ Im } \psi' = 0 \quad \text{or} \quad y = 0 \tag{6.46}$$

Define a new function η such that

$$\eta = z \, \phi'' + \psi' \tag{6.47}$$

Then eq. (6.46) is equivalent to

$$\text{Im } \eta = 0 \quad \text{or} \quad y = 0 \tag{6.48}$$

Using eq. (6.47) in (6.44) gives

$$\sigma_{yy} - \sigma_{xx} + 2 \, i \, \tau_{xy} = 2 \left[(\bar{z}-z)\phi'' + \eta \right] \tag{6.49}$$

Then from eqs. (6.49), (6.43) and (6.45) one can write

$$\sigma_{xx} = 2 \text{ Re } \phi' - 2y \text{ Im } \phi'' - \text{Re } \eta \tag{6.50}$$

$$\sigma_{yy} = 2 \text{ Re } \phi' + 2y \text{ Im } \phi'' + \text{Re } \eta \tag{6.51}$$

$$\tau_{xy} = - 2y \text{ Re } \phi'' + \text{Im } \eta \tag{6.52}$$

If $\eta = 0$ for all z then eqs. (6.50) – (6.52) reduce to the original Westergaard eqs. (6.21) – (6.23). A more useful set of equations is obtained by imposing the symmetry condition

$$\text{Im } \eta = 0 \quad \text{on} \quad y = 0 \tag{6.46bis}$$

and letting

$$2\phi' = Z - \eta \tag{6.53}$$

For this case eqs. (6.50) – (6.52) can be written in terms of familiar Westergaard stress function Z and the new stress function η

$$\sigma_{xx} = \text{Re } Z - y \text{ Im } Z' + y \text{ Im } \eta' - 2 \text{ Re } \eta \tag{6.54}$$

$$\sigma_{yy} = \text{Re } Z + y \text{ Im } Z' - y \text{ Im } \eta' \tag{6.55}$$

$$\tau_{xy} = y \text{ Re } Z' + y \text{ Re } \eta' + \text{Im } \eta \tag{6.56}$$

The function Z must satisfy $\text{Re } Z = 0$ over the crack and the function η must satisfy eq. (6.46). A series function can be used to represent η of the form

$$\eta = \sum_{m=0}^{M} \alpha_m z^m \tag{6.57}$$

Recalling

$$Z = K/\sqrt{2\pi r} \tag{6.35bis}$$

gives the two stress functions required in eqs. (6.54) - (6.56) to
completely determine the stress field near the neighborhood of a crack.

This generalized approach will be developed in a method to determine
the stress intensity factor K from photoelastic isochromatic fringe patterns.

References

1. Griffith, A.A., "The Phenomena of Rupture and Flow in Solids", Phil.
    Trans. of Royal Soc. of London, Vol. 221, (1921), pp. 163-198.
2. Irwin, G.R. and Kies, J., "Fracturing and Fracture Dynamics", Welding
    Jnl. Research Supplement, Feb. 1952.
3. Irwin, G.R., "Fracture", Handbuch der Physik, Vol 6, (1958), pp.
    551-590.
4. Westergaard, H.M., "Bearing Pressures and Cracks", Transactions ASME,
    Vol. 61, (1939), pp. A49-A53.
5. Sanford, R.J., "A Critical Re-examination of the Westergaard Method
    for Solving Opening-Mode Crack Problems", Mechanics Research
    Communications, Vol. 6, (1979), pp. 289-294.
6. Sih, G.C., "On the Westergaard Method of Crack Analysis", Int'l. J.
    Fracture Mechanics, Vol. 2, (1966), pp. 628-631.

CHAPTER 7

PHOTOELASTIC DATA ANALYSIS METHODS FOR FRACTURE

7.1 Introduction

Post[1] and Post and Wells[2] in the early 50's were the first
investigators to show the application of photoelasticity to fracture
mechanics. Irwin[3] in a discussion to Ref. 2 showed that the stress-
intensity factor K could be determined from a single isochromatic-fringe
loop at the tip of the crack. With the Irwin method, the stress-
intensity factor K and a uniformly distributed stress σ_{ox} are functions
of the radius r_m of the fringe loop and the θ_m the angle associated with
the tilt of the loop as defined in Fig. 7.1.

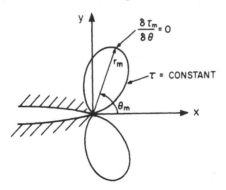

Fig. 7.1 Geometry of the Isochromatic-fringe Loop at the Crack Tip

Since this work in the early 50's, Bradley and Kobayashi[4] and C.W.
Smith[5] have modified Irwin's method. Bradley and Kobayashi redefined
the relationship between σ_{ox} and K which resulted in a simplified relation for K in terms of the fringe-loop parameters r_m and θ_m. Kobayashi
and Bradley then introduced a differencing technique involving measurement of θ and r on two fringe loops in an attempt to avoid difficulties
associated with the singular terms in Irwin's solution.

Smith omitted one term in one of Irwin's equations and measured the position r of the fringe loop at $\theta = \pi/2$ (along a line perpendicular to the crack extension). Smith then employed a differencing technique similar to that of Bradley and Kobayashi to uncouple K and σ_{ox} in the analysis. The data are then statistically conditioned using results from several pairs of fringes to improve the accuracy of the K determination.

The objective of this lecture is to review Irwin's methods of analysis indicating the effect of errors in the measurement of both r_m and θ_m. Finally a new method will be introduced which provides an approach to minimize the effects of inherent error in the measurement of r and θ on the determination of K.

7.2   Irwin's Two-Parameter Deterministic Methods of Analysis

Irwin[3,6] showed that the stress-intensity factor K could be determined from the isochromatic-fringe loop at the tip of the crack illustrated in Fig 7.1. Irwin began by writing the cartesian components of stress σ_x, σ_y and τ_{xy} in the local neighborhood (r $\ll$ a) of the crack tip as:

$$\sigma_x = \frac{K}{\sqrt{2\pi r}} \cos \frac{\theta}{2} \left(1 - \sin \frac{\theta}{2} \sin \frac{3\theta}{2}\right) - \sigma_{ox}$$

$$\sigma_y = \frac{K}{\sqrt{2\pi r}} \cos \frac{\theta}{2} \left(1 + \sin \frac{\theta}{2} \sin \frac{3\theta}{2}\right) \qquad (7.1)$$

$$\tau_{xy} = \frac{K}{\sqrt{2\pi r}} \sin \frac{\theta}{2} \cos \frac{\theta}{2} \cos \frac{3\theta}{2}$$

This representation (except for the σ_{ox} term) is the exact solution for the case of the semi-infinite crack subject to biaxial loading. The σ_{ox} term was subtracted from the expression for σ_x to provide another degree of freedom so that the analytically determined fringe loop can be brought into closer correspondence with the experimentally observed loop.

The maximum shear stress τ_m is expressed in terms of the cartesian stress components as:

$$(2\tau_m)^2 = (\sigma_y - \sigma_x)^2 + (2\tau_{xy})^2 \tag{7.2}$$

From eqs. (7.1) and (7.2), it is apparent that

$$(2\tau_m)^2 = \frac{K^2}{2\pi r} \sin^2\theta + \frac{2\sigma_{ox}K}{\sqrt{2\pi r}} \sin\theta\sin\frac{3\theta}{2} + \sigma_{ox}^2 \tag{7.3}$$

Next, Irwin observed the geometry of the fringe loops and noted that:

$$\frac{\partial\tau_m}{\partial\theta} = 0 \tag{7.4}$$

at the extreme position on the fringe loop where $r = r_m$ and $\theta = \theta_m$.

Differentiating eq. (7.3) with respect to θ and using eq. (7.4) gives

$$\sigma_{ox} = \frac{-K}{\sqrt{2\pi r_m}} \frac{\sin\theta_m \cos\theta_m}{(\cos\theta_m\sin\frac{3\theta_m}{2} + \frac{3}{2}\sin\theta_m\cos\frac{3\theta_m}{2})} \tag{7.5}$$

The two unknown parameters K and σ_{ox} are determined from the complete solution of eqs. (7.3) and (7.5) as:

$$\sigma_{ox} = \frac{-2\tau_m \cos\theta_m}{\cos(3\theta_m/2)[\cos^2\theta_m + (9/4)\sin^2\theta_m]^{1/2}} \tag{7.6}$$

and

$$K = \frac{2\tau_m\sqrt{2\pi r_m}}{\sin\theta_m} \left[1 + \left(\frac{2}{3\tan\theta_m}\right)^2\right]^{-1/2} \left(1 + \frac{2\tan\frac{3\theta_m}{2}}{3\tan\theta_m}\right) \tag{7.7}$$

To obtain K and σ_{ox} measurements of r_m and θ_m are made to define the maximum point on a fringe loop of order N. The value of τ_m used in eq. (7.7) is given by:

$$\tau_m = \frac{Nf_\sigma}{2h} \tag{7.8}$$

where f_σ is the material-fringe value

   h is the model thickness

Dally and Etheridge[7] have examined the range of applicability of Irwin's two parameter method and have shown that the relative error $\Delta K/K$ is a function of tilt angle for the fringe loop as indicated in Fig. 7.2. Inspection of Fig. 7.2 shows that the two-parameter method of Irwin will predict K to within ±5 percent providing $73 < \theta_m < 139$ deg. and to within ±2 percent providing $73.5 < \theta_m < 134$ deg. Outside these ±5 percent, the error increase rapidly and the two-parameter method is not applicable.

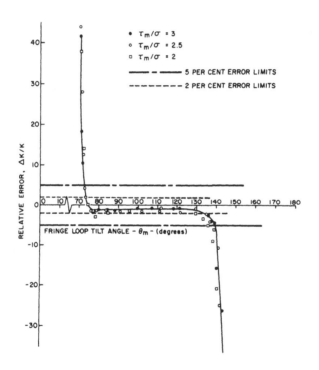

Fig. 7.2 Error as a Function of Tilt Angle θ_m for Irwin's Method.

The error in K is dependent upon r_m/a and in the analysis which produced the results depicted in Fig. 7.2 the value of $r_m/a < 0.03$ was used for θ_m near 130 deg. and the value $r_m/a < 0.065$ was used for θ_m near 73 deg. The smaller the value of r_m/a consistent with the precise measurement of both θ_m and r_m gives more accurate predictions of K provided $r/h > 0.5$ to insure plane stress conditions.

7.3 Multi-Parameter Methods

Two-parameter models are not adequate when the crack tip approaches a boundary on a load point since the state of stress in the measurement region deviates from that implied by the central crack problem. A more general theory[8] is required to accurately express the stress field in the measurement region and to accurately determine K.

Recall the discussion of Section 6.5 and the generalized Westergaard equations where the stress near the crack tip was given by eqs. (6.54), (6.55) and (6.56). The stress functions Z and η are expressed as:

$$Z = \sum_{n=0}^{N} \frac{A_n}{(n-1/2)} z^{(n-1/2)} \qquad (7.9)$$

and

$$\eta = \sum_{m=0}^{M} \alpha_m z^m \qquad (7.10)$$

Note from eq. (7.9) that the stress intensity factor K is written in this notation as

$$K = -2\sqrt{2\pi} \, A_o \qquad (7.11)$$

Next, substitute eq. (7.8) into (7.2) to obtain

$$\left(\frac{Nf_\sigma}{2h} \right)^2 = D^2 + T^2 \qquad (7.12)$$

where

$$D^2 = \left(\frac{\sigma_{yy} - \sigma_{xx}}{2} \right)^2$$

$$T^2 = \tau_{xy}^2$$

From eqs. (6.54), (6.55) and (6.56) into eq. (7.12) one can write

$$D = y \text{ Im } Z' - y \text{ Im } \eta' + \text{Re } \eta \tag{7.13}$$

$$T = - y \text{ Re } Z' + y \text{ Re } \eta' + \text{Im } \eta \tag{7.14}$$

Substituting eqs. (7.9) and (7.10) into (7.13) and (7.14) and relating the complex variable $z = re^{i\theta}$ to crack tip co-ordinates the functions D and T become

$$D = \sum_{n=0}^{N} A_n r^{(n-1/2)} \sin\theta\sin(n-1/2)\theta$$

$$+ \sum_{m=0}^{M} \alpha_m r^m [\cos(m\theta) - m\sin\theta\sin(m-1)\theta] \tag{7.15}$$

$$T = \sum_{n=0}^{M} (-A_n) r^{(n-1/2)} \sin\theta\cos(n-3/2)\theta$$

$$+ \sum_{m=0}^{M} \alpha_m r^m [\sin(m\theta) + m\sin\theta\cos(m-1)\theta] \tag{7.16}$$

Combining eqs. (7.15) and (7.16) with eq. (7.12) gives the general solution for the isochromatic pattern around the crack tip for any size region. The size of the region (i.e., r/a) and the required accuracy determines the number of terms which must be retained in the series expansion of Z and η. The constants A_0, A_1, A_2 ... and α_0, α_1, α_3 ... most be determined from photoelastic data measured in the region of the crack tip. As the form of the equations are involved specialized computational procedures are necessary for solutions yielding accurate results for A_0, A_1, A_2 ... and α_0, α_1, α_2 ... .

7.4 Over Deterministic Solution of Multi-Parameter Constants

Sanford and Dally[9] showed that a non-linear least squares method could be utilized to solve for the constants which occur in the multi-parameter representation of the stress field at a crack tip. This

approach utilizes a combination of the least squares method and the
Newton-Raphson method to give an iterative procedure which can be imple-
mented with a relatively simple numerical routine on a computer.

Consider a set of functions g_k of the form

$$g_k(A_0, A_1, \ldots, A_N, \alpha_0, \alpha_1, \ldots, \alpha_M) = 0 \tag{7.17}$$

where $k = 1, 2, \ldots, L$ $(L>N+M)$.

Taking the Taylor's series expansion of eq. (7.17) yields:

$$(g_k)_{i+1} = (g_k)_i + \left[\frac{\partial g_k}{\partial A_0}\right]_i \Delta A_0 + \left[\frac{\partial g_k}{\partial A_1}\right]_i \Delta A_1 + \ldots + \left[\frac{\partial g_k}{\partial A_N}\right]_i \Delta A_N$$

$$\tag{7.18}$$

$$+ \left[\frac{\partial g_k}{\partial \alpha_0}\right]_i \Delta\alpha_0 + \left[\frac{\partial g}{\partial \alpha_1}\right]_i \Delta\alpha_1 + \ldots + \left[\frac{\partial g_k}{\partial \alpha_M}\right]_i \Delta\alpha_M$$

where i refers to the ith iteration step, and ΔA_0, ΔA_1, $\ldots \Delta A_N$, $\Delta\alpha_1$, $\ldots$
$\Delta\alpha_M$ are corrections to the previous estimates of A_0, A_1, $\ldots A_N$, α_0, α_1,
$\ldots\alpha_M$, respectively.

Recognizing from eq. (7.17) that it is necessary for $(g_k)_{i+1} = 0$,
yields an iterative equation of the form:

$$\left[\frac{\partial g_k}{\partial A_0}\right] \Delta A_0 + \left[\frac{\partial g_k}{\partial A_1}\right] \Delta A_1 + \ldots + \left[\frac{\partial g_k}{\partial A_N}\right] \Delta A_N +$$

$$\tag{7.19}$$

$$\left[\frac{\partial g_k}{\partial \alpha_0}\right] \Delta\alpha_0 + \left[\frac{\partial g_k}{\partial \alpha_1}\right] \Delta\alpha_1 + \ldots + \left[\frac{\partial g_k}{\partial \alpha_M}\right] \Delta\alpha_M = -(g_k)_i$$

In matrix notation eq. (7.19) becomes:

$$[g] = [c] [\delta] \tag{7.20}$$

where

$$[g] = \begin{bmatrix} -g_1 \\ -g_2 \\ \cdot \\ \cdot \\ \cdot \\ -g_L \end{bmatrix} \quad ;$$

$$[c] = \begin{bmatrix} \dfrac{\partial g_1}{\partial A_0} & \cdots & \dfrac{\partial g_1}{\partial A_N} & \dfrac{\partial g_1}{\partial \alpha_0} & \cdots & \dfrac{\partial g_1}{\partial \alpha_M} \\ \cdot & & \cdot & \cdot & & \cdot \\ \cdot & & \cdot & \cdot & & \cdot \\ \cdot & & \cdot & \cdot & & \cdot \\ \dfrac{\partial g_L}{\partial A_0} & \cdots & \dfrac{\partial g_L}{\partial A_N} & \dfrac{\partial g_L}{\partial \alpha_0} & \cdots & \dfrac{\partial g_L}{\partial \alpha_M} \end{bmatrix} \quad ;$$

$$[\delta] = \begin{bmatrix} \Delta A_0 \\ \cdot \\ \cdot \\ \cdot \\ \Delta A_N \\ \Delta \alpha_0 \\ \cdot \\ \cdot \\ \cdot \\ \Delta \alpha_M \end{bmatrix}$$

Since matrix [c] is not square (L>M+N) eq. (7.20) has no unique solution. However, it can be shown that a solution in the least squares sense[10] can be obtained from an auxiliary equation of the form:

$$[\delta] = [d]^{-1}[c]^T[g] \qquad (7.21)$$

where

$$[d] = [c]^T[c]$$

and

$$[c]^T = \text{transpose of } [c]$$

To apply this solution scheme to the isochromatic equation, eq.
(7.17) is recast in the form:

$$g_k = D_k^2 + T_k^2 - (\frac{N_k f}{2t}\sigma)^2 = 0 \qquad (7.22)$$

where k refers to the value of the function evaluated at a point in the
field (r_k, θ_k) at which the fringe order is N_k.

Note that the column elements of the matrix [c] are of the form:

$$\frac{\partial g_k}{\partial A_0} = 2D_k(\frac{\partial D}{\partial A_0})_k + 2T_k(\frac{\partial T}{\partial A_0})_k$$

.
.
.

$$\frac{\partial g_k}{\partial \alpha_0} = 2D_k(\frac{\partial D}{\partial \alpha_0})_k + 2T_k(\frac{\partial T}{\partial \alpha_0})_k$$

.
.
.

Thus, the procedure for determining the best fit values of the
constants consists of the following steps:

1.   from the fringe pattern select a sufficiently large set of data
      points (r_k, θ_k, N_k) over the region to be characterized.
2.   assume initial values for $A_0, A_1, \cdots, A_N, \alpha_0, \alpha_1, \cdots \alpha_m$,
3.   compute the elements of the matrices [g] and [c] for each data
      point,
4.   compute [δ] from eq. (7.20),
5.   revise the estimates of the unknowns, i.e.,

$$(A_0)_{i+1} = (A_0)_i + \Delta A_0$$

$$(A_1)_{i+1} = (A_1)_i + \Delta A_1$$

.   .
.   .
.   .

$$(\alpha_M)_{i+1} = (\alpha_M)_i + \Delta\alpha_m \quad,$$

6.   repeat steps (3), (4), and (5) until $[\delta]$ becomes acceptably
      small.

As an example of the application of the method developed above, consider the fringe pattern shown in Fig. 7.3. From this pattern 60 data
points were selected and used with the least-squares method to compute
the best fit values of the first three terms of the series representation
of each of the functions Z and η, i.e., a six-parameter isochromatic
model. The computed values of these constants are given in Table 7.1.
Using the values from this table, the theoretical isochromatic pattern
was plotted as shown in Fig. 7.4.

TABLE 7.1

Coefficients of Six-parameter Model

A_0	-303.7 psi-$\sqrt{in}$
A_1	-382.4 psi/$\sqrt{in}$
A_2	-870.0 psi-$(in)^{-3/2}$
α_0	-33.0 psi
α_1	-211.1 psi/in
α_2	-114.3 psi/in^2
$f_\sigma = 134$ psi-in/fringe	
$h = 0.5$ in	

7.5 Conclusions – Multi-Parameter Methods

The complexity of the state of stress at the crack tip can be treated
by using the series expansion for Z and η. The numerical approach
offered with a combination of the Newton-Raphson method and the least

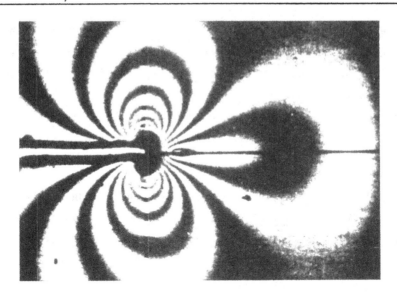

Fig. 7.3 Fringe Pattern at the Crack Tip Used to Provide 60 Data Points
in a Six Parameter Analysis.

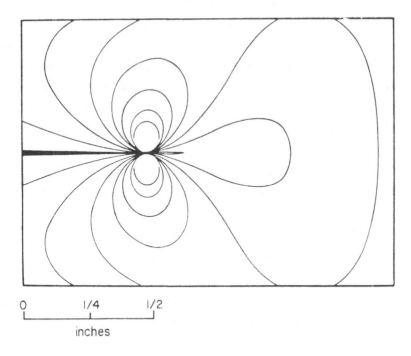

Fig. 7.4 Theoretical Isochromatic Pattern Resulting from the Six
Parameter Fit to the Experimental Data.

squares method provides solutions for K and other constants with surprising ease. Comparisons of the theoretical solutions and experimental solutions are essential to test the adequacy of the fit of the solution to problem at hand.

Reference 9 treats a mixed mode case which should be used when the fringe pattern is not symmetrical with respect to the x axis.

References

1. Post, D., "Photoelastic Stress Analysis for an Edge Crack in a Tensile Field", Proc. of SESA, Vol. 12, No. 1, (1954), pp. 99-116.

2. Wells, A. and Post, D., "The Dynamic Stress Distribution Surrounding a Running Crack - A Photoelastic Analysis", Proc. of SESA, Vol. 16, No. 1, (1958), pp. 69-92.

3. Irwin, G.R., Dist. of Ref. 2, Proc. of SESA, Vol. 16, No. 1, (1958), pp. 93-96.

4. Bradley, W.B., "A Photoelastic Investigation of Dynamic Brittle Fracture", Ph.D. Thesis, University of Washington (1969).

5. Schroedl, M.A. and Smith, C.W., "Local Stress Near Deep Surface Flaws Under Cylindrical Bonding Fields", Progress in Flaw Growth and Fracture Toughness Testing, ASTM STP 536, ATM, (1973), pp. 45-63.

6. Irwin, G.R., "Analysis of Stresses and Strains Near the End of a Crack Traversing a Plate", J. Appl. Mech, Vol. 24, No. 3, (Sept. 1957).

7. Etheridge, J.M. and Dally, J.W., "A Critical Review of Methods for Determining Stress Intensity Factors from Isochromatic Fringes", Expl. Mechanics, Vol. 17, No. 7, (1977), pp. 248-254.

8. Sanford, R.J. et al, "A Photoelastic Study of the Influence of Non-Singular Stresses in Fracture Test Specimens", University of Maryland Report, College Park, MD, March 1981.

9. Sanford, R.J. and Dally, J.W., "A General Method for Determining Mixed Mode Stress Intensity Factors from Isochromatic Fringe Patterns", Eng. Fracture Mechanics, Vol. 11, (1979), pp. 621-633.

10. Sanford, R.J., "Application of the Least Squares Method to Photoelastic Analysis", Experimental Mechanics, Vo. 20, (1980), pp. 192-197.

CHAPTER 8

CHARACTERIZATION OF DYNAMIC CRACK PROPAGATION

8.1   Introduction

Since the pioneering paper by A. A. Griffith[1] in 1921 considerable research[2,3,4] has been directed toward an understanding of the stability of a crack tip in a flawed structure. Significant progress has been made and the concepts of fracture mechanics are commonly used today in the design of most high performance structures. The procedure involves determining the stress intensity factor K associated with a flawed structure. The value of K depends upon the geometry of the crack and the body and the loading of the structure. Well established methods (analytical, numerical and experimental) have been developed for determining K and this portion of the design procedure can usually be accomplished with relative ease.

The value of the stress intensity factor is compared with the appropriate fracture initiation toughness K_{Ic} which characterizes the material and crack initiation is predicted when

$$K \geq K_{Ic} \qquad\qquad (8.1)$$

The adequacy of the prediction depends on the accuracy of the fracture initiation toughness K_{Ic}. Since K_{Ic} is dependent on material, body thickness, temperature and a host of other environmental parameters[4], its determination requires considerable care and skill in the materials laboratory. In most instances K and K_{Ic} can be determined with sufficient accuracy to permit a close prediction of the load which will produce crack initiation. At this point most engineers lose interest because crack initiation is synonymous with failure.

The behavior of a rapidly propagating crack after initiation has

received considerably less attention. While the topic is of less engineering significance than crack initiation, it is vitally important in
dynamic fracture control, fragmentation, and crack arrest methods in
large structures. The topic is difficult and the traditional approaches
- analytical, numerical and experimental - yield limited results and then
only after major investment of time and energy on the part of the
investigator.

The very high velocity of the propagating crack coupled with its
unpredictable path essentially dictates the use of whole field optical
methods of stress analysis to study experimentally its behavior. This
paper will review one experimental approach to the study of dynamic fracture phenomena, namely photoelasticity. The other experimental approach,
the method of caustics[5,6], which has yielded significant results will
be covered by Dr. Kalthoff in another part of this book.

8.2 Application of Dynamic Photoelasticity to Dynamic Fracture

Fundamental experimental studies of the behavior of propagating
cracks have been based on optical methods of stress analysis together
with high speed photographic recording techniques. Since the behavior of
the propagating crack is controlled by the highly localized stress singularity at the crack tip and the propagation path is often unpredictable,
experimental studies have been limited to the whole-field and high-
resolution optical methods. Of the optical methods available to the
experimentalist, which include photoelasticity, caustics, moire,
holographic interferometry and speckle interferometry, it appears that
photoelasticity and the method of caustics are the most suitable at this
time.

Beginning with the work of Schardin[7,8] and including the contrib-
utions of researchers (A. S. Kobayashi[9], P. S. Theocaris[5], J. F.
Kalthoff[6] and J.W. Dally[10] experimental studies of propagating
cracks have been conducted with a multiple-spark, high-speed photographic
system to record the image of the propagating crack. The multiple-spark
high-speed photographic system, which has been described in Chapter 1,
provides a set of isochromatic fringe patterns, as illustrated in Fig.

Fig. 8.1   A Sequence of 16 Dynamic Isochromatic Fringe Patterns
           Associated with High Velocity Crack Propagation Across a
           Center-Pin Loaded SEN Specimen.

8.1, which cover the entire crack propagation event from initiation to finalization of rupture. The detailed features of the state of stress at the propagating crack tip are clearly depicted in the enlargement shown in Fig. 8.2.

Fig. 8.2   Isochromatic Fringe Pattern Showing the Detail
of the Dynamic Stresses at the Tip of the Propagating Crack

The isochromatic fringe pattern in the field adjacent to the crack tip provides the data necessary to determine the instantaneous stress intensity factor K_{dyn} and the location of the crack tip as a function of time. It is necessary to use analytical procedures similar to those discussed in Chapter 7 to determine the value of K, and to make corrections when necessary to account for the effect of crack tip velocity.

The study of the effect of crack velocity on the stress-intensity factor has been limited to the opening mode. The results of the early work by A. S. Kobayashi[11] and by Etheridge[12] indicated that the velocity effect was small. However, more recent work by Rossmanith and Irwin[13] shows that significant differences in K occur for the higher crack velocity, and the method of analysis described in Chapter 7 must be modified to account for the effect of velocity of the crack tip.

The method for extracting K_I from the isochromatic data involves the use of the general dynamic solution[14] for the state of stress near the crack tip where the stresses are expressed in a power series containing $r^{-1/2}$, r^0, $r^{1/2}$, r, $r^{3/2}$ and r^2. Studies[15,16] show that the higher order terms are necessary to reduce the errors in determining K_I.

The dynamic higher order solution for the stress field is given by[14]

$$\sigma_x = \frac{1 + \lambda_2^2}{4\lambda_1\lambda_2 - (1+\lambda_2^2)^2} \left\{ (1+2\lambda_1^2 - \lambda_2^2)\text{Re}Z_1 - \frac{4\lambda_1\lambda_2}{1 + \lambda_2^2} \text{Re}Z_2 \right\}$$

$$+ \frac{1}{\lambda_1^2 - \lambda_2^2} \left\{ (1+2\lambda_1^2 - \lambda_2^2)\text{Re}Y_1 - (1+\lambda_2^2)\text{Re}Y_2 \right\} \tag{8.2}$$

$$\sigma_y = \frac{1 + \lambda_2^2}{4\lambda_1\lambda_2 - (1+\lambda_2^2)^2} \left\{ - (1+\lambda_1^2)\text{Re}Z_1 + \frac{4\lambda_1\lambda_2}{1 + \lambda_2^2} \text{Re}Z_2 \right\}$$

$$+ \frac{1 + \lambda_2^2}{\lambda_1^2 - \lambda_2^2} \left\{ \text{Re} Y_2 - \text{Re}Y_1 \right\} \tag{8.3}$$

$$\tau_{xy} = \frac{2\lambda_1(1 + \lambda_2^2)}{4\lambda_1\lambda_2 - (1+\lambda_2^2)^2} \left\{ \text{Im}Z_2 - \text{Im}Z_1 \right\}$$

$$+ \frac{1}{\lambda_1^2 - \lambda_2^2} \left\{ \frac{(1 + \lambda_2^2)^2}{2\lambda_2} \text{Im}Y_2 - 2\lambda_1\text{Im}Y_1 \right\} \tag{8.4}$$

where

$$Z_j = Z(z_j) = \sum_{n=0}^{n=N} C_n z_j^{n-1/2} \quad , \quad j = 1,2 \tag{8.5}$$

$$Y_j = Y(z_j) = \sum_{m=0}^{m=M} D_m z_j^m \quad , \quad j = 1,2 \tag{8.6}$$

$$z_j = x + i\lambda_j y \quad , \quad j = 1,2 \tag{8.7}$$

$$\lambda_j = 1 - (a/c_j)^2 \quad , \quad j = 1,2 \tag{8.8}$$

where $\overset{.}{a}$ is the crack velocity

c_1 is the plate wave speed (plane stress for thin plate analysis)

c_2 is the shear wave speed.

These stresses are used with the stress optic relationship

$$(2\tau_m)^2 = (\frac{Nf_\sigma}{h})^2 = (\sigma_x - \sigma_y)^2 + (2\tau_{xy})^2 \tag{8.9}$$

which yields a system of non-linear equations to be solved for the unknown coefficients C_n and D_m by the overdeterministic method of Sanford and Dally[16]. This method utilizes many data points (40 to 60) where N, r and θ are experimental quantities taken from the field of data provided in the isochromatic pattern. The use of redundant data points permit a least squares analysis which reduces the effect of errors in determining r and θ for specific values of N.

8.3 K-á Relation for the Characterization of Dynamic Fracture

8.3.1 Homalite 100

The dynamic behavior of a propagating crack can be characterized over the entire range from crack arrest to crack branching from the relationship between the instantaneous stress-intensity factor and the crack-tip velocity $\dot{a}$. The K-á relation represents a set of material properties. Of this set of material properties, the K associated with $\dot{a}$ = 0 and $\dot{a} = \dot{a}_{max}$ are of particular importance. With $\dot{a}$ = 0, the instantaneous stress-intensity factor defines K_{Im}, the arrest toughness of the model material. With $\dot{a} = \dot{a}_{max}$, the instantaneous stress-intensity factor defined K_{Ib}, the branching toughness. Monitoring two different fracture experiments run-arrest and run-branch with dynamic photoelasticity permits determination of the entire K-á relation and the specification of both K_{Im} and K_{Ib}.

Initial photoelastic studies[10] were performed with Homalite 100, a brittle transparent polyester. Large single-edge notched specimens were used so that any stress waves generated by the formation of fracture surfaces would disperse and attenuate before reflection back into the local

region near the crack tip. The SEN specimens were loaded with central
pins to provide a K field which increased with increasing crack length.
The specimens were also loaded by applying parting forces on the crack
line which produced a K field which decreased with increasing crack
length. Photographs of the isochromatic-fringe patterns were recorded
with a Cranz-Schardin camera and the instantaneous stress-intensity factor was determined by using procedures which have been described in previous sections.

The results obtained for the K-$\overset{\circ}{a}$ relation for Homalite 100 are presented in Fig. 8.3. The results show that there is a minimum value of K
below which the crack cannot propagate. This minimum value denoted as
K_{Im} = 380 psi$\sqrt{}$in. (0.418 MPa$\sqrt{}$m) represents the arrest toughness of
Homalite 100. The arrest toughness K_{Im} is slightly less than the initiation toughness K_{Ic}, since K_{Im}/K_{Ic} = 0.94.

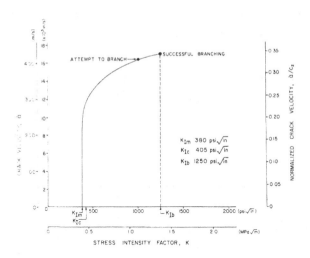

Fig. 8.3 Crack Velocity a as a Function of K_{dyn} (Homalite 100).

Small increases in K above this minimum level result in very sharp
increases in crack velocity until $\overset{\circ}{a}$ = 8000 in/s (200 m/s) is attained. A
transition region exists for velocities between 8000 and 15000 in./s (200
to 318 m/s) where the stress-intensity factor must be increased by a factor of nearly 2 to produce this increase in velocity. For higher velocities, 15,000 < a < 17,000 in./s (381 < a < 432 m/s) very significant

increases in K are required to drive the crack at these slightly larger
velocities. The value of K must be increased by an additional 1.5
K_{Im} to achieve the final increment of 2000 m/s (50 m/s) in crack velocity. The fracture surface is extremely rough in this high velocity
region of crack propagation and the large increase in energy required to
drive the higher velocity cracks is dissipated by production of many
small fractures in the region near the crack.

The ratio $K_{Ib}/K_{Im} = 3.3$ gives the range of K associated with crack
propagation in Homalite 100. The crack propagates with relatively low
energy $\mathscr{G} = 0.23$ in.-lb/in.^{2} (40.3 J/m^2) at low velocity. However, crack
propagation near the terminal velocity requires an order-of-magnitude
more energy with $\mathscr{G} = 2.48$ in.-lb/m^2 (435 J/m^2).

8.3.2   Influence of Specimen Geometry

The K-$\overset{\bullet}{a}$ relationship can be determined from any fracture-type specimen by using dynamic photoelasticity providing that the methods used to
determine K from the isochromatic loops are valid. Studies by Irwin, et
al[18-21] have been conducted with six different types of fracture specimens, which include the single-edge notched (SEN), compact tension (CT),
rectangular double-cantilever beam (RDCB), contoured double-cantilever
beam (CDCB) with wedge and machine loading and a ring segment, to determine if the K-$\overset{\bullet}{a}$ relationship is independent of specimen size, configuration and loading system. The results obtained in this very
extensive experimental program are shown in Fig. 8.4. Considering that
these results were (1) obtained with two different shipments of Homalite
100, (2) over a three-year period, (3) involved four different investigators and (4) with experimental errors ranging from ±10 percent on individual values of K to about ±3 to 5 percent on average values, the
agreement for the K-$\overset{\bullet}{a}$ relations is satisfactory. Based on these findings, it is believed that the vertical stem segment of the K-$\overset{\bullet}{a}$ relation
and the arrest toughness K_{Im} are independent of specimen size, configuration and loading system. More extensive studies pertaining to the
horizontal portion of the K-$\overset{\bullet}{a}$ relationship must be performed. There is
some evidence that the size of the fracture process zone and the

branching toughness K_{Ib} is influenced by σ_{ox}. Since σ_{ox} is strongly
dependent upon specimen geometry, it is anticipated that K_{Ib} may depend
upon specimen geometry and the type of loading.

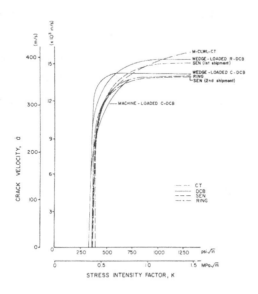

Fig. 8.4 Experimental Results Showing that the Vertical
Stem of the K-å Curve is Independent of the Specimen Geometry

8.3.3 Epoxies

The K-å relation has been determined for three different epoxies by
T. Kobayashi[22]. The results are quite similar to those obtained for
Homalite 100 in that the K-å relation is a Γ-shaped curve with a minimum
value of K (i.e., the arrest toughness K_{Im}) occurring at å = 0. Also,
results for all three epoxies exhibit a nearly vertical stem indicating
very large increases in velocity for small increases in K above K_{Im}.
Near the terminal velocity ($\mathring{a}/c_2 = 0.30$), very large increases in K are
required for small increases in velocity.

A complete K-å curve for a tough epoxy is shown in Fig. 8.5. The
epoxy was a casting prepared from Epon 828, cured with a polyoxypropyle-
namine and an accelerator. The blending provides a tough but somewhat

viscous epoxy. The results presented in Fig. 8.5 show two branches in
the K-á relation - one corresponding to higher energy and the other with
lower energy fracture. The higher energy branch is usually associated
with the accelerating crack and the lower energy branch is usually
associated with decelerating cracks. Cross over from the high energy
branch to the low energy branch occurs as the crack slows prior to
arrest. Further studies will be necessary to explain this phenomenon;
however, it is similar to behavior in linear polymers such as polystyrene
and PMMA where crazing produces two branches in the á-K relation.

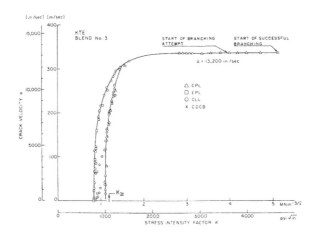

Fig. 8.5 The K-á Relation for a Toughened Epoxy Showing a Double Stem.

8.4 Conclusions

Significant progress has been made in developing analysis techniques
to determine K from the isochromatic fringe loops near the tip of a
crack. The higher parameter theory used with the Newton-Raphson method
provides an over deterministic approach which uses the full-field data
from the isochromatic pattern to accurately predict K even when the crack
is located near the boundary of the model.

Analysis methods have been extended to the dynamic problem where the effects of crack velocity on K have been established. For the higher crack velocities this effect is significant and corrections which lower the value of K by 10 to 30 percent must be applied.

Dynamic photoelasticity has been applied to characterize the behavior of a propagating crack in Homalite 100. It was shown that the fracture behavior could be described by a K-$\dot{a}$ relation and that two material properties K_{Im} the arrest toughness and K_{Ib} the branching toughness could be determined. The uniqueness of K_{Im} was established; however, much more work remains to be done to adequately characterize the instability associated with crack branching. The effect of the adjacent field stress σ_{ox} must be thoroughly explored.

References

1. Griffith, A.A., "The Phenomena of Rupture and Flow in Solids", Phil. Trans. Roy. Soc. of London, Vol. A221, (1921), pp. 163-197.
2. Irwin, G.R., "Fracture", Handbuch der Physik, Vol. IV, Springer, Berlin, (1958), pp. 558-590.
3. Liebowitz, H., "Fracture", Academic Press, New York-London, Seven volumes (1968).
4. Fracture Toughness Testing and Its Application, ASTM STP No. 381, (1965), I-409.
5. Theocaris, P.S., "Caustics for the Determination of Singularities in Cracked Plates", Proc. of IUTAM Sym., Optical Methods in Solid Mechanics, Poitiers, France, Sept. 10-14, 1979.
6. Kalthoff, J.F. and Beinert, J., "Experimental Determination of Dynamic Stress Intensity Factors by the Method of Shadow Patterns", in Sih., G.C. (ed.) Mechanics of Fracture, Vol. 7, (1980).
7. Cranz, C. and Schardin, H., "Kinematographic auf ru hendem Film und mit extrem hoher Bildefrequenz", Zeits. f. Phys., Vol. 56, (1929), p. 147.
8. Schardin, H., "Velocity Effects in Fracture", in Averbach, B.L., et.al. (eds.), Fracture (Technology Press of Massachusetts Institute of Technology and John Wiley & Sons, Inc., New York), (1959), pp. 297-330.
9. Kobayashi, A.S. and Bradley, W.B., "An Investigation of Propagating Cracks by Dynamic Photoelasticity", Experimental Mechanics, Vol. 10 No. 3, (1970), pp. 106-113.
10. Dally, J.W., "Dynamic Photoelastic Studies of Fracture", Experimental Mechanics, Vol. 19, No. 10, (1979), pp. 349-361.
11. Kobayashi, A.S. and Mall, S., "Dynamic Fracture Toughness of Homalite 100", Experimental Mechanics, Vol. 18, No. 1, (1978), pp. 11-18.
12. Etheridge, J.M., "Determination of the Stress Intensity Factor K from Isochromatic Fringe Loops", Ph.D. Thesis, University of Maryland

(1976).

13. Rossmanith, H.P. and Irwin, G.R., "Analysis of Dynamic Isochromatic Crack-Tip Stress Patterns", University of Maryland Report (1979).

14. Irwin, G.R., "Constant Speed Semi-Infinite Tensile Crack Opened by a Line Force P, at a Distance, b, from the Leading Edge of the Crack Tip", Lehigh University Lecture Notes (1968).

15. Sanford, R. J., Chona, R., Fourney, W. L. and Irwin, G.R., "A Photoelastic Study of the Influence of Non-singular Stresses in Fracture Test Specimens, University of Maryland Report, March, 1981.

16. Cottron, M. and Lagarde, A., SM Archives, Vol. 7, Issue 1, pp. 1-18, 1982.

17. Sanford, R. J. amd Dally, J. W., Engrg. Fract. Mech., Vol. 11, pp. 621-633, 1979.

18. Irwin, G.R., et al., "A Photoelastic Study of the Dynamic Fracture Behavior of Homalite 100", U.S. NRC Report NUREG-75-107, Univ. of Maryland (1975).

19. Irwin, G.R., et al., "A Photoelastic Characterization of Dynamic Fracture", U.S. NRC Report NUREG-0072, Univ. of Maryland (1976).

20. Irwin, G.R., et al., "Photoelastic Studies of Crack Propagation and Arrest", U.S. NRC Report NUREG-0342, Univ. of Maryland (1977).

21. Irwin, G.R., et al., "Photoelastic Studies of Crack Propagation and Arrest", U.S. NRC Report NUREG/CR-0542, Univ. of Maryland (1978).

22. Kobayashi, T. and Dally, J.W., "A System of Modified Epoxies for Dynamic Photoelastic Studies of Fracture", Experimental Mechanics, Vol. 17, No. 10, (1977), pp. 367-374.

CHAPTER 9

CONTROL OF EXPLOSIVELY INDUCED FRACTURE

9.1 Introduction

Excavation in hard rock is usually accomplished with a drill and
blast procedure where a hole is drilled in the rock, packed with high
explosive, primed, stemmed and then detonated. The detonation pressures
are extremely high and an extensive amount of energy is dissipated at the
bore hole. The energy is expended in crushing the adjacent rock, by producing stress waves and in creating a dense radial crack pattern about
the hole. These radial cracks arrest quickly and only about 8 to 12 randomly oriented cracks extend for any significant distance from the bore
hole.

Since this process dissipates much of the energy in crushing the rock
at the bore hole, the resulting crack pattern is randomly oriented and
very little control of the fracture process is achieved. Where control
of the fracture process is important, the conventional drill and blast
procedure has been modified. Pre-splitting, post-splitting, and smooth
blasting procedures have been developed which to some degree control the
fracture process.

In pre-splitting, a row of closely spaced and highly charged holes
are detonated simultaneously. The resulting stress waves overlap and
reinforce to produce cracking in the region between the holes. Highly
charged holes, of course, produce extensive cracking at the bore hole and
degrade the quality of the wall of an excavation. Also, simultaneous
detonation results in excessively high ground shocks when excavations are
made in populated urban areas.

Post-splitting is almost identical except that the holes are fired
after the central core of the excavation has been fragmented. Post-
splitting is usually employed in highly stressed rock where the residual

stress system diverts the crack. Removal of the central core alters the residual stress distribution and the cracks can be directed along the contour of the opening.

In smooth blasting, the holes are drilled on very close centers and cushioned charges are used. As control is obtained by close spacing of the holes, delays can be used and the ground shock reduced. Smooth blasting gives satisfactory results when enough holes are drilled and when the charge is properly cushioned; however, the number of holes which must be drilled and loaded increases the cost of the excavation.

An alternative procedure is described where a high degree of fracture control can be achieved while minimizing the ground shock associated with detonation of the charge. Fracture control implies that control be exercised over all three phases of the fracture process which include crack initiation, crack propagation, and crack arrest. Control of crack initiation involves specifying the number of cracks to be initiated and the location of the initiation sites on the wall of the bore hole. Control of the propagation phase involves orienting the cracks (usually in the radial direction) and providing a stress field which will produce the stress intensity required to maintain sufficiently high crack velocity. Finally, control of crack arrest involves maintaining a stress intensity which is sufficiently large to avoid crack arrest until the crack has achieved its specified length. If all three aspects of the fracture process can be controlled, then a blasting round can be designed for optimum performance in a construction process.

9.2 Fracture Plane Control

Excavation for underground construction requires close control of the contour of the opening, minimal damage to the remaining rock wall and reduced ground vibration from the blast. These objectives can be met by using a modified drill and blast process where the bore hole is notched and is loaded with a very light and cushioned charge.

The concept of using longitudinal notches on the side of a bore hole is not new. Langefors and Kihlstrom[1] in 1963 indicate notches can be used to initiate cracks and control the fracture plane. Indeed, notching was described by Foster[2] in 1905, as a method of promoting fracture.

Notching is an effective means for concentrating stress at the bore
hole and insuring that cracks initiate at the notch location. However,
Ladegaard-Pederson et al[3], showed cracks will start at other locations
about the bore hole when the detonation pressure is sufficiently high and
the beneficial effect of the notch is nullified. There is a pressure
range which must be achieved to control initiation. If the pressure is
too low, the cracks will not initiate even at the notches. When the
pressure is too high, cracks will form at the natural flaws on the side
of the bore hole.

If the residual stress is small, as is usually the case in near surface excavations, then the cracks will propagate in a radial direction.
Control of the crack path along radial lines is achieved if crack
branching can be prevented. Crack branching will occur for two reasons.
First, if the crack intersects a large flaw in the rock structure, the
flaw can arrest, divert or bifuricate the crack. Flaws present a serious
problem to fracture plane control and dictate that the center to center
distance between holes must be less than the flaw spacing. The second
reason for crack branching is over-driving the crack. If the strain
energy available is sufficiently large, the crack will branch.

Crack length is controlled by preventing premature arrest of the
crack. Kobayashi and Dally[4] have shown that cracks move in relatively
high velocity if the stress intensity factor at the crack tip is greater
than a critical arrest toughness K_{Im}. The stress intensity factor is
determined by the pressure of the gas in the bore hole. If the gas is
confined to the bore hole, the stress intensity factor K decreases with
increasing crack length until $K < K_{Im}$ and the crack arrests. However,
if the gas flows into the opening crack and pressurizes the fracture surface, the stress intensity at the crack tip increases with increasing
crack length and the cracks arrest only when the gas supply is depleted
due to the increase in the volume of the cavity. Plewman[5] has conducted tests with notched bore holes and concludes that the crack length
to bore hole diameter D can be as large as s/D = 50, providing the wall
of the hole is not crushed by over-pressure.

9.3 Control of Crack Initiation

When the stress intensity factor K associated with a notch or a natural flaw exceeds the initiation toughness K_{Ic} of rock, cracks will initiate. Ouchterlony[6] has shown that the pressure required to initiate shallow flaws on a bore hole is given by

$$p_c = K_{Ic}/2.24 \sqrt{\pi a} \qquad\qquad (9.1)$$

where a is the length of the notch of flaw.

The results from eq. (9.1) can be used to compute the pressure required to initiate cracks at the tip of sharp notches on the side of the bore hole as shown in Fig. 9.1. Two facts are evident from this figure. First, the bore hole pressures required to initiate cracks at the notches are quite low even for very shallow notches. Second, cracks can be initiated at notches with low pressures for all types of rocks. The rock property which is important is the fracture toughness K_{Ic}. Measurements of the fracture toughness of rock is relatively new; however, some limited data exists which indicates that the value of K_{Ic} is between about 0.2 to 17 ksi-$\sqrt{in}$ (.18-15 MNm$^{-3/2}$) for most of the commonly excavated rock.

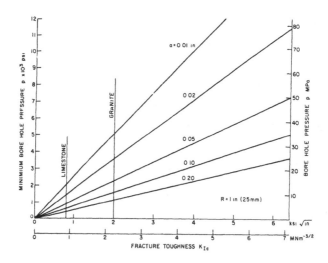

Fig. 9.1 Pressure Required to Initiate Cracks at the Bore Hole

The next factor which must be considered in control of rock initiation is over-pressure in the bore hole. The pressures given in Fig. 9.1 are minimum pressures which should be exceeded to overcome the effect of notch end bluntness and to provide additional energy to drive the crack. The amount of over-pressure which can be tolerated can also be determined from eq. (9.1). In this instance, the flaw length a_f is equated to the natural flaw size. These natural flaws occur in large numbers randomly distributed about the periphery of the wall of the bore hole. Little is known about the distribution of the size of these natural flaws; however, photomicrographs by Bienawski, 1967, suggest that many grain boundaries represent small cracks. Thus, it appears reasonable as a first approximation to equate the natural flaw length a_f to the grain size of the rock.

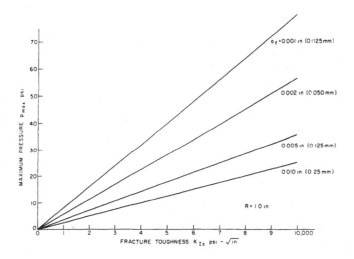

Fig. 9.2   Pressure Required to Initiate Small
Natural Flaws at the Walls of the Bore Hole.

Taking a_f ranging from 0.001 in (0.025 mm) to 0.01 in. (0.25mm) and using eq. (9.1) gives the maximum pressure which can be tolerated in the bore hole before producing a large number of random cracks. This pressure as a function of flaw size and fracture toughness is shown in Fig. 9.2. These results show that the very fine grain rock materials with a_f = 0.001 in. (0.025 mm) support much higher pressures prior to random crack initiation than the coarse grain materials where large size natural flaws are encountered.

The pressure range in which crack initiation can be controlled will depend on three factors - the fracture toughness K_{Ic} of the rock, the natural flaw size a_f, and the depth a of the side notches.

A typical example for a granite with K_{Ic} = 2 ksi$\sqrt{}$in. (0.18 MNm$^{-3/2}$) showing the allowable pressure range for several sets of operating conditions is presented in Table 9.1.

TABLE 9.1
Pressure Range for Controlling Crack
Initiation with Side Grooving

Operating Conditions		a_f	a	p_{max}	p_{min}	$\dfrac{p_{max}}{p_{min}}$
Rock Grain Size	Notch Size	mm	mm	MPa	MPa	
Very fine	deep	0.025	5.0	110	7.6	14.5
fine	medium	0.050	2.50	76	11.0	6.9
medium	medium	0.125	2.50	48	11.0	4.4
coarse	shallow	0.250	1.25	34	15.9	2.2

Inspection of these results shows a very wide range of pressure, 1100 to 16,000 psi (7.6 to 110 MPa) which gives satisfactory control of crack initiation when deep grooves are used in very fine grain material. However, when relatively shallow grooves are used in coarse grain rock, the range of pressure 2300 to 5000 psi (15.9 to 34 MPa) is much narrower.

The very low pressures required suggest the use of a low explosives where pressure could be controlled more closely and where hole stemming can be maintained for a longer period of time after detonation. Black powder, smokeless power, and nitro-cotton all detonate slowly and complete the conversion of gaseous products in milliseconds rather than the microseconds associated with high explosives. The pressures generated are usually less than 50,000 psi (345 MPa) and can be controlled by adjusting the weight of the charge per unit length of bore hole and the diameter of the hole.

9.4 Control of Fracture Plane Orientation

The cracks initiated at the notches will begin to propagate at high
velocity in the radial direction. Fundamental research by Kobayashi and
Dally[4,8], on the dynamic behavior of propagating cracks is relatively
recent and studies on rock or ceramic materials by Bienawski[9], and
Chubb and Congleton[10], are quite limited. Nevertheless, it appears
reasonable to characterize the dynamic behavior of a crack with a relation between the crack velocity $\dot{a}$ and the instantaneous stress intensity
factor K by a curve of the form shown in Fig. 9.3.

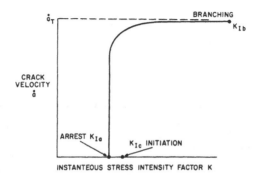

Fig. 9.3 Stress Intensity Factor as a Function of Crack Velocity

There are several features of this relation which should be emphasized. First, the curve has a very steep initial portion which implies
that a crack, when initiated at $K > K_{Ic}$, begins to propagate at a high
velocity. Further increases in K above this initiation value produce
little change in the velocity since it is near its terminal value. The
terminal velocity $\dot{a}_T$ in norite has been measured by Bienawski[9] at
73,800 in/sec (1875 m/sec) or 51 percent of the shear wave velocity c_2.
The terminal velocity of cracks propagating in plate glass has been
determined by Doll[11] as 59,000 in/sec (1500 m/sec) or 47 percent of
shear wave velocity.

Continued increases in K will eventually produce branching when K

K_{Ib} where sufficient energy is available to drive two or more cracks.
The branching phenomena is particularly important in control of the
fracture plane because bifurication results in cracks deviating from the
control plane. Consequently, the instantaneous stress intensity factor
should be limited to $K < K_{Ib}$ to avoid branching.

It is helpful to examine the K field associated with a pair of
diametrically opposed cracks in a bore hole. It is important to
distinguish if the cracks are open so that gas penetration is obtained or
if the cracks are sealed. If the cracks are sealed by small rock particles, the pressure required to drive the crack at a constant K value
increases markedly with crack length a. Ouchterlony[6] has shown for
long cracks that the pressure -K relation is

$$p = \frac{K\sqrt{\pi(R+a)}}{2R} \qquad (9.2)$$

The pressure required to drive a sealed crack at a constant K value
increases as the crack grows as illustrated in Fig. 9.4.

If the cracks remain open as the gas in the bore hole expands, then
the cracks are pressurized and this is a more favorable condition for
crack extension. For an open crack, the pressure required to propagate a
pair of diametrical opposite cracks with a constant stress intensity factor K is given by:

$$p = K/\sqrt{\pi(R+a)} \qquad (9.3)$$

This result indicates the pressure required to drive the crack at
constant K decreases as the crack extends.

A comparison of the pressure required to drive open or sealed cracks
at a constant stress intensity factor K as presented in Fig. 9.4 shows
the marked difference between the two cases. At the boundary of the hole
after crack initiation, the pressure is identical; however, for cracks 5R
long the pressure required to drive sealed cracks is 8 times more than

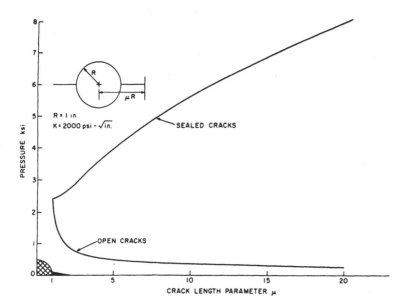

Fig. 9.4   Pressure Required to Propagate Two cracks at a Constant K.

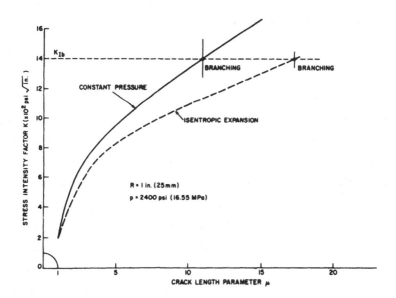

Fig. 9.5   Increase in K During Crack Extension Under Constant Pressure
and Under Pressure Decay Due to Isentropic Expansion.

open cracks and for cracks 20R long the pressure ratio is nearly 32.

It is possible to demonstrate the occurrance of branching with eq.
(9.3). To simplify the discussion, consider the pressure constant during
the crack propagation period and note the increase in K with crack length
as shown in Fig. 9.5.

When $K > K_{Ib}$ branching will occur and the crack will bifuricate and
deviate out of the control plane. In the example illustrated in Fig.
9.5, K_{Ib} was arbitrarily set at 14,000 psi-$\sqrt{in}$ (12.6 $MNm^{-3/2}$) and
branching occurred at $\mu = 11$ or a crack length of 10R.

It is evident that the crack can be driven along the control plane as
long as $K_{Im} < K < K_{Ib}$. Constant bore hole pressure during the propagation period (0.5 to 1.5 msec) is not recommended since K would increase
beyond K_{Ib} and the crack would eventually branch. In practice, the
pressure decays during the propagation period due to isentropic expansion
of the gas with increase in size of the fracture cavity and due to gas
leakage. The effect of pressure reduction is to decrease the rate of
increase of K with crack extension, thus permitting growth to greater
distances before the onset of branching as is illustrated in the example
presented in Fig. 9.5.

9.5 Flaw Induced Branching

Branching occurs in a homogeneous flaw-free material when the instantaneous stress intensity factor K exceeds the branching toughness K_{Ib} for
that material. However, if the crack encounters a flaw as it propagates,
then branching can occur at much lower K values. The flaw is in effect a
pre-existing crack and the energy required to form the primary (driven)
crack in the local neighborhood of the flaw is greatly reduced.

Flaw orientation with respect to the driven crack is also important.
Flaws perpendicular to the direction of the driven crack appear to
inhibit crack propagation. Flaws parallel to the driven crack assist
propagation and while branches are often produced at these aligned flaws,
these branches usually arrest before deviating any significant distance
from the fracture plane. Flaws making an angle from 30 to 60 degrees
with the driven crack produce extensive branching and cause considerable

difficulty in control of the fracture plane.

The role of small natural flaws in rock on the control of the fracture plane does not appear to be important if the stress wave is
suppressed by using small highly cushioned charges.

In laboratory tests, with a fine grain pink westerly granite with
relatively light explosive charges, flaw induced branching has not been
observed. An illustration of a clean fracture plane produced by
controlled blasting in laboratory tests with pink westerly granite is
shown in Figure 9.6.

Field tests in small limestone boulders and in large sandstone
boulders showed that the fracture plane could be closely controlled and
that branching due to natural flaws could be suppressed by using relatively light charges. Results obtained in these field tests are
illustrated in Fig. 9.7.

Limited experience in field and laboratory tests indicates that small
flaws will not introduce branching with subsequent loss of control of the
fracture plane. However, this is not the case for large flaws such as
joints. When the driven crack reaches a joint, it will usually turn and
run along the joint cleaving the poorly bonded joint plane. Since extension of the fracture control plane across a joint plane is not possible,
it is necessary to use two bore holes with fracture control planes that
meet at the joint.

9.6 Control of Crack Length

Previous discussion indicated that cracks would propagate at relatively high velocity provided the stress intensity factor K was maintained
at a level higher than the arrest toughness for the rock. It was also
shown that K increases with crack length, provided that the bore hole
pressure could be maintained and that gas could flow freely into the
cracks. Thus, it appears that the cracks can be extended over large
distances and fracture control can be exercised with widely spaced bore
holes.

Plewman[5] indicates that the cracks can be driven 25 diameters to
either side of the bore hole giving a bore hole spacing of 50D. This is

Fig. 9.6   Fracture Plane Control Tests in Pink Westerly Granite.

Fig. 9.7   Fracture Plane Control in Limestone Boulders

reasonable, providing the pressure can be maintained over the crack propagation period. Since the cracks propagate at high velocity, the time
of confinement of the gas pressure is short (about 1 msec) and pressure
reduction due to thermal losses will be small. Also, Dally and
Fourney[12] showed that the pressure reduction due to cavity expansion
resulting from extension of two diametrically opposed cracks will not
cause crack arrest.

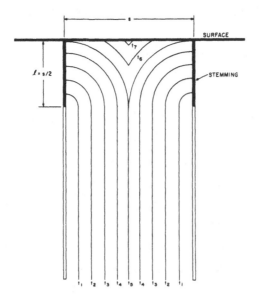

Fig. 9.8 Development of the Fracture Control Plane
Showing the Length of Stemming Required to Avoid Surface Venting

It appears then that the length of the cracks which can be achieved
prior to their arrest will be limited by pressure loss due to venting.
Venting will occur with the loss of the stemming or when the cracks
intersect the free surface of the rock. The time required to blow the
stemming from a bore hole can easily be controlled by adjusting the
length ℓ of the stemming column. For instance, with a bore hole pressure
of 10,000 psi, which is about as high as would be used in controlled
blasting, a stemming column of 12 in. (0.3 m) provides 2 ms of propagation time.

Venting by intersection of the cracks with the surface is more likely to occur. As the cracks are driven from the bore holes, they propagate in both the radial and longitudinal directions as illustrated in Fig. 9.8. It is evident by tracing the crack fronts at intervals of time during the dynamic event, that surface venting will occur prior to complete formation of the fracture plane unless the stemming length $\ell = s/2$. As spacings of s from 40 to 50D are suggested, the stemming length will be 20 to 25D, which is much longer than that required to avoid blowing the stemming during the crack propagating period.

9.7 Summary

Fracture plane control can be achieved by using a modified drill and blast process where the bore hole is notched and loaded with a very light and cushioned charge. Notching enables control of the initiation site of the cracks which produce the specified fracture plane and reduces the bore hole pressures required for initiation.

Fracture mechanics concepts using crack initiation toughness indicate that the bore hole pressures required to initiate cracks at notches range from about 1000 to 5000 psi (7 to 35 MPa) for most common rock types. Higher pressures are possible, but pressures in excess of 5000–15,000 psi (35–175 MPa) will cause crack initiation at small natural flaws on the bore hole wall. Highly cushioned charges of a low explosive or a pro-pellant are recommended for producing these bore hole pressures.

The fracture plane is formed in about one ms due to the very high crack propagation velocities. The fracture plane is produced by radially outgoing cracks which will not branch provided the pressure in the bore-hole is not excessive and the branching toughness of the rock material is not exceeded. Field experience indicates that small natural flaws will not produce branching; however, larger flaws such as joints will cause the cracks to deviate from the control plane.

Providing the gas flows into the cracks, the fracture plane can be extended over a considerable distance (s = 50D). Crack arrest occurs due to venting of the pressure when the cracks intersect the surface. Premature arrest can be avoided by using stemming columns of length $\ell = s/2 = 25D$.

The drill and notch procedure has several advantages in construction blasting. First, the drilling costs should be reduced. Langefors, and Kihlstrom[1] recommend s/D ratios of 16 and 8 for smooth blasting and presplitting respectively. Increasing the s/D ratio to say 40 will reduce the number of bore holes required on the contour by a factor of 2.5 to 5.

Second, there will be a cost savings since the relatively low density charges specified with notched bore holes require less explosive than the more highly loaded smooth blasting or presplitting rounds. Also, as low explosives may be employed instead of high explosives further cost reductions for explosives may be achieved.

Third, the control of the fracture plane should reduce the possibility of overbreak and underbreak, and the costs associated with scaling and/or forms and concrete should be greatly reduced.

Fourth, relatively few cracks will be produced in the wall remaining after excavation improving its strength and stability and minimizing the need for auxiliary support such as rock bolts, shot concrete, and frames.

Finally, the reduced and highly cushioned charges which can be employed to presplit an excavation will greatly reduce the ground vibration and will alleviate the number and frequency of complaints when blasting in heavily populated urban areas.

References

1. Langefors, U. and Kihlström, B., Rock Blasting, John Wiley and Son, (1963), pp. 300-301.
2. Foster, Clement LeNeue, A Treatise of Ore and Stone Mining, Charles Griffin & Co., (1905).
3. Ladergaard, Peterson, A., Fourney, W.L. and Dally, J.W., "Investigation of Presplitting and Smooth Blasting Techniques in Construction Blasting", Univ. of Maryland Report to the National Science Foundation, (1974).
4. Kobayashi, T. and Dally, J.W., "The Relation Between Crack Velocity and Stress Intensity Factor in Birefringent Polymers", ASTM Special Technical Publication, Vol. 627, (1977), pp. 257-273.
5. Plewman, R.P., "An Exercise in Post-Plitting at Vlakfaitem Gold Mining Co., Ltd.", Papers and Discussion of Assoc. of Mine Managers of So. Africa, (1968-69), pp. 62-81.
6. Ouchterlony, Finn., "Analys av Spanning-stillstandet Krmg Nagra Olika Geometrier Med Radiellt Riktade Sprickor I Ett Oandlight Plant Medium Under Inverkan av Expansionskrafter", Swedish Detonic

Research Foundation Report, DS 1972: Vol. 11, (1972).

7. Bienawski, Z.T., "Mechanism of Brittle Fracture of Rock, Part II, Experimental Studies", Int. J. of Rock Mech. and Min. Sci., Vol. 14, No. 4, (1967), pp. 407-423.

8. Kobayashi, T. and Dally, J.W., "A System of Modified Epoxies for Dynamic Studies of Fracture", Experimental Mechanics, Vol. 17, No. 10, (1977), pp. 367-374.

9. Bienawski, Z.T., "Fracture Dynamics of Rock", International Journal of Fracture Mechanics, Vol. 4, No. 4, (1968), pp. 415-430.

10. Chubb, J.P. and Congleton, J., "Crack Velocity due to Combined Tensile and Impact Loading", Philosophical Magazine, Vol. 28, No. 5, (1973), pp. 1987-1097.

11. Doll, W., "Investigation of the Crack Branching Energy", International Journal of Fracture, Vol. 11, (1975), p. 184.

12. Dally, J.W. and Fourney, W.L., "Fracture Control in Construction Blasting", Univ. of Maryland Report to the National Science Foundation, (1976).

CHAPTER 10

CRACK INITIATION AND CRACK ARREST

10.1 Introduction

Most of the studies in fracture mechanics pertain to the determination of a static stress intensity factor which exists at the tip of
the crack immediately prior to crack initiation. Some experimental work
by Kobayashi[1,2] and Dally[3,4] have shown characteristic features of
crack propagation and arrest. In particular it has been shown that crack
velocity a depends upon the instantaneous stress intensity factor $K(t)$
and the K-å curve represents material properties which characterize dynamic crack behavior. This form of the characterization of the dynamic
crack behavior has been questioned by Knaus and Ravi-Chandar[5] since
their experiments show that $K(t)$ varies while crack velocity remains
constant (see Fig. 10.1). Close inspection of Fig. 10.1 shows extremely
small variations in $K(t)$ at a constant velocity on the vertical stem of
the curve and large variations of $K(t)$ with crack tip velocity on the
horizontal reach of the curve. From an engineering viewpoint it appears
reasonable to consider the vertical stem unique and to consider K_{Im} the
arrest fracture toughness a material property independent of specimen
size and specimen type.

The high energy crack propagation where $a/c_2 > 0.25$ shows considerable variation in experimental results by both Dally[6] and
Knauss[5]. It appears that characterization of this region of the K-å
relation will require the introduction of additional physical parameters
to adequately explain the crack propagation behavior. This topic will be
considered in the final chapter (11).

In this chapter attention will focus on low energy propagation where
the K-a curve appears to be unique. Specifically three topics will be
addressed

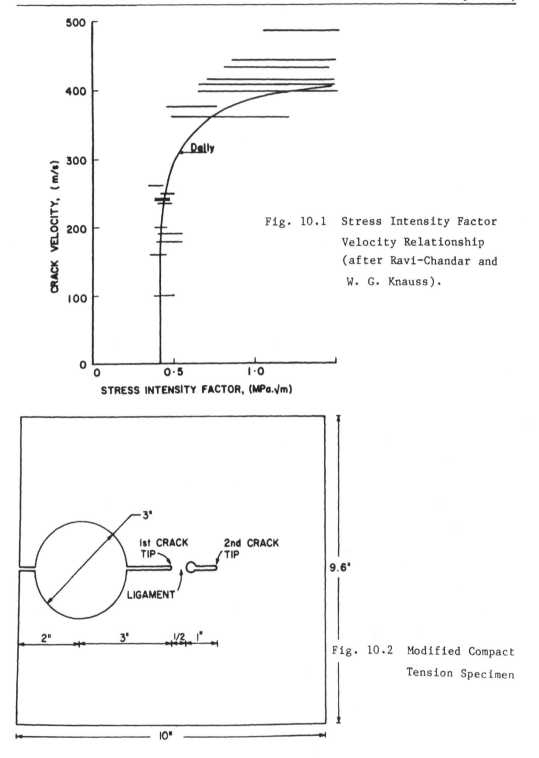

Fig. 10.1 Stress Intensity Factor
Velocity Relationship
(after Ravi-Chandar and
W. G. Knauss).

Fig. 10.2 Modified Compact
Tension Specimen

(1)   Initiation which is the transition from a static K to the dynamic K.

(2)   Abrupt arrest where the crack is caused to arrest either by a sudden change in either K(t) or by a sudden change in the arrest toughness K_{Im}.

(3)   Gradual arrest where the crack velocity slowly decreases toward zero in a brittle material (Homalite 100) or in a slightly viscoelastic material (modified epoxy).

The results obtained from experiments designed to provide data on these three topics clearly provide insight regarding crack acceleration, crack deceleration and some effects of viscoelasticity on arrest behavior.

10.2 Crack Initiation

It is impossible in the ordinary fracture experiment with a SEN, DCB or MCT type specimen to obtain data or crack initiation. In these experiments the crack is started and 5 to 10 µs is required to initiate the high speed photographic system used to record the movement and stresses associated with a high-speed crack. To alleviate this experimental difficulty a special photoelastic model was designed[7]. This model was a modified compact tension specimen machined from 12.7 mm thick sheets of Homalite 100 as illustrated in Fig. 10.2. A near crack line load was applied with a split D type fixture which was inserted in the 3 in. (75 mm) diameter hole. The split D fixture was forced apart with a transverse wedge to give a specified value of K_Q (static stress intensity factor) at the first blunt crack tip. The crack was initiated at the first crack tip by drawing a sharp knife edge across the tip of the blunted crack.

The crack initiated and propagated across the ligament cutting through a strip of silver conducting paint before coming to arrest at the small hole which terminates the ligament. Cutting the silver conducting paint provides an electrical pulse which is used to initiate the Cranz-Schardin camera used to record the dynamic event.

After the crack arrests at the hole, the value of K at the tip of the second crack begins to increase until it becomes sufficiently large to reinitiation the second crack. The bluntness of the second crack is controlled so that the crack remains at arrest for a relatively long time (approximately 200 μs) and reinitiation occurs at high values of K_Q.

The Cranz-Schardin camera is operated with an inter frame time of 5 μs (i.e., 200,000 frames/sec) and initiated to record the increase in K at the stationary crack prior to initiation and the dynamic value of K(t) immediately after initiation.

Five different experiments were conducted with K_Q ranging from 1.48 to 1.14 MPa√m. The instantaneous values of the stress intensity factor were determined from the dynamic isochromatic fringe loops by using the multiple point over deterministic method developed by Sanford and Dally[7]. A dynamic correction following the procedure developed by Irwin and Rossmanith[8] was made to account for the effect of velocity on K(t).

Typical results showing the variation in K with time and the position of the crack tip with time are shown in Fig. 10.3. These results show that K increases monotonically at the arrest crack until K_Q becomes large enough to produce initiation at the second crack tip. After initiation it is not possible to determine K from the isochromatic fringe loops until the shear stress wave has cleared the near field region (r = 5 mm and t = 4 μs). In the first frame after the shear wave has cleared the near field region (about 8 to 10 μs after initiation), the value of K(t) was determined. The decrease in K is quite rapid, for example K decreases from K_Q = 1.29 MPa√m to K(t) = 0.81 MPa√m in 8 μs.

The graph of crack position as a function of time indicates that the crack propagated at essentially constant velocity (342.9 m/s) after initiation. By interpolating from the stationary crack and from the straight line with a slope of $\overset{\circ}{a}$, a close estimate (i.e., ±0.2 μs) of the initiation time t_i = 34.7 μs can be made. The crack was observed at t = 36.5 μs at a position a = 0.62 mm propagating at the velocity $\dot{a}$ = 342.9 m/s. If it is assumed that the acceleration of the crack is uniform over the initiating time Δt = 36.5 − 34.7 = 1.8 μs, then the acceleration of the crack at initiation may be estimated from $\ddot{a} = \dot{a}/\Delta t$ =

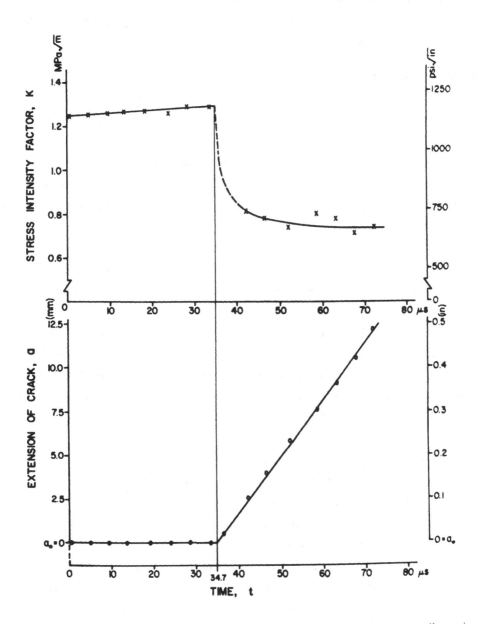

Fig. 10.3   Stress Intensity Factor and Crack Tip Position as a Function
             of Time t.

2×10^7 g's. Since the initiating time Δt may be less than the 1.8 µs measured in this experiment, it is probable that crack tip acceleration in Homalite 100 is even higher.

10.3 Abrupt Arrest

In order to study abrupt arrest a duplex specimen was designed using the familiar modified compact tension specimen as illustrated in Fig. 10.4. The starter section and arrest sections were bonded together with a structural epoxy adhesive known as EA9410 and manufactured by Hysol Corp. This two part adhesive system is formulated to give a high shear strength (34.5 MPa) with a room temperature cure. While K_{Ic} of the adhesive has not been determined, it is estimated as 1.4 $MNm^{-3/2}$. The thickness of the adhesive joint was varied from 0.025 to 0.36 mm. The surfaces of the specimen were faced and then polished to provide for a uniform and smooth transition from the starter section to the arrest section.

These experimental results were reproduced in several tests where the initial K_Q value was sufficiently high to produce reinitiation of the arrested crack at the adhesive joint. In other experiments with lower K_Q values the crack arrested at the adhesive joint and K oscillated but did not become large enough to reinitiate the crack.

In six of these experiments, the crack propagated to the adhesive joint and abruptly arrested. The stress intensity factor increased until the crack reinitiated in the adhesive joint and propagated into the arrest section of the duplex specimen. The results for model No. 178 presented in Fig. 10.5 show the position of the crack tip as a function of time. It is evident that the crack in the starter section of the duplex specimen propagates at nearly constant velocity, 358 m/sec ($\dot{a}/c_2$ = 0.29) until it is abruptly arrested at the adhesive joint. The crack remains at arrest for a pause period t_p which in this instance was 128 µsec prior to reinitiation. Upon reinitiation in the epoxy material of the arrest section, the velocity is initially 237 m/sec ($\dot{a}/c_2$ = 0.21). As the crack extends sufficiently far into the arrest section, the stress intensity factor K decreases resulting in a progressive loss of velocity until final crack arrest occurs.

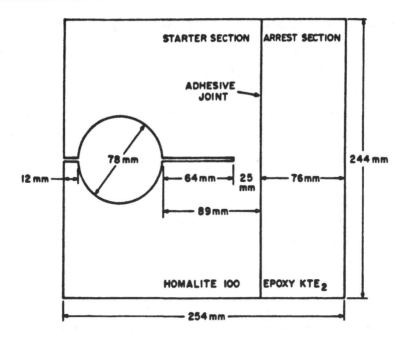

Fig. 10.4   Geometry of Duplex Specimen.

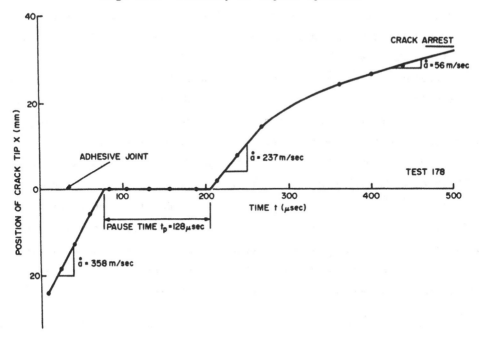

Fig. 10.5   Position-time Function Describing Crack Propagation Across the
            Duplex Specimen.

The arrest of the crack tip at the joint of the duplex specimen was extremely abrupt. Deceleration of the crack tip was estimated based on $\Delta\dot{a}/\Delta t$ as 4×10^7 by using results similar to those shown in Fig. 10.5.

The appearance of the crack tip in the adhesive joint for two specimens was determined by taking photomicrographs. The procedure was to polish the surface of the specimen in the local neighborhood of the adhesive joint where the crack had arrested. Photomicrographs were made using obliquely reflected light and a magnification ratio of 100. The results obtained are illustrated in Fig. 10.6a and 10.6b. In Fig. 10.6a the crack is shown to penetrate a distance of about 0.02 mm into the 0.36 mm thick adhesive joint. Ahead of the crack tip is a light region approximately circular in shape. This light region is due to crazing or microvoids which have developed in the fracture affected zone in front of the crack. The reflected light is intensified due to the presence of the craze and/or microvoids producing a light region on the photomicrograph. It is interesting to note that the diameter of the fracture affected zone is only about one-half the thickness of the adhesive joint. Thus it appears that arrest would have occurred under similar K conditions for a much thinner adhesive joint.

The photomicrograph for crack arrest in a thinner adhesive joint is presented in Fig. 10.6b. In this instance, the crack has propagated only about 0.01 mm into the joint which was 0.15 mm thick. The light region, representing the fracture affected zone, approximately 0.14 mm in diameter, extends across the unfractured portion of the adhesive joint.

These experiments show that a crack arrests abruptly when encountering another material where $K_{Im} > K(t)$. The deceleration of the crack tip is of the order of 4×10^7 g's. The extension of the crack into the tougher material is small which is consistent with the high deceleration rates. Upon arrest the value of K(t) oscillates as indicated in Fig. 10.7 and the crack can reinitiate if the oscillation carries $K(t) > K_{Im}$ for the second material.

10.4 Fracture Behavior in a Tough Epoxy

Kobayashi and Dally[3] developed a relatively tough photoelastic

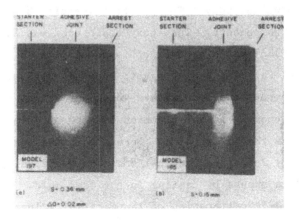

Fig. 10.6   Photomicrographs of the Arrested Crack in the Adhesive Joint

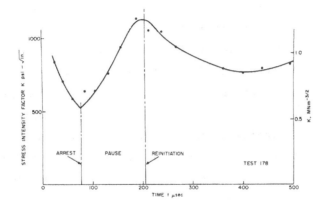

Fig. 10.7   Stress Intensity Factor as a Function of Time in a Specimen with Reinitiation

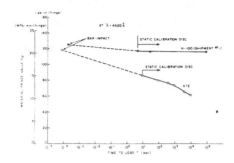

Fig. 10.8   Material Fringe Value as a Function of Loading Time for

Homalite 100 and KTE Epoxy (λ - 4920Å).

polymer by modifying an epoxy. The material identified as KTE Blend No.
3 contained three components. The epoxy resin used was Epon 828, a low
molecular weight (380) epichlorohydrin/bisphenol A resin manufactured by
the Shell Chemical Co. The curing agent, known as Jeffamine D-400,
which is a polyoxypropyleneamine with a molecular weight of 400 was used
to polymerize the epoxy. As the reaction is slow an accelerator A-398
was added to reduce the curing time. Both the curing agent and accelerator are produced by Jefferson Chemical Co. The modified epoxy called
KTE was prepared in sheets by casting in a closed mold.

Static and dynamic calibration of the material fringe value f_σ which
is shown in Fig. 10.8 and Table 10.1 for both Homalite 100 and KTE indicates the viscoelastic behavior of the two materials.

Dynamic photoelastic experiments were conducted with CPL, EPL and CLL
type SEN specimens fabricated from the KTE material. These experiments
provided the data for K-á curve for KTE presented in Fig. 8.5.

There are three features which aid in characterizing the dynamic
crack propagation in this material. First, there is a terminal velocity
of about 13000 in./s (330 m/s) which is 0.30 times the shear wave velocity. This terminal velocity is achieved when K = 1600 psi√in (1.76
$MNm^{-3/2}$), and further increases in K do not appear to drive the crack at
higher velocities. As K is increased in this region it does result in
rougher fracture surfaces, atempts to branch and finally successful
branching occurs at K = 4650 psi√in. (5.12 $MNm^{-3/2}$).

Second, there appears to be a minimum value of K below which cracks
cannot propagate. This minimum value is denoted as the arrest toughness
K_{Im} = 770 psi√in. (0.85 $MNm^{-3/2}$) which is about one half the value of K
associated with terminal velocity and about one sixth the value associated with successful branching.

TABLE 10.1

Mechanical and Optical Properties of Homalite 100 and KTE Epoxy

	Homalite 100	KTE Epoxy
c_1(in/s);(m/s)	84700;2150	77000;1970
c_2(in/s);(m/s)	48600;1230	44700;1130
E_s(psi);(GPa)	560000;3.86	437000;3.01
E_d(psi);(GPa)	700000;4.82	573000;3.86
μ_d(psi);(GPa)	267000;1.84	214000;1.47
ν_d...;...	0.31	0.34
ρ(lb-s^2/in.^{4});Kg-s^2/m^4)	0.000112;122	0.000107;117
G_{Ic}lb/in.;J/m^2	0.33;57.8	2.65;464
K_{Ic}psi$\sqrt{\text{in.}}$;MNm$^{-1/2}$	405;0.45	1072;1.18
K_{Im}psi$\sqrt{\text{in.}}$;MNm$^{-1/2}$	380;0.42	770;0.85
$f_{\sigma s}$psi-in.fringe;NM-mm/fringe[a]	121;21.1	85;14.9
$f_{\sigma d}$psi-in./fringe;MN-mm/fringe[b]	125;21.9	118;20.7

[a]Static - 10 s after application of load.

[b]Dynamic at a loading time of 200 μs both for ⋏ = 4920 Å for spark gap
light source with Kodak 4135 film and type K-2 filters.

Third, in the velocity range from near zero to about 12000 in./s
(305 m/s), there appear to be two branches in the å-K relation. The
lower energy branch is associated with decelerating cracks, and the
higher energy branch is associated with both accelerating and decelerating cracks. Crossover from the higher energy to the lower energy
branch was observed for both the EPL and CLL specimens when the initial
crack velocity was below 12000 in./s (305 m/s). From these limited observations, it appears that it is not possible for a crack to propagate on

the lower energy branch without having first propagated for some distance
on the higher energy branch. Further studies will be necessary to
explain this phenomenon; however, it is similar to the behavior observed
in linear polymers such as polystyrene and polymethylmethacrylate (PMMA)
where crazing causes two branches for the slow growth region of the $\dot{a}$-K
relationship.

10.5 Summary

The arrest toughness K_{Im} can be defined from the $\dot{a}$-K relation for
elastic and viscoelastic materials. Cracks arrest when $K(t) < K_{Im}$ and
cracks will reinitiate if $K(t)$ oscillates above K_{Im} in the post arrest
period.

Crack initiation is sudden when $K_Q > K_{Im}$. High velocities are
achieved almost instantaneously because of two factors. Firstly, the $\dot{a}$-K
relation exhibits a nearly vertical stem and value of $K(t)$ slightly
larger than K_{Im} are associated with a > 0.2 c_2. Secondly, the acceleration of the crack tip is of the order of 10^7 g's or higher. This very
high acceleration permits cracks to achieve velocities of 0.1 to 0.3
c_2 in a few μs or less.

Crack arrest can also be abrupt when the crack encounters a second
material which is tougher. If the arrest toughness of the second
material $K_{Im} > K(t)$, the crack arrests almost instantaneously with deceleration of the order of 10^7 g's. Arrest is possible with very thin
joints and arrest is permanent if subsequent oscillations of $K(t)$ do not
exceed K_{Im}. However, if $K(t)$ in the post arrest phase does oscillate to
a higher value than K_{Im} the crack will reinitiate.

References

1. Bradley, W.B. and Kobayashi, A.K., "An Investigation of Propagating
   Cracks by Dynamic Photoelasticity", Expl. Mech., Vol. 10, No. 3,
   (1970), pp. 106-113.
2. Kobayashi, A.S., Wade, B.G. and Bradley, W.B., "Fracture Dynamics of
   Homalite 100", Deformation and Fracture of High Polymers, H.H.
   Kausch, J.A. Hassel, R.I. Jaffee eds., Plenum Press, New York,
   (1973), pp. 487-500.
3. Kobayashi, T. and Dally, J.W., "The Relation Between Crack Velocity
   and the Stress Intensity Factor in Birefringent Polymers", ASTM STP
   627, (1977), pp. 257-273.
4. Irwin, G.R., Dally, J.W., Kobayashi, T., Fourney, W.L., Etheridge,

M.J. and Rossmanith, H.P., "On the Determination of the a-K Relationship for Birefringent Polymers", Expl. Mech., Vol. 19, No. 4, (1979), pp. 27-33N.

5. Ravi-Chandar, K. and Knauss, W.G., "Process Controlling the Dynamic Fracture of Brittle Solids", Workshop on Dynamic Fracture, California Institute of Technology, (1983), pp. 119-128.

6. Dally, J.W., "Dynamic Photoelastic Studies of Fracture", Experimental Mechanics, Vol. 19, No. 10, (1979), pp. 349-361.

7. Sanford, R.J. and Dally, J.W., "A General Method for Determining Mixed-Mode Stress Intensity Factors from Isochromatic Fringe Patterns", Jrnl. of Engr. Fract. Mech., Vol. 11, (1979), pp. 621-633.

8. Rossmanith, H.P. and Irwin, G.R., "Analysis of Dynamic Isochromatic Crack-Tip Stress Patterns", Univ. of Maryland Report, 1979.

CHAPTER 11

CRACK PROPAGATION AND BRANCHING

11.1 Introduction

The behavior of a propagating crack can be classified in four different phases which include:

1. Initiation and subsequent acceleration of the crack tip.

2. Propagation in constant, increasing and decreasing K fields.

3. Crack tip deceleration and subsequent arrest.

4. Terminal velocity of the crack tip and subsequent branching.

Phases 1, 2 and 3 have been covered in previous chapters. This chapter will show several different experiments conducted to study terminal velocity and crack branching.

Theoretical work by Yoffe[1] indicated changes in the stress field at the crack tip with the velocity of crack propagation. The circumferential stress $\sigma_{\theta\theta}$ is a maximum at $\theta = 0$ for low velocity crack propagation; however, $\sigma_{\theta\theta}$ is a maximum at $\theta = 60°$ when $\dot{a}/c_2 = 0.6$. This work implied a terminal velocity of $\dot{a}_T = 0.6\ c_2$ with a branching angle of $120°$.

Experimental work by Dally[2], A.S. Kobayashi[3], Ravi-Chandar and Knauss[4], Fourney, et al[5] and Shukla and Anand[6] will be examined to describe the branching process. While these experiments assist in the understanding of the branching process much remains to accurately and completely characterize this process. The conclusive series of experiments which completely describes branching have not yet been conducted.

11.2 Branching in Homalite 100

Experiments were conducted with large single edge notched (SEN) specimens of Homalite 100 to obtain data in the high velocity region of the $K-\dot{a}$ field. The results shown in Fig. 11.1 indicate attempts to branch and successful branching with $K_{Ib} = 1250$ psi$\sqrt{in}$ at $\dot{a}_T = 430$ m/s. The

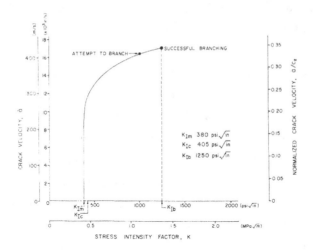

Fig. 11.1   Crack Velocity a as a Function of the Instantaneous Stress
Intensity Factor K(t) for Homalite 100.

Fig. 11.2   Isochromatic Crack Pattern showing Characteristic Loops for a
Single Crack and Two Unsuccessful Surface Branches.

branching process does not occur suddenly but instead it occurs after several unsuccessful attempts to branch as the crack propagates in the high velocity region.

Examination of the fracture surfaces gives some evidence for the changes in K required to drive the crack at higher velocities. For $\dot{a}$ < 200 m/s the fracture surface is mirror smooth, indicating the fracture process zone is relatively small and microcracking off the main crack path is essentially non-existent. In the transition region 200 < $\dot{a}$ < 380 m/s, the fracture surfaces show some roughness going from a frosted appearance to appreciable surface roughening. This progressively increasing roughness indicates that the fracture process zone is increasing in size and that microcracking off of the main crack path is occurring. Some of the microcracks intersect the main crack path and produce the irregularities noted as surface roughness. For the very high velocity region, the fracture process zone is enlarged further and the microcracks off the main crack path coalese to form small surface cracks which are unsuccessful branches as noted in Fig. 11.2.

When K becomes sufficiently large, the fracture process zone is enlarged to the point where the microcracks form a collection of off-line branches which are sufficiently long and propagate at a high enough velo-city that they are not unloaded by the main crack. At this point, K has exceeded the branching toughness of the material and the crack branches, as illustrated in Fig. 11.3, at a terminal velocity $\dot{a}_T$ = 430 m/s. The branching toughness of Homalite 100 K_{Ib} = 1250 psi$\sqrt{in}$ (1.38 MPa $\sqrt{m}$) at the terminal ratio of $\dot{a}/c_2$ = 0.35.

The ratio of K_{Ib}/K_{Im} = 3.3 gives the range of K associated with crack propagation in Homalite 100. In the low velocity region $\dot{a}$ < 200 m/s, the crack propagates with low fracture energy with G = 40.3 J/m ; however, as the velocity increases the energy consumed by the propagating crack increases by more than an order of magnitude with G = 435 J/m at branching.

Kobayashi[3] has also measured K_{Ib} for Homalite 100 in similar experiments and obtained values for K_{Ib} in the range of 1.54 to 1.98 MPa$\sqrt{m}$. These results compare closely with those reported by Dally.

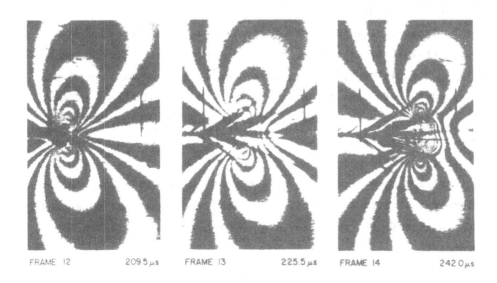

FRAME 12   209.5 μs   FRAME 13   225.5 μs   FRAME 14   242.0 μs

Fig. 11.3   Isochromatic-fringe Pattern Representing Crack Propagation
            During the Branching Period.

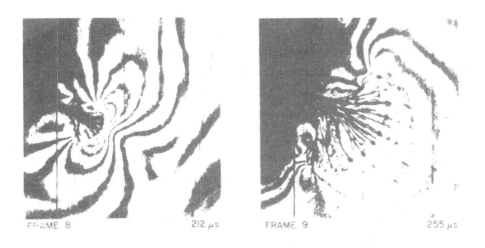

FRAME 8   212 μs   FRAME 9   255 μs

Fig. 11.4   Branching Pattern Showing the Formation of a Candelabra Where
            15 Separate Cracks are Produced.

11.3 Branching Criteria

The horizontal segment of the K-å relation is terminated when K
increases until it becomes large enough to cause crack bifurcation or
branching. With large K, the fracture process zone at the tip of the
crack expands in size and the crack advances by a coalescence of many
off-line microcracks which occur in the process zone. The fan-shaped
front of the propagating crack, presented in Fig. 11.3, illustrates the
mechanism of crack advance with high K conditions. When $K = K_{Ib}$, the
fracture process zone becomes large enough to form a collection of off-
line microcracks which are sufficiently long and are extending at a sufficiently high velocity that they are not unloaded by the extension of
the main crack.

Comparisons of the branching toughness measured for Homalite 100 was
made with an equation derived by Congleton[7] for K_{Ib}/K_{Im}.

$$K_{Ib}/K_{Im} = \pi \left[\frac{n}{2(1-\nu^2)} \right]^{1/2} \tag{11.1}$$

where n is a factor describing the position of the 'advance' microcracks
relative to the main crack. Velocity considerations[8] indicate n > 2 to
prevent unloading of the microcracks by extension of the main crack.
Substitution of n = 2 into eq. (11.1) gives $K_{Ib}/K_{Im} = 3.3$ which
corresponds closely to the experimental measurements that varied from 3.3
to 3.8 for Homalite 100.

The substitution of n = 2 into eq. (11.1) is somewhat arbitrary and n
should be considered as an open parameter which characterizes the size of
the fracture-process zone at the instant of branching. In experiments
with Homalite 100, using a single-edge-notched (SEN) specimen, the
branching occurred by trifurcation. In reality, the trifurcation is produced by a coalescence of multiple bifurcations and indicated in Figs.
11.3b and 11.3c. If the branching instability can be suppressed, K can
increase beyond K_{Ib} (as measured with an SEN specimen) and the fracture-
process zone is greatly enlarged. When the instability finally occurs,

the number of branches formed can be quite large as illustrated in Fig. 11.4. The crack pattern in Fig. 11.4 is for an explosively loaded model where a large compressive stress exists parallel to the crack. The compressive σ_{ox} suppresses branching and permits the cracks to propagate with a super velocity and extremely high values of K. As σ_{ox} decreases with distance of crack extension instability occurs at high K forming a candelabra with 15 separate branches.

When the specimen geometry and loading produces compressive σ_{ox} suppressing the branching instability, the value of K_{Ib}/K_{Im} is much larger than 3.3 to 3.8 and n is much larger than 2. It may be feasible to modify eq. (11.1), to read

$$K_{Ib}/K_{Im} = C(k-1)^m \tag{11.2}$$

where k is the number of branches formed, and C and m are material constants. This modification of eq. (11.1) accounts for the enlargement of the fracture-process zone and utilizes the number of successful branches which can be measured much more easily than the parameter n.

11.4 Super-velocity Crack Propagation

It has been shown theoretically[1] that the maximum crack velocity which can be achieved is a = c_R where c_R is the Rayleigh wave velocity. For most materials, this maximum is not achieved because branching occurs and the energy driving the crack is divided. For Homalite 100, branching occurs with the SEN specimen when $\dot{a}/c_2$ = 0.31-0.35. The velocity at branching is often called the terminal velocity $\dot{a}_T$ and it is generally believed that $\dot{a}_T$ is the maximum which can be achieved for a given material. However, if branching can be suppressed, K can be increased to levels greater than K_{Ib}, and $\dot{a} > \dot{a}_T$ can be observed.

If a specimen is loaded explosively with a small charge of PETN placed in a central through hole, several radial cracks are produced which propagate away from the hole. A typical curve showing the position a of one of these cracks as a function of time is presented in Fig. 11.5. Differentiation of these a-t data shows that the initial velocity of the

crack was 28,600 in./s (726 m/s) which corresponds to $\dot{a}/c_2$ = 0.59. The
crack velocity decreases as the crack extends into the field with $\dot{a}$ =
23,800 in./s (604 m/s) just prior to branching. The stress intensity
factor at this instant was estimated at K_{Ib} = 4900 psi√in. (5.30 MPa√m)
or K_{Ib}/K_{Im} = 12. Six branches were produced when the crack became
unstable indicating that m = 1 and C = 2.4 in eq. (11.2). Of course,
these results are preliminary and further branching experiments should be
conducted to verify the adequacy of eq. (11.2)

    Similar results on super velocity cracks were obtained by Fourney, et
al[5] in Homalite 100 models loaded explosively. Crack velocities in
excess of 625 m/s were recorded. Reflected P waves from the boundary of
the model interacted with the propagating crack and induced high energy
branching.

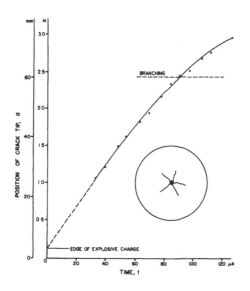

Fig. 11.5 Crack Position as a function of Time for a Super Velocity Crack

11.5 Effects of Non-singular Terms on Crack Branching

    From the inception of fracture mechanics, Irwin's basic concepts per-
taining to fracture initiation involved only a single parameter K for prediction. Studies of low velocity crack propagation and crack arrest

indicate that the process can be characterized with K(t). However, for high velocity crack propagation there is a great amount of experimental evidence that the crack tip is not a point but a region. The size of the region depicting the fracture process zone grows larger as the velocity increases and as the branching process develops. Since the branching phenomena encompasses a region of material, the singularity term K in the stress field equation can not by itself characterize the branching process. Higher order terms will be required.

Fourney, et al[4] have proposed a representation of the high velocity region of the K-å field illustrated in Fig. 11.6. Several curves are represented on the K-å field depending upon the value and sign of σ_{ox}.

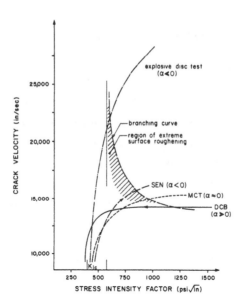

Fig. 11.6 Schematic Representation of the a-K
Showing the Effects of σ_{ox} on Crack Branching.

Kobayshi and Ramulu[9] have developed another criteria for branching based on K_{Ib} and a critical characteristic distance $r_o = r_c$ which is a material property. They define

$$r_o = \frac{1}{128\pi} \left[\frac{K_I}{\sigma_{ox}} f(c, c_1, c_2) \right]^2 \qquad (11.3)$$

$$\theta_c = \cos^{-1} \left\{ f \left. \frac{\sigma_{ox}}{K_I} \right|^2, r_o \right\} \tag{11.4}$$

In experiments conducted by Kobayashi and Ramulu with SEN specimens fabricated from Homalite 100 they observed $K_{Ib} = 2.04$ MPa$\sqrt{m}$, $r_c = 1.3$ mm and $\theta_c = 28°$ (where θ_c is half the branching angle).

Anand and Shukla[6] have recently completed a series of experiments conducted with cross type fracture experiment capable of supporting biaxial loads to produce remote stresses σ_{xx} and σ_{yy}. In these experiments the biaxial remote loading σ_{xx}/σ_{yy} was varied from −0.51 to +1.72. (Note: the σ_{xx} stress is in the direction parallel to the crack direction and the minus sign indicates compression stresses.) Measurements of $\dot{a}_T$, θ_c and K_{Ib} which were made during these experiments are reported in Table 11.1.

The results indicate that the biaxiality of the loading does not appear to markedly affect $\dot{a}_T$ and K_{Ib}. The variations in $\dot{a}_T$ are less than 4 percent of the average value of $\dot{a}_T$ and the variations in K_{Ib} are at most 10 percent of the average $K_{Ib} = 2.01$ MPa$\sqrt{m}$. The remote biaxiality ratio does markedly affect the branching angle as the angle increases significantly with the tensile biaxiality ratio.

Anand and Shukla have also examined the validity of eqs. (11.3) and (11.4). Their results indicate r_c ranges from 1.1 to 8.1 mm which differs significantly from the constant value of 1.3 mm which Kobayashi and

TABLE 11.1

Influence of Remote Load Biaxiality on
Branching Characteristics a_T, θ_c, and K_{Ib}

Remote Load Ratio σ_{xx}/σ_{yy}	Terminal Velocity $\dot{a}_T$ (m/s)	Branching Angle θ_c (deg)	Branching Toughness K_{Ib} MPa√m
-0.51	372	25	1.91
-0.26	385	29	2.01
0	400	23	--
+0.35	375	30	2.2
+1.03	380	45	2.0
+1.72	400	73	1.91

Ramulu considered as a material property. Also, the branching angle θ_c depends strongly on the sign of σ_{ox}/K_I which is in disagreement with eq. (11.4).

11.6 Conclusions

The understanding of the branching process is in its early stages. Significant evidence has been collected which shows that the fracture process zone increases in size as K(t) increases. When the process zone becomes sufficiently large cracks are formed by coalesence of microvoids at locations off of the crack line and at orientations making an angle with the crack line. Many of these off line cracks are initiated; however, they quickly arrest because the fracture process zone which is dominated by the primary crack moves out of the region of the initiating crack. The initiating crack is unloaded and it arrests. This formation of voids, coalesence into a crack, minor extension and arrest is repeated thousands of times as the primary crack propagates at nearly the terminal velocity. This repeated and unsuccessful branching accounts for the very rough fracture surface texture associated with high velocity crack propagation.

Branching or more accurately bifurication occurs when a new crack is initiated far enough ahead of the primary crack that the subsequent movement of the fracture process zone increases the size and velocity of the newly initiated crack. It then becomes the dominant crack as it unloads the primary crack often causing it to arrest.

While the branching process has been observed, the factors which control the process and the predictive equation for branching have not been established. For elastic and brittle materials it appears that one will need at least three parameters K_I, σ_{ox} and a material constant similar to Kobayashi's r_c, to predict branching. For ductile materials with elastic plastic behavior, the situation is more complex because both viscoelastic and plastic behavior must be taken into account.

References
1. Yoffee, E.H. "The Moving Griffith Crack, Philosophical Magazine, Series 7, Vol. 42, (1951), p. 739.
2. Dally, J.W., "Dynamic Photoelastic Studies of Fracture", Experimental Mechanics, Vol. 19, No. 10, (1979), pp. 349-361.
3. Kobayashi, A.S., et al, "Crack Branching in Homalite 100 Sheets", Eng. Fracture Mechanics, Vol. 6, (1974), pp. 81-92.
4. Ravi-Chandar, K. and Knauss, W.G., "Processes Controlling the Dynamic Fracture of Brittle Solids", Workshop on Dynamic Fracture, California Institute of Technology, (1983), pp. 119-128.
5. Fourney, W.L., Chona, R. and Sanford, R.J., "Dynamic Crack Growth in Polymers", Workshop on Dynamic Fracture, California Institute of Technology, (1983), pp. 75-86.
6. Anand, S. and Shukla, A., "High Speed Crack Propagation and Branching Under Uniaxial and Biaxial Loading", paper to be presented at 17th National Fracture Symposium, Albany, N.Y., August 1984.
7. Congleton, J., "Practical Applications of Crack-Branching Measurements", Proc. of Intnl. Conf. on Dynamic Crack Propagation, Ed. G.C. Sih, (1972), pp. 427-438.
8. Congleton, J. and Petch, N.J., "Crack Branching", Phil. Mag., Vol. 16, (1967), pp. 749-760.
9. Ramulu, M. et al, "Dynamic Crack Branching - A Photoelastic Evaluation", ONR Technical Report No. UWA/DME/TR-82/43.

THE SHADOW OPTICAL METHOD OF CAUSTICS

Jörg F. Kalthoff
Fraunhofer-Institut für Werkstoffmechanik
Freiburg, Federal Republic of Germany

LIST OF SYMBOLS

a	Crack length
a,b	Elasto-optical constants
A,B	Material constants in Maxwell-Neumann's law
α	Velocity dependent factor
$a_{2,3,4...}$	Coeffients of higher order terms in crack tip stress distribution
c	Shadow optical constant
c_o	Sound wave speed
c_1	Longitudinal wave speed
c_2	Transverse wave speed
C	Compliance of specimen
d	1. Specimen thickness
	2. Distance between two cracks in double crack configuration
d_{eff}	Effective thickness of the specimen
D	Characteristic length parameter for caustic evaluation
$D_{o,i}$	Outer, inner characteristic length parameter
$D_{max,min}$	Maximum, minimum length parameter of mixed mode caustics
e	Index characterizing elastic behavior
E	Young's modulus
ϵ	Strain
f	Numerical factor for caustic evaluation
$f_{o,i}$	Numerical factor for evaluating outer, inner caustic
g	Numerical factor for K_I-determination from mixed mode caustics
G	1. Function
	2. Lamé's constant, $G = E/2(1+\nu)$
H	Height of specimen
I_n	Numerical Factor in the HRR stress field equations
J	J-Integral
K	Stress intensity factor
n	1. Refractive index
	2. Strain hardening coefficient

ν	Poisson's ratio
0	Origin of coordinate system
P	Edge load, unit N/m
p	Index characterizing plastic behavior
p,q	Biaxial stresses in y,x-direction
R	Radius of circular hole
r,φ	Polar coordinate system in object plane (specimen)
r',φ'	Polar coordinate system in image (reference) plane
$\bar{r},\bar{\varphi}$	Polar coordinate system at the tip of a moving crack
r_o	Radius of initial curve
r_{pl}	Radius of plastically deformed region around the crack tip
r_{ps}	Smallest radius around the center of stress concentration outside which a state of plane stress exists
ρ	1. Density of the material
	2. Notch tip radius
	3. Wedge tip radius
S	Support span
s	Optical path length
σ	Normal stress
σ_o	Tensile yield stress
t	Time
λ	Coefficient of anisotropy
τ	1. Shear stress
	2. Period of the oscillation of impacted specimen
μ	Ratio of the mode II to mode I stress intensity factor, K_{II}/K_I
u,v,w	Displacements in x,y,z-direction
W	Width of the specimen
x,y	Cartesian coordinates system in object plane (specimen)
x',y'	Cartesian coordinates system in image (reference) plane
$\bar{x},\bar{y}$	Cartesian coordinates system at the tip of a moving crack
z	Direction of optical axis
z_o	Distance between object plane (specimen) and image (reference) plane

1. INTRODUCTION

The shadow optical method of caustics is a relatively new experimental technique in stress strain analysis. It was introduced by Manogg[1,2] in 1964. The method is sensitive to stress gradients and therefore is an appropriate tool for quantifiying stress concentration problems. Manogg originally used the method for investigating crack tip stress intensifications. The technique was extended later by Theocaris[3-5], Rosakis[6,7], and the author and his colleagues[8-11] to different conditions of loading, material behavior, in static as well as dynamic situations. Shadow optical images of test specimens under loading in general are characterized by very simple geometric patterns which can be easily evaluated. Because of the simplicity of shadow patterns, the method can also be successfully applied for investigating rather complex phenomena, for example transient problems. Despite the complexity which may be inherent in the problems to be investigated the clearness of the generated recordings allows the derivation of reliable informative data. The author and his colleagues have applied the caustic technique to investigate various problems of practical interest in the field of fracture dynamics, in particular to the behavior of propagating and subsequently arresting cracks and the behavior of cracks under different impact loading conditions.

This article illustrates the basic physical principle of the shadow optical method of caustics and presents the mathematical description of the imaging process. Caustic evaluation formulas are presented for various stress strain concentration problems. The experimental techniques for the generation and the recording of shadow patterns under static and dynamic situations are discussed with special emphasis on laboratory test conditions. The applicability of the technique for investigating problems of practical relevance is demonstrated by several examples.

2. PHYSICAL PRINCIPLE

Stresses in a solid alter the optical properties of the solid. Tensile stresses reduce the thickness of the body due to Poisson's effects and they also reduce the refractive index of the material, since the material becomes optically less dense. The reverse situation applies for compressive stresses. These changes in the optical properties are utilized in the shadow optical method of caustics to make stress distributions in solids visible.

2.1 Model Considerations for Shadow Optical Images

The principle of the shadow optical imaging process is illustrated in Figure 1. Considered is a plate the middle section of which is subjected to linearly increasing compressive stresses. According to this stress distribution the thickness of the plate and the refractive index of the material linearly increase in this section of the plate. Both effects have the same influence on the deflection of light rays. The two effects are here combined and substituted in a simple model having the same net influence on the light deflection, i.e. a prism. The plate is illuminated from the left side by a parallel incident light beam. The stress free parts of the plate are traversed by the light rays without deflection, in the middle section of the plate, however, the light rays are deflected (in this simplified model by the same angle). Consequently the light distribution in an arbitrary plane (image plane or reference plane) behind the specimen (i.e. at the right side of the specimen) is not uniform any more. Certain areas are not hit by light rays and thus appear dark, whereas other areas are hit by light rays twice and thus appear as areas of increased brightness. The light distribution can be made visible on a screen. The resulting pattern represents a quantitative description of the stress distribution in the plate.

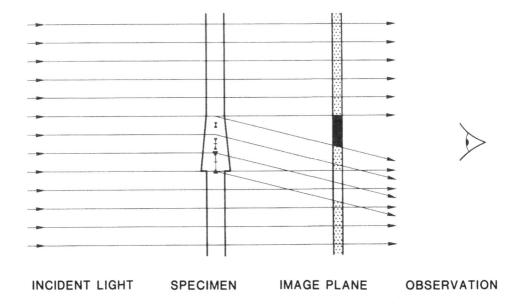

INCIDENT LIGHT   SPECIMEN   IMAGE PLANE   OBSERVATION

Fig. 1   Principle of shadow optical imaging process (schematically)

Shadow optical light distributions can be observed in different ways, in transmission or in reflection arrangements, as real or as virtual images. The different possibilities of observation and the resulting light distributions are explained in Figure 2. Two characteristic examples are considered which illustrate typical but simplified compressive and tensile stress concentration problems. A plate in its middle section is subjected to a linearly increasing and subsequently decreasing distribution of compressive stresses (upper diagrams in Figure 2) or it is subjected to equivalent tensile stresses (lower diagrams in Figure 2). The specimens are again illuminated from the left side by a parallel incident light beam. The transmission arrangements are shown on the right side of the figure, the reflection arrangements on the left side.

The light configurations 2b and 4b in Figure 2 represent the previously discussed simple case of a transmission arrangement with the real image observed on the right side of the specimen (see Fig. 1 for comparison). The observation direction is opposite to the direction of the illuminating light beam. With the compressive stress distribution the light rays are refracted towards the center line, thus creating a central region of increased brightness surrounded by an area of darkness (configuration 2b). With the tensile stress distribution, the reverse situation is obtained, i.e. a central area of darkness surrounded by an area of increased brightness (configuration 4b).

Light rays which are reflected at the front side of the plate, i.e. the side facing the light source, also form real shadow optical images, but on the left side of the specimen (configurations 1a and 3a in Figure 2). The observation direction now is the same as the direction of the illuminating light beam. In the reflection arrangement the compressive stress model forms a central area of darkness (configuration 1a) whereas the tensile stress model leads to a central region of brightness (configuration 3a). The characteristic features of light intensity distributions are thus reversed (i.e., dark and bright regions are ex-

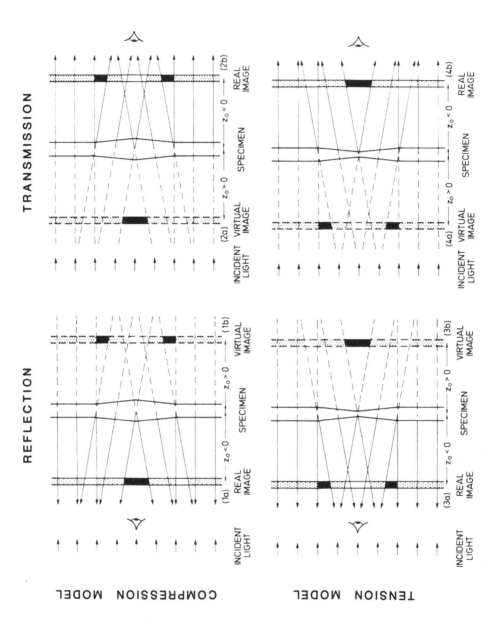

Fig. 2   Shadow optical light distributions for different registration arrangements (schematically)

changed) if the observation mode is changed from transmission to reflection.

In addition to real shadow optical images, which can be directly made visible on a screen, virtual images can be observed as well. These virtual images are obtained on the opposite side of the specimen where the real images are observed, i.e. on the left side of the specimen in transmission arrangements (configurations 2a and 4a in Figure 2) and on the right side of the specimen in reflection arrangements (configurations 1b and 3b in Figure 2). The light ray tracing in the virtual image space behind the specimen is obtained from a backward extrapolation of the refracted light rays. (For simplicity this schematic representation assumes that the light deflection in transmission is quantitatively the same as in reflection, which is of course not true in reality). As can be easily seen the characteristic features of the resulting light intensity distributions are now reversed (exchange of bright and dark regions) if virtual instead of real images are considered.

The following crude rules are useful for a preliminary tentative interpretation of shadow optical images:

o In transmission arrangements compressive (tensile) stress concentrations have an effect similar to convex (concave) lenses, thus generating areas of central brightness (darkness) in real shadow optical images.

o In reflection arrangements compressive (tensile) stress concentrations act similar to convex (concave) mirrors, thus generating areas of central darkness (brightness) in real shadow optical images.

o The characteristic features of shadow optical light distributions are reversed if virtual instead of real images are considered.

Consequently, depending on the observation mode it is possible to obtain different shadow optical images for the same stress distribution or similar (sometimes even identical) shadow optical images for differ-

ent stress distributions. Therefore, quantitative evaluations require an
exact knowledge of the particular observation mode. For the mathematical
description of shadow optical images, the following definition is made:

o   The distance between the reference plane and the specimen is de-
         fined to be negative (positive) if the reference plane with regard
         to the observation direction is located ahead of (behind) the spe-
         cimen (see Fig. 2).

2.2 Generation of Caustics by Stress Concentrations

The models for stress concentration problems considered in the examples previously discussed are of necessity oversimplified. Thus, the
deduced light distribution rules discussed there can only represent crude
hints for the interpretation of shadow optical images. The stress distributions for real stress concentration problems in general are not given
by linear relationships but by more complicated functions. In particular,
the stresses increase steeper than linearly when the center of the stress
concentration field (e.g. the tip of a crack) is approached. Figure 3
shows the distribution of light rays in a transmission arrangement for
the case of a crack under tensile (mode I) loading. The closer to the
crack tip a light ray traverses the specimen the larger its deflection.
Consequently, for light rays with decreasing distance to the center point
(i.e. the crack tip in the object plane) the image points first also
approach the corresponding center point in the image plane. Then the tendency is reversed. For light rays which are traversing the specimen even
closer to the crack tip the image points move away from the central point
again. The shadow optical image, therefore, exhibits a sharp boundary
line between the area of darkness and the surrounding area of light concentration. This boundary line between the two areas is called the caustic curve (caustic = greek for focal line).

These particular light rays which directly hit the caustic curve in
the image plane are of special importance; they are called the "initial"

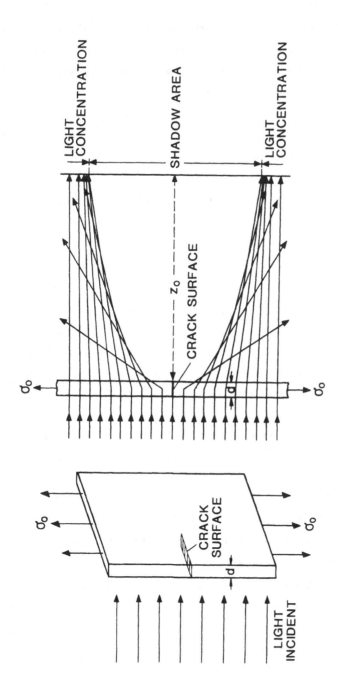

Fig. 3   Light ray distribution for a realistic stress concentration
problem (schematically)

light rays. The locus of all points where the initial light rays traverse the object plane (specimen) is called the initial curve. Light rays traversing the object plane at points that, when compared with the initial curve, are closer to (or farther away from) the crack tip are deflected steeper (or shallower) than the initial light rays. Consequently in both cases the image points lie outside the caustic curve. The caustic curve in the image plane is the direct image of the initial curve in the object plane. In the mathematical sense of the theory of imaging this mapping process of the initial curve onto the caustic curve is in general not reversible and does not represent a one-to-one-correspondence.

Figure 4 shows examples of experimentally observed shadow optical images and the resulting caustic curves for various stress concentration problems. These are, a 60°-notch subjected to compressive and tensile loading (Fig. 4a), a compressive edge load acting on a half plane (Fig. 4b), and a crack subjected to tensile (mode I) loading (Fig. 4c). The shadow optical pictures have been photographed under different observation modes specified in the figure.

The derived interpretation rules of shadow optical light distributions are verified by these examples. For example, the virtual shadow pattern of the notch under compression is identical to the real pattern of the notch under tension (Fig. 4a). The same is true for the real pattern of the notch under compression and the virtual pattern of the notch under tension. The real shadow pattern of the notch under tension and the virtual pattern for the compressive edge load show distinct regions of central darkness, whereas the real patterns of the notch under compression and of the compressive edge load exhibit central regions of increased brightness (Figs. 4a and 4b). The shape of the real crack tip caustic observed in transmission is the same as that of the virtual crack tip caustic observed in reflection (Fig. 4c).

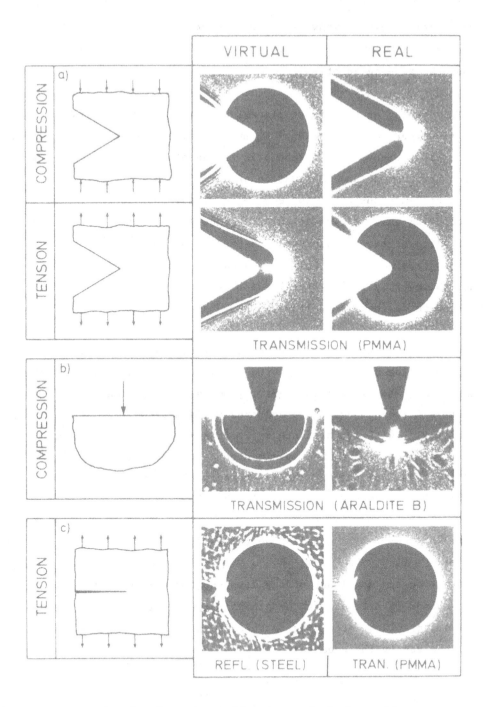

Fig. 4   Experimentally observed shadow patterns

3. QUANTITATIVE DESCRIPTION OF SHADOW OPTICAL IMAGES

3.1 General Mapping Equations

The mapping of the object plane E (specimen) onto a shadow optical image plane (or reference plane) E' by a parallel incident light beam is considered in Figure 5. The schematic representation applies for a transmission arrangement and observation of a real shadow optical image. A notched specimen subjected to tensile loading is considered in the figure. The following quantitative description, however, applies quite generally for any kind of stress concentration problem or observation mode. The cartesian (x,y) and polar (r,φ) coordinate systems are used in the object plane, while (x',y') and (r',φ') are used in the image plane. Tensile stresses are defined as positive.

A light ray is considered which traverses the object plane E at the point P with the distance $\vec{r}$ from the origin of the coordinate system. In an unloaded specimen this light ray would not experience any light deflection in the object plane and thus would hit the image plane E' at the point P_{nd}, with the distance $\vec{r}_{nd} = \vec{r}$ from the origin. However, the stress distribution in a specimen subjected to loading would deflect the light ray. Consequently the light ray will hit the image plane E' displaced by the vector $\vec{w}$ at the point $P'(\vec{r'})$ with

$$\vec{r'} = \vec{r} + \vec{w} \tag{1}$$

The direction and magnitude of the displacement vector $\vec{w}$ are controlled by the change in optical path length Δs which the light ray experiences in the object plane. According to the eikonal theory (see e.q. Reference 12) $\vec{w}$ is given by the equation

$$\vec{w} = - z_0 \overrightarrow{grad} \, \Delta s(r,\varphi). \tag{2}$$

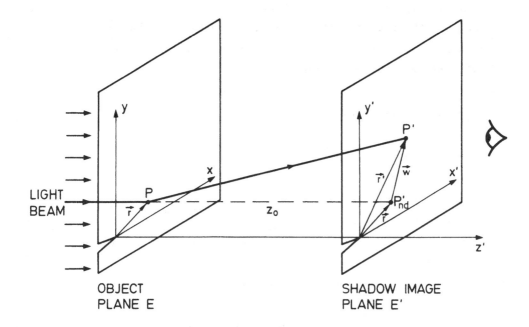

Fig. 5   Light ray deflection

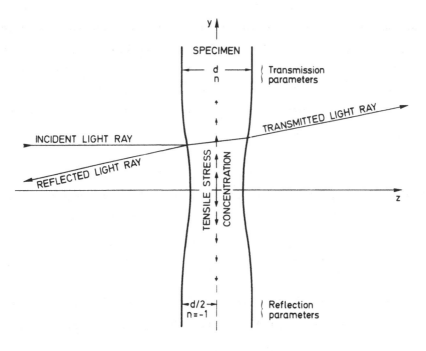

Fig. 6   Relevant parameters for transmisssion and reflection of
          light rays

The change in the optical path length Δs results from changes in the thickness of the plate and in the refractive index of the material. For a plate with plane parallel surfaces

$$\Delta s = (n-1) \ \Delta d_{eff} + d_{eff} \ \Delta n \tag{3}$$

where s = optical path length
             d_{eff} = effective thickness of the plate
             n = refractive index $\left.\right\}$ for transparent specimens
             d_{eff} = d
and n = -1 $\left.\right\}$ for non-transparent specimens
             d_{eff} = d/2   in reflection
with d = actual thickness of the specimen.

In transmission, the surface deformations at both sides of the speci-men and the change in the refractive index along the total thickness of the specimen contribute to the change in the optical path length. In re-flection only the deformation of the illuminated front surface determines the change in the optical path length (see Fig. 6). Thus, the reflection case is obtained from the more general transmission case by formally set-ting the refractive index $n = -1$ (change in the direction of the light rays and $\Delta n = 0$) and by using an effective thickness $d_{eff} = d/2$ (con-sideration of the deformation at one surface only).

Furthermore, changes Δn in the refractive index due to the princi-pal stresses σ_1, σ_2, σ_3 are described by Maxwell-Neumann's law. For the general case of a birefringend, transparent material

$$\Delta n_1 = A\sigma_1 + B(\sigma_2 + \sigma_3), \qquad \Delta n_2 = A\sigma_2 + B(\sigma_1 + \sigma_3) \tag{4}$$

where A, and B are material constants. For optically isotropic, non-bire-fringent materials $A = B$ and $\Delta n_1 = \Delta n_2 = \Delta n$. For reflection $A = B = 0$.

Changes Δd_{eff} due to the prevailing stresses are described by Hooke's law

$$\Delta d_{eff} = \left[\frac{1}{E} \sigma_3 - \frac{\nu}{E} (\sigma_1 + \sigma_2) \right] d_{eff} \qquad (5)$$

with $\sigma_3 = 0$ for plane stress, and $\Delta d_{eff} = 0$ for plane strain,

E = Young's modulus, ν = Poisson's ratio.

With equations (4) and (5) equation (3) can be written as

$$\Delta s_1 = d_{eff}(a\sigma_1 + b\sigma_2), \qquad \Delta s_2 = d_{eff}(a\sigma_2 + b\sigma_1) \qquad (6)$$

with $a = A - (n-1)\nu/E$, $b = B - (n-1)\nu/E$ for plane stress,

and $a = B - \nu B$, $b = B + \nu B$ for plane strain.

Equation (6) can be rearranged in a more convenient form in terms of the sum and the difference of the principal stresses as

$$\Delta s_{1/2} = c\, d_{eff}\left[(\sigma_1 + \sigma_2) \pm \lambda (\sigma_1 - \sigma_2) \right] \qquad (7)$$

with $c = \dfrac{A+B}{2} - (n-1)\nu/E, \quad \lambda = \dfrac{2(A-B)}{A+B - 2(n-1)\nu/E}$ for plane stress,

and $c = \dfrac{A+B}{2} - \nu B \quad , \quad \lambda = \dfrac{2(A-B)}{A+B + \nu B}$ for plane strain.

The constant c describes the change in the optical path length obtained with a specific material for a certain stress situation. Thus the constant c is a quantitative measure of the resulting shadow optical effect and, therefore, is called the "shadow optical constant". Influences on the change of the optical path length due to anisotropy effects of the material ($A \neq B$) are described by the coefficient λ. Numerical values for the material constants A, B, n, and the deduced shadow optical constant c and the anisotropy coefficient λ are given for different materials in Table 1.

TABLE 1 - Constants for Caustic Evaluation

Material	Elastic Constants		General Optical Constants			Shadow Optical Constants				Effective Thickness
						for Plane Stress		for Plane Strain		
	Young's Modulus MN/m^2	Poisson's Ratio	Refractive Index	A m^2/N	B m^2/N	c m^2/N	λ	c m^2/N	λ	d_{eff}
TRANSMISSION ($z_0 < 0$)										
Optically Anisotropic										
Araldite B	3660[*]	0.392[*]	1.592	-0.056×10^{-10}	-0.620×10^{-10}	-0.970×10^{-10}	-0.288	-0.580×10^{-10}	-0.482	d
CR - 39	2580	0.443	1.504	-0.160×10^{-10}	-0.520×10^{-10}	-1.200×10^{-10}	-0.148	-0.560×10^{-10}	-0.317	d
Plate Glass	73900	0.231	1.517	$+0.0032 \times 10^{-10}$	-0.025×10^{-10}	-0.027×10^{-10}	-0.519	-0.017×10^{-10}	-0.849	d
Homalite 100	4820[*]	0.310[*]	1.561	-0.444×10^{-10}	-0.672×10^{-10}	-0.920×10^{-10}	-0.121	-0.767×10^{-10}	-0.149	d
Optically Isotropic										
PMMA	3240	0.350	1.491	-0.530×10^{-10}	-0.570×10^{-10}	-1.080×10^{-10}	~ 0	-0.750×10^{-10}	~ 0	d
REFLECTION ($z_0 > 0$)										
All materials	E	v	-1	0	0	2v/E	0	-	-	d/2

[*]) dynamic values

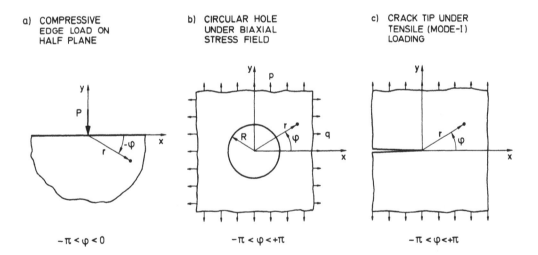

a) COMPRESSIVE EDGE LOAD ON HALF PLANE

$-\pi < \varphi < 0$

b) CIRCULAR HOLE UNDER BIAXIAL STRESS FIELD

$-\pi < \varphi < +\pi$

c) CRACK TIP UNDER TENSILE (MODE-I) LOADING

$-\pi < \varphi < +\pi$

Fig. 7   Typical stress concentration problems

Equations (1), (2), and (7) describe the mapping of the object plane onto the shadow optical image plane for arbitrary stress distributions $\sigma_{1,2}(r, \varphi)$. The specific mapping equation for a special stress concentration problem is obtained by inserting into the general equation (7) the individual formula for the stress distribution of the particular stress concentration problem considered.

3.2 Mapping Equations and Caustics for Specific Examples

Three stress concentration problems will be discussed simultaneously in a comparative manner. These are: a) a compressive edge load P acting on a half plane, b) a circular hole in a plate subjected to a biaxial stress field p,q, and c) a crack in a plate under tensile load with mode I stress intensity factor K_I*. Graphical presentations of the three problems considered are given in Figure 7.

The linear elastic stress concentration fields for the three examples are given by the following equations

a)
$$\sigma_r = \frac{2P}{\pi} \frac{\sin\varphi}{r}$$

$$\sigma_\varphi = 0$$

$$\tau_{r\varphi} = 0$$

b)
$$\sigma_r = \frac{p+q}{2}\left(1 - \frac{R^2}{r^2}\right) - \frac{p-q}{2}\left(1 - \frac{R^2}{r^2} + 3\frac{R^4}{r^4}\right) \cos 2\varphi$$

$$\sigma_\varphi = \frac{p+q}{2}\left(1 + \frac{R^2}{r^2}\right) + \frac{p-q}{2}\left(1 + 3\frac{R^4}{r^4}\right) \sin 2\varphi \qquad (8)$$

$$\tau_{r\varphi} = \frac{p-q}{2}\left(1 + 2\frac{R^2}{r^2} - 3\frac{R^4}{r^4}\right) \sin 2\varphi$$

---

* For the definition of the stress intensity factor K see textbooks on fracture mechanics, e.g. Reference 13.

c)
$$\sigma_r = \frac{K_I}{\sqrt{2\pi r}} \frac{1}{4} (5 \cos \frac{1}{2}\varphi - \cos \frac{3}{2}\varphi)$$

$$\sigma_\varphi = \frac{K_I}{\sqrt{2\pi r}} \frac{1}{4} (3 \cos \frac{1}{2}\varphi + \cos \frac{3}{2}\varphi)$$

$$\tau_{r\varphi} = \frac{K_I}{\sqrt{2\pi r}} \frac{1}{4} (\sin \frac{1}{2}\varphi + \sin \frac{3}{2}\varphi)$$

With these stress distributions and Equation (7) (for $\lambda = o$)* the mapping equations (1), (2) become

a)
$$x' = r \cos \varphi + \frac{2P}{\pi} z_o c d_{eff} r^{-2} \sin 2\varphi$$

$$y' = r \sin \varphi - \frac{2P}{\pi} z_o c d_{eff} r^{-2} \cos 2\varphi$$

b)
$$x' = r \cos \varphi + 4 z_o c d_{eff} R^2 (p-q) r^{-3} \cos 3\varphi$$

$$y' = r \sin \varphi + 4 z_o c d_{eff} R^2 (p-q) r^{-3} \sin 3\varphi$$

(9)

c)
$$x' = r \cos \varphi + \frac{K_I}{\sqrt{2\pi}} z_o c d_{eff} r^{-3/2} \cos \frac{3}{2} \varphi$$

$$y' = r \sin \varphi + \frac{K_I}{\sqrt{2\pi}} z_o c d_{eff} r^{-3/2} \sin \frac{3}{2} \varphi$$

The complete family of the deflected light rays forms a shadow space behind the object plane. Its surface is an envelope to the light rays and is called the caustic surface. The intersection of this surface with the image plane forms the caustic curve. This caustic curve is a multivalued, singular solution of equations (1) and (2), i.e. the mapping of points along the caustic curve is not reversible. Thus a necessary and sufficient condition for the existence of the caustic curve is obtained

---

* For simplicity only the isotropic case is considered in this context.

if the Jacobian of equations (1), (2) vanishes, i.e.

$$\frac{\partial x'}{\partial r}\frac{\partial y'}{\partial \varphi} - \frac{\partial x'}{\partial \varphi}\frac{\partial y'}{\partial r} = 0 \tag{10}$$

The coordinates r, φ of points P which fulfill equation (10) form the initial curve in the object plane and the mapping of this initial curve onto the image plane is the caustic curve.

Consequently, application of equation (10) to the mapping equations (9) gives the equation of the initial curves

a) $\quad r = \left[\frac{4}{\pi} |z_o| |c| d_{eff} P\right]^{1/3} \equiv r_o$

b) $\quad r = \left[12 |z_o| |c| d_{eff} R^2 (p-q)\right]^{1/4} \equiv r_o \tag{11}$

c) $\quad r = \left[\frac{3}{2} \frac{K_I}{\sqrt{2\pi}} |z_o| |c| d_{eff}\right]^{2/5} \equiv r_o$

For the three examples considered the initial curves are circles around the center point of stress concentration with fixed radii r_o. With the mapping equations (9) the caustic curves are finally obtained as images of the initial curves, equations (11), and are given as

a) $\quad x' = r_o (\cos \varphi + \text{sgn}(z_o c) \frac{1}{2} \sin 2\varphi)$

$\qquad\qquad\qquad\qquad\qquad\qquad$ for $\quad -\pi < \varphi < 0$

$\quad y' = r_o (\sin \varphi - \text{sgn}(z_o c) \frac{1}{2} \cos 2\varphi)$

b) $\quad x' = r_o (\cos \varphi + \text{sgn}(z_o c) \frac{1}{3} \cos 3\varphi)$

$\qquad\qquad\qquad\qquad\qquad\qquad$ for $\quad -\pi < \varphi < +\pi \quad$ (12)

$\quad y' = r_o (\sin \varphi + \text{sgn}(z_o c) \frac{1}{3} \sin 3\varphi)$

c) $\quad x' = r_o (\cos \varphi + \text{sgn}(z_o c) \frac{2}{3} \cos \frac{3}{2}\varphi)$

$\qquad\qquad\qquad\qquad\qquad\qquad$ for $\quad -\pi < \varphi < +\pi$

$\quad y' = r_o (\sin \varphi + \text{sgn}(z_o c) \frac{2}{3} \sin \frac{3}{2}\varphi)$

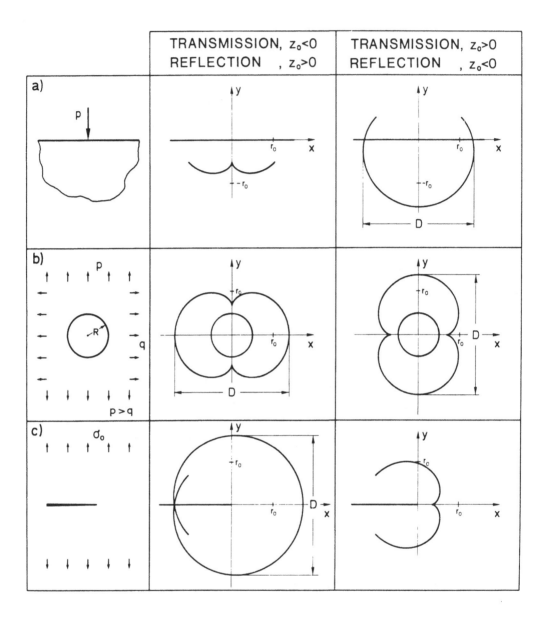

Fig. 8   Caustic curves for the considered stress concentration
          problems

Mathematically the caustic curves are generalized epicycloids. The caustics are graphically shown in Figure 8 for positive and negative signs of the distance z_o between the specimen and reference plane.

An illustrative picture of the complete distribution of deflected light rays in the image plane is obtained in a simple manner by considering the mapping of rays which traverse the object plane along lines φ = const. The images of those lines obtained with equations (9) for the three examples considered are shown in Figure 9. The caustic curves appear as envelopes to the obtained families of image lines (see Fig. 8 for comparison).

For the quantitative evaluation of caustics a length parameter between characteristic points on the caustic curve is defined, e.g. the distances D given in Figure 8. These distances are related to the radii of the initial curves by the equations

a) $\qquad\qquad D = 2.6\ r_o$

b) $\qquad\qquad D = 2.67\ r_o$ $\qquad\qquad\qquad\qquad\qquad\qquad\qquad$ (13)

c) $\qquad\qquad D = 3.17\ r_o$

With equations (11) and (13) a quantitative formula is obtained in each case relating the size of the shadow optical pattern with the generating load parameter

$$\text{a)} \quad P = \frac{\pi}{4\ (2.6)^3\ z_o\ c\ d_{eff}}\ D^3$$

$$\text{b)} \quad p\text{-}q = \frac{1}{12\ (2.67)^4\ z_o\ c\ d_{eff}\ R^2}\ D^4 \qquad\qquad (14)$$

$$\text{c)} \quad K_I = \frac{2\ \sqrt{2\pi}}{3\ (3.17)^{5/2}\ z_o\ c\ d_{eff}}\ D^{5/2}$$

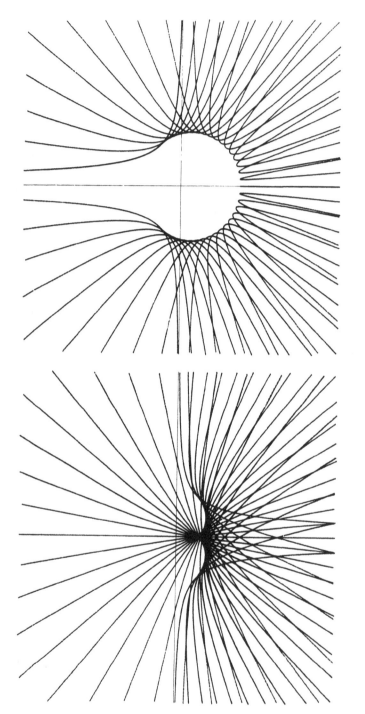

COMPRESSIVE EDGE LOAD ON HALF PLANE

Fig. 9a   Shadow optical light distribution

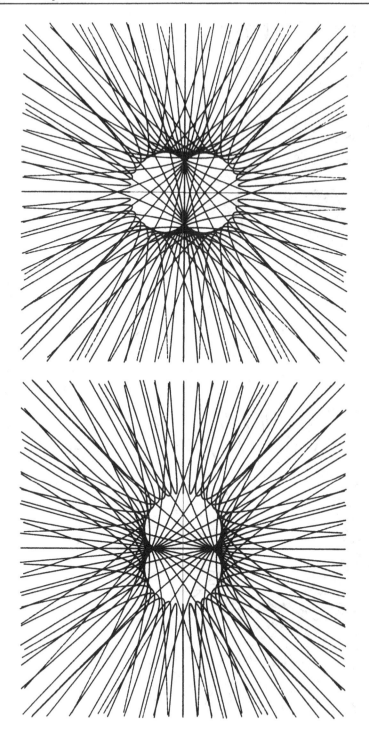

CIRCULAR HOLE UNDER BIAXIAL STRESSES

Fig. 9b   Shadow optical light distribution

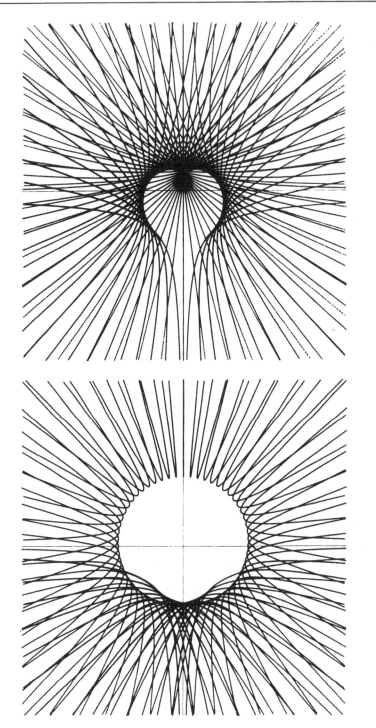

CRACK UNDER MODE–I LOADING

Fig. 9c   Shadow optical light distribution

Thus, with equation (14) and the distance D measured in an experimentally obtained caustic the generating load parameter can be determined, i.e. the magnitude of the compressive edge load P, or the difference of the biaxial stresses p-q, or the crack tip stress intensity factor K_I.

It is also possible to use other characteristic length parameters of the caustic curves than those defined in Figure 8, or to use caustic curves obtained in reference planes of opposite sign. However, in order to obtain evaluation formulas which are reliable and sufficiently accurate, it is advantageous to evaluate that caustic and to select that length parameter between two characteristic points on the caustic which represents the largest value. Furthermore, both characteristic points should be points on the caustic curve itself. The use of a characteristic point which does not lie on the caustic, e.g. a point which is related to the boundary of the specimen, should be avoided. Due to the shadow optical imaging process such a point is not sharply reproduced and thus cannot be located with sufficient accuracy. Also some points which lie on the caustic curve may first appear as suitable characteristic points but a more detailed consideration shows that they are not appropriate, e.g. the cusps of the caustic curves in Figure 8. The plots of the light ray distributions (Fig. 9) indicate that these points can be experimentally located only with reduced accuracy due to the focal-point-character of the cusps. The characteristic length parameters D defined in Figures 8a,b,c have been selected on the basis of the above discussed criteria.

With transparent materials that are optically anisotropic ($\lambda \neq 0$) the caustic curve splits up into a double caustic. Figure 10 shows crack tip caustics for the optically anisotropic material Araldite B in comparison to the corresponding single caustic. Since the coefficient of anisotropy λ depends on the state of stress ($|\lambda|$ is larger for plane stress than for plane strain, see Table 1) the split-up of the double caustic is larger for plane stress then for plane strain.

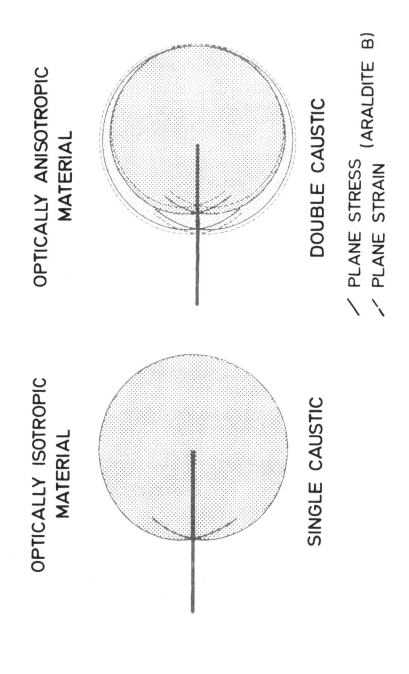

Fig. 10   Single and double caustic

The outer as well as the inner caustic can be used for determining the stress intensity factor. The evaluation formula for the crack tip stress intensity factor then is

$$K_I = \frac{2\sqrt{2\pi}}{3\ f_{o,i}^{5/2}\ z_o\ c\ d_{eff}}\ D_{o,i}^{5/2} \tag{15}$$

with $D_{o,i}$ = outer, inner diameter of the crack tip caustic

$f_{o,i}$ = numerical factor for evaluation of the outer, inner caustic.

Numerical values of the factors $f_{o,i}$ which lie around the value 3.17 for the single caustic are given in Figure 11 as a function of the aniso-tropy coefficient λ. Anisotropic evaluation formulas for the other two examples considered (see Fig. 7) can be derived in an analogous manner.

4. CRACK TIP CAUSTICS

Most applications of the shadow optical method of caustics have so far been in the field of fracture mechanics. Consequently the theory of caustics around crack tips has been developed to the greatest degree. Some of the results that are of more general interest and that can be transferred to other caustic problems as well are presented in this Chapter.

4.1 Mode I, Mode II, and Mode III Caustics

Three possible loading conditions exist for cracks (see Fig. 12), i.e. tensile loading (mode I), in-plane shear loading (mode II), and anti-plane shear loading (mode III). Any arbitrary loading condition can be represented by a superposition of these three fundamental types of loading.

Since both in-plane and out-of-plane loading states are discussed it is necessary to consider the complete stress and displacement fields in all three directions.[13,14] These are given by

$$\text{Mode I} \qquad \sigma_r = \frac{K_I}{\sqrt{2\pi r}} \; \frac{1}{4} \left\{ 5 \cos \tfrac{1}{2}\varphi - \cos \tfrac{3}{2}\varphi \right\}$$

$$\sigma_\varphi = \frac{K_I}{\sqrt{2\pi r}} \; \frac{1}{4} \left\{ 3 \cos \tfrac{1}{2}\varphi - \cos \tfrac{3}{2}\varphi \right\}$$

$$\tau_{r\varphi} = \frac{K_I}{\sqrt{2\pi r}} \; \frac{1}{4} \left\{ \sin \tfrac{1}{2}\varphi + \sin \tfrac{3}{2}\varphi \right\}$$

$$\sigma_z = \nu \, (\sigma_r + \sigma_\varphi), \quad \tau_{rz} = \tau_{\varphi z} = 0$$

$$u = \frac{K_I}{G} \; \sqrt{\frac{r}{2\pi}} \left\{ \cos \tfrac{1}{2}\varphi \, (1-2\nu + \sin^2 \tfrac{1}{2}\varphi) \right\}$$

$$v = \frac{K_I}{G} \; \sqrt{\frac{r}{2\pi}} \left\{ \sin \tfrac{1}{2}\varphi \, (2-2\nu - \cos^2 \tfrac{1}{2}\varphi) \right\}$$

$$w = 0$$

Mode II $\quad \sigma_r = \dfrac{K_{II}}{\sqrt{2\pi r}} \dfrac{1}{4} \left\{ -5 \sin \dfrac{1}{2}\varphi + 3 \sin \dfrac{3}{2}\varphi \right\}$

$\quad\quad\quad\quad \sigma_\varphi = \dfrac{K_{II}}{\sqrt{2\pi r}} \dfrac{1}{4} \left\{ -3 \sin \dfrac{1}{2}\varphi - 3 \sin \dfrac{3}{2}\varphi \right\}$

$\quad\quad\quad\quad \tau_{r\varphi} = \dfrac{K_{II}}{\sqrt{2\pi r}} \dfrac{1}{4} \left\{ \cos \dfrac{1}{2}\varphi + 3 \cos \dfrac{3}{2}\varphi \right\}$

$\quad\quad\quad\quad \sigma_z = \nu\,(\sigma_r + \sigma_\varphi), \quad \tau_{rz} = \tau_{\varphi z} = 0 \quad\quad\quad (16)$

$\quad\quad\quad\quad u = \dfrac{K_{II}}{G} \sqrt{\dfrac{r}{2\pi}} \left\{ \sin \dfrac{1}{2}\varphi \left(2-2\nu + \cos^2 \dfrac{1}{2}\varphi \right) \right\}$

$\quad\quad\quad\quad v = \dfrac{K_{II}}{G} \sqrt{\dfrac{r}{2\pi}} \left\{ \cos \dfrac{1}{2}\varphi \left(-1+2\nu + \sin^2 \dfrac{1}{2}\varphi \right) \right\}$

$\quad\quad\quad\quad w = 0$

Mode III $\quad \tau_{xz} = -\dfrac{K_{III}}{\sqrt{2\pi r}} \sin \dfrac{1}{2}\varphi$

$\quad\quad\quad\quad \tau_{yz} = +\dfrac{K_{III}}{\sqrt{2\pi r}} \cos \dfrac{1}{2}\varphi$

$\quad\quad\quad\quad w = \dfrac{K_{III}}{G} \sqrt{\dfrac{2r}{\pi}} \sin \dfrac{1}{2}\varphi$

where G is the shear modulus of the material,

$$G = \frac{E}{2(1+\nu)}$$

u, v, and w are the displacements in x, y, and z direction and K_I, K_{II}, K_{III} are the stress intensity factors for mode I, mode II, and mode III, respectively.

Since the equations for the stress fields for the mode I and the mode II cases are similar in form the procedure for obtaining mode II caustics is

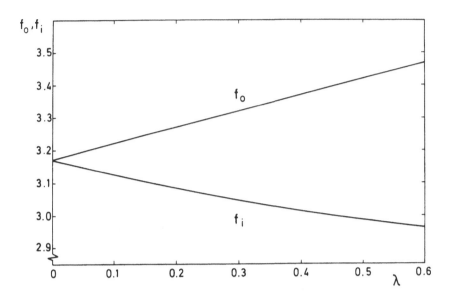

Fig. 11   Numerical factors for evaluating anisotropic crack tip caustics

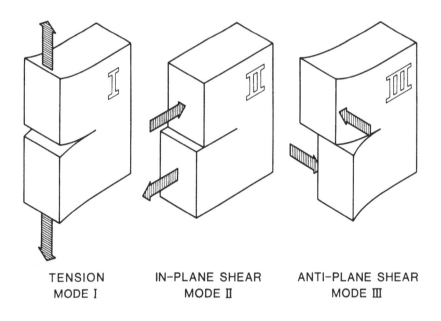

TENSION   IN-PLANE SHEAR   ANTI-PLANE SHEAR
MODE I   MODE II   MODE III

Fig. 12   Fundamental crack tip loading conditions

analogous to the one described in the previous Chapter for the mode I
case. For mode III, however, the in-plane stresses and displacements are
zero and changes in the optical path length according to equation (7)
would be zero. Physically, the out-of-plane loading does not produce
changes in the thickness of the plate or (for the transmission case) in
the refractive index of the material. Thus, a shadow optical effect on
the basis of the previously discussed considerations does not result. The
out-of-plane displacement w(x,y), however, can be utilized for generating
a shadow optical image, but only in reflection. The change in optical
path length in this case is simply given by the equation

$$\Delta s = -2w(x,y) \tag{17}$$

A shadow optical effect due to the out-of-plane displacement w does not
result for transmission arrangements, since the surfaces of the plate although nonplanar remain parallel. Consequently light rays traversing the
plate are not deflected but only slightly displaced.

The mapping equations for the three fracture modes, therefore, are

Mode I $x' = r \cos\varphi + \dfrac{K_I}{\sqrt{2\pi}} z_o \, c \, d_{eff} \, r^{-3/2} \cos\dfrac{3}{2}\varphi$

$\qquad\quad y' = r \sin\varphi + \dfrac{K_I}{\sqrt{2\pi}} z_o \, c \, d_{eff} \, r^{-3/2} \sin\dfrac{3}{2}\varphi$

Mode II $x' = r \cos\varphi - \dfrac{K_{II}}{\sqrt{2\pi}} z_o \, c \, d_{eff} \, r^{-3/2} \sin\dfrac{3}{2}\varphi$

$\qquad\quad y' = r \sin\varphi + \dfrac{K_{II}}{\sqrt{2\pi}} z_o \, c \, d_{eff} \, r^{-3/2} \cos\dfrac{3}{2}\varphi$ $\tag{18}$

Mode III $x' = r \cos\varphi + 2\dfrac{K_{III}}{\sqrt{2\pi}} \dfrac{z_o}{G} r^{-1/2} \sin\dfrac{1}{2}\varphi$

$\qquad\quad y' = r \sin\varphi - 2\dfrac{K_{III}}{\sqrt{2\pi}} \dfrac{z_o}{G} r^{-1/2} \cos\dfrac{1}{2}\varphi$

For simplicity only the isotropic case is considered in this context. Application of equation (10) to these mapping equation gives the equation of the initial curves

$$\text{Mode I} \qquad r = \left[\frac{3}{2} \frac{K_I}{\sqrt{2\pi}} |z_o| |c| \, d_{eff} \right]^{2/5} \equiv r_o$$

$$\text{Mode II} \qquad r = \left[\frac{3}{2} \frac{K_{II}}{\sqrt{2\pi}} |z_o| |c| \, d_{eff} \right]^{2/5} \equiv r_o \qquad (19)$$

$$\text{Mode III} \qquad r = \left[\frac{K_{III}}{\sqrt{2\pi}} \frac{|z_o|}{G} \right]^{2/3} \equiv r_o$$

As in the examples previously discussed, the initial curves are circles around the origin with fixed radius r_o. With the mapping equations (18) the caustic curves are obtained as images of the initial curves, equations (19). The caustic equations are

$$\text{Mode I} \qquad x' = r_o \left(\cos \varphi + sgn(z_o c) \, \frac{2}{3} \cos \frac{3}{2}\varphi \right)$$

$$y' = r_o \left(\sin \varphi + sgn(z_o c) \, \frac{2}{3} \sin \frac{3}{2}\varphi \right)$$

$$\text{Mode II} \qquad x' = r_o \left(\cos \varphi + sgn(z_o c) \, \frac{2}{3} \sin \frac{3}{2}\varphi \right)$$

$$\qquad\qquad\qquad\qquad\qquad\qquad\qquad\qquad\qquad\qquad\qquad (20)$$

$$y' = r_o \left(\sin \varphi + sgn(z_o c) \, \frac{2}{3} \cos \frac{3}{2}\varphi \right)$$

$$\text{Mode III} \qquad x' = r_o \left(\cos \varphi + sgn(z_o) \, 2 \sin \frac{1}{2}\varphi \right)$$

$$y' = r_o \left(\sin \varphi - sgn(z_o) \, 2 \cos \frac{1}{2}\varphi \right)$$

Graphical presentations of caustic curves for the three fracture modes and the different possible observation modes are given in Figure 13.

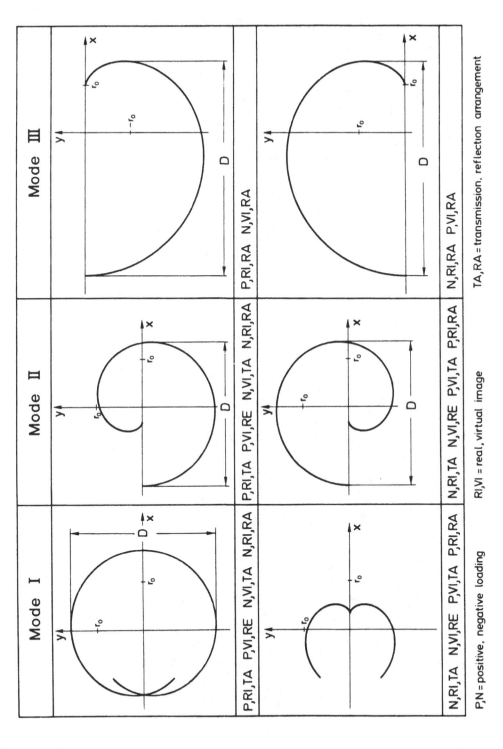

Fig. 13   Crack tip caustic curves for mode I, mode II, and mode III loading

P,N = positive, negative loading   RI,VI = real, virtual image   TA,RA = transmission, reflection arrangement

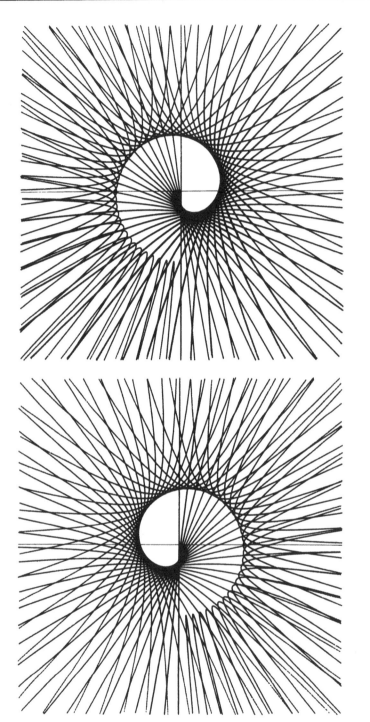

CRACK UNDER MODE–II LOADING

Fig. 14a   Shadow optical light distribution

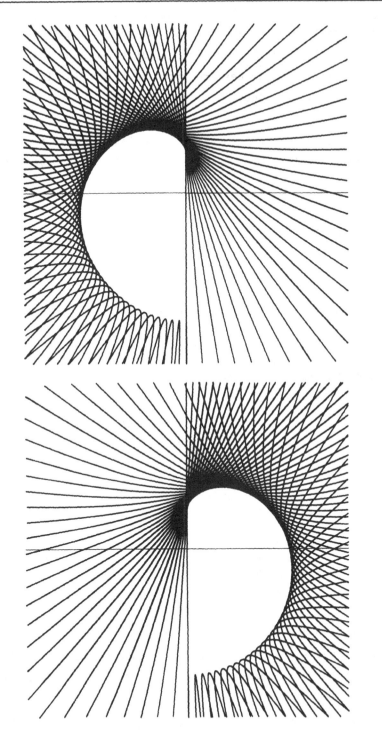

CRACK UNDER MODE-III LOADING

Fig. 14b   Shadow optical light distribution

In contrast to the mode I caustics the mode II and the mode III caustics are asymmetric. Graphical presentations of light distributions for the mode II and the mode III shadow images are given in Figure 14. For the mode I light distributions see Figure 9.

It can be easily seen that the characteristic length parameters D defined in Figure 13 are related to the radii of the initial curves by the equations

Mode I $D = 3.17\ r_o$

Mode II $D = 3.02\ r_o$ (21)

Mode III $D = 4.5\ r_o$ (for reflection only).

Consequently with equations (19) and (20) the evaluation formulas for the three fracture modes become

$$\text{Mode I}\qquad K_I = \frac{2\ \sqrt{2\pi}}{3\ (3.17)^{5/2}\ z_o\ c\ d_{eff}}\ D^{5/2}$$

$$\text{Mode II}\qquad K_{II} = \frac{2\ \sqrt{2\pi}}{3\ (3.02)^{5/2}\ z_o\ c\ d_{eff}}\ D^{5/2} \qquad (22)$$

$$\text{Mode III}\qquad K_{III} = \frac{\sqrt{2\pi}\ G}{(4.5)^{3/2}\ z_o}\ D^{3/2}$$

Results for cracks under mixed mode I mode II loading are obtained by a superposition of the stress field equations and consequently of the mapping equations for the pure mode I and mode II cases. Caustic curves

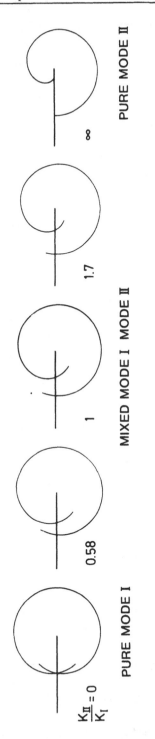

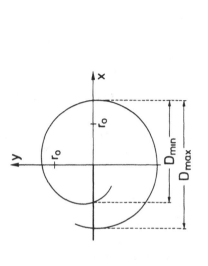

Fig. 15   Mixed mode crack tip caustics

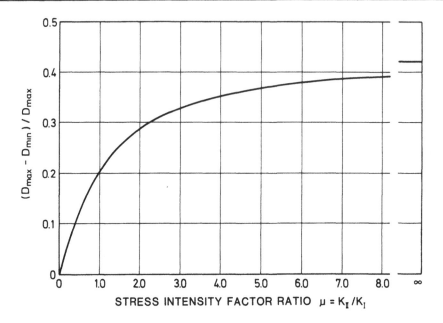

Fig. 16   Determination of the stress intensity factor ratio μ from
           mixed mode caustics

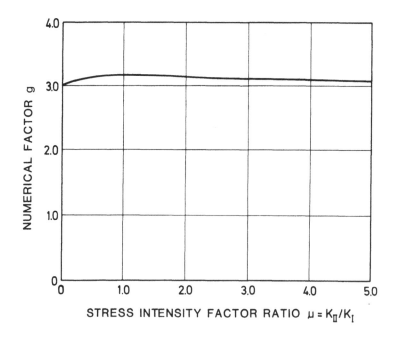

Fig. 17   Numerical factor for stress-intensity-factor-determination
           from mixed mode caustics

for different ratios of the mode II to the mode I stress intensity factor $\mu = K_{II}/K_I$ are shown in Figure 15a. Depending on the K_{II}/K_I ratio all intermediate stages between the pure mode I and the pure mode II caustics are possible.

From the two diameters, D_{max} and D_{min}, defined in Figure 15b, the two stress intensity factors K_I and K_{II} are determined.[15,16] With Figure 16, resulting from the equations (20) for the complete mode I mode II caustic curve, the stress intensity factor ratio $\mu = K_{II}/K_I$ is first determined from the measured value $(D_{max}-D_{min})/D_{max}$. According to Figure 17 this μ-value then defines the value of the numerical factor g, describing the relationship between the characteristic length parameter D_{max} and the radius r_o of the initial curve, $D_{max} = g\, r_o$. Thus, the absolute value of the mode I stress intensity factor K_I can be determined from the measured distance D_{max} as

$$K_I = \frac{2\sqrt{2\pi}}{3\, g^{5/2}\, z_o\, c\, d_{eff}}\, D_{max}^{5/2}$$

and   (23)

$$K_{II} = \mu\, K_I$$

4.2 Dynamic Caustics

Two cases of dynamic fracture problems are considered. These are: stationary cracks under dynamic loading and propagating cracks under stationary external loading.

For a stationary crack subjected to a dynamically applied load the near field stress distribution is the same as for a statically loaded crack.[17] The stress intensity factor, however, becomes a function of

time. Consequently, since the dependences from the radial distance r and
the angle φ are identical for the dynamic and the static case, the results presented in Section 4.1 also apply to dynamically loaded cracks
with the static stress intensity factors K_I, K_{II}, or K_{III} replaced
by $K_I(t)$, $K_{II}(t)$, or $K_{III}(t)$.

For a propagating crack the near field stress distribution differs
from that for a stationary crack due to inertia effects (see e.g. References 18,19).

In a coordinate system $(\bar{x},\bar{y})$ or $(\bar{r},\bar{\varphi})$ with its origin in the moving
crack tip (see Fig. 18), i.e.

$$
\begin{aligned}
x &= a(t) + \bar{x} = a(t) + \bar{r}\cos\bar{\varphi} \\
y &= \bar{y} = \bar{r}\sin\bar{\varphi}
\end{aligned}
\tag{24}
$$

the near field stress distribution for a propagating crack with an instantaneous crack velocity v is given as

$$
\sigma_x = \frac{K_I}{\sqrt{2\pi\bar{r}}}\,\frac{1+\alpha_2^2}{4\alpha_1\alpha_2-(1+\alpha_2^2)^2}\left[(1+2\alpha_1^2-\alpha_2^2)\,p(\bar{\varphi},\alpha_1) - \frac{4\alpha_1\alpha_2}{1+\alpha_2^2}\,p(\bar{\varphi},\alpha_2)\right]
$$

$$
\sigma_y = \frac{K_I}{\sqrt{2\pi\bar{r}}}\,\frac{1+\alpha_2^2}{4\alpha_1\alpha_2-(1+\alpha_2^2)^2}\left[-(1+\alpha_2^2)\,p(\bar{\varphi},\alpha_1) + \frac{4\alpha_1\alpha_2}{1+\alpha_2^2}\,p(\bar{\varphi},\alpha_2)\right]
\tag{25}
$$

$$
\tau_{xy} = \frac{K_I}{\sqrt{2\pi\bar{r}}}\,\frac{1+\alpha_2^2}{4\alpha_1\alpha_2-(1+\alpha_2^2)^2}\,\alpha_1\left[q(\bar{\varphi},\alpha_1) - q(\bar{\varphi},\alpha_2)\right]
$$

where $p(\bar{\varphi},\alpha_j) = \left[\cos\bar{\varphi} + (\cos^2\bar{\varphi} + \alpha_j^2 \sin^2\bar{\varphi})^{1/2}\right]^{1/2} / (\cos^2\bar{\varphi} + \alpha_j^2 \sin^2\bar{\varphi})^{1/2}$

$q(\bar{\varphi},\alpha_j) = \left[-\cos\bar{\varphi} + (\cos^2\bar{\varphi} + \alpha_j^2 \sin^2\bar{\varphi})^{1/2}\right]^{1/2} / (\cos^2\bar{\varphi} + \alpha_j^2 \sin^2\bar{\varphi})^{1/2}$

and $\qquad \alpha_j = (1-v^2/c_j^2)^{1/2}, \qquad j = 1,2 \qquad\qquad\qquad (27)$

with

$$c_1 = \sqrt{\frac{E}{\rho}} \sqrt{\frac{1-v}{(1+v)(1-2v)}} \qquad \text{longitudinal wave speed}$$

$$(28)$$

$$c_2 = \sqrt{\frac{E}{\rho}} \sqrt{\frac{1}{2(1+v)}} \qquad \text{transverse wave speed.}$$

With this stress distribution the mapping equations with regard to the moving coordinate systems are

$$\bar{x}' = \bar{r}\cos\bar{\varphi} + \frac{K_I}{\sqrt{2\pi}} z_o c\, d_{eff}\, \bar{r}^{-3/2}\, F^{-1}\, G_1(\alpha_1,\bar{\varphi})$$

$$(29)$$

$$\bar{y}' = \bar{r}\sin\bar{\varphi} + \frac{K_I}{\sqrt{2\pi}} z_o c\, d_{eff}\, \bar{r}^{-3/2}\, F^{-1}\, G_2(\alpha_1,\bar{\varphi})$$

where

$$F = \frac{4\alpha_1\alpha_2 - (1+\alpha_2^2)^2}{(\alpha_1^2 - \alpha_2^2)(1+\alpha_2^2)} \qquad\qquad\qquad (30)$$

and

$$G_1(\alpha_1,\bar{\varphi}) = \frac{-1}{\sqrt{2}}(g^{1/2} + \cos\bar{\varphi})^{-1/2}(g^{-1/2} - g^{-1}\cos\bar{\varphi} - 2g^{-3/2}\cos^2\bar{\varphi})$$

$$(31)$$

$$G_2(\alpha_1,\bar{\varphi}) = \frac{1}{\sqrt{2}}(g^{1/2} + \cos\bar{\varphi})^{-1/2}(\alpha_1^2 g^{-1}\sin\bar{\varphi} + \alpha_1^2 g^{-3/2}\sin 2\bar{\varphi})$$

with $\qquad\qquad\qquad g = 1 + (\alpha_1^2-1)\sin^{-2}\bar{\varphi}. \qquad\qquad (32)$

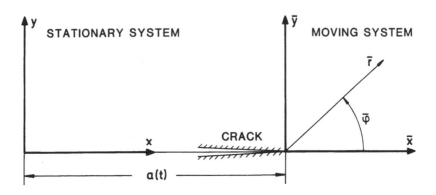

Fig. 18   Coordinate systems at the tip of a moving crack

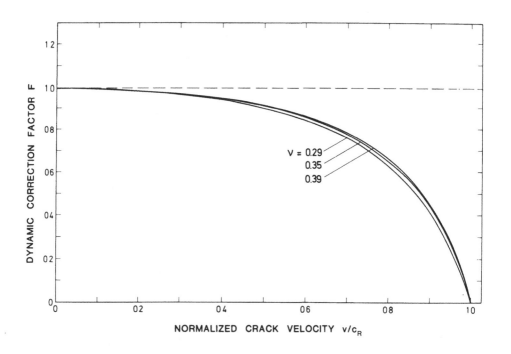

Fig. 19   Correction factor for evaluating dynamic crack tip caustics

It can be easily shown by numerical calculations (see more detailed discussion in Reference 9) that the influences of α_1 and α_2 on the functions $G_1(\alpha_1, \bar{\varphi})$ and $G_2(\alpha_1, \bar{\varphi})$ are negligibly small for all crack velocities of practical relevance (i.e. $v < 0.3c_1$). In particular, the errors made by neglecting the α_1- and α_2-dependences on G_1 and G_2 are mall in comparison to those on the factor F. Thus with an accuracy sufficient for engineering purposes these functions can be approximated by

$$G_1(\alpha_1, \bar{\varphi}) \approx \cos \frac{3}{2}\bar{\varphi}$$

$$G_2(\alpha_1, \bar{\varphi}) \approx \sin \frac{3}{2}\bar{\varphi} \tag{33}$$

With these approximations the equation for the initial curve is obtained in a manner analogous to the static considerations (see Chapter 3, eq. (11c)) as

$$\bar{r} = \left[\frac{3}{2} \frac{K_1}{2\pi} |z_o| |c| d_{eff} F^{-1} \right]^{2/5} \equiv \bar{r}_o \tag{34}$$

and consequently the evaluation formula becomes

$$K_I = \frac{2\sqrt{2\pi}\ F}{3\ (3.17)^{3/2}\ z_o\ c\ d_{eff}} D^{5/2} \tag{35}$$

Thus, the stress intensity evaluation formula for a propagating crack is the same as that for a stationary crack (eq. (22)) except for a correction factor F(v). This factor accounts for velocity effects on the r- and φ-distribution of the dynamic stress field for a propagating crack. The factor F as a function of crack velocity v is given by Figure 19. F is less than 1 but for practically relevant crack velocties it is nearly equal to 1.

Figure 20 shows crack tip caustics for different crack velocities but with a fixed stress intensity factor. The crack velocities are norma-

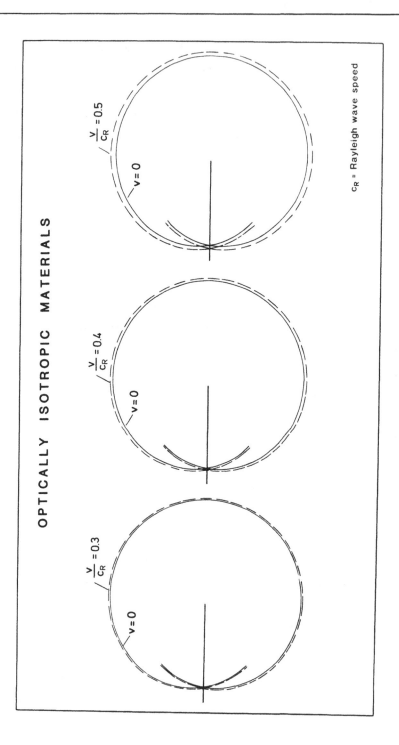

Fig. 20   Caustics for propagating cracks

lized by the Rayleigh wave speed $c_R \approx (0.862 + 1.14\nu)/(\sqrt{2}(1+\nu)^{3/2}) \sqrt{E/\rho}$.
The dynamic caustics are shown in comparison to the corresponding caustic
for a stationary crack. For all crack velocities the shape of the dynamic
caustic is practically the same as for the stationary caustic, but the
size of the caustic increases slightly with increasing velocity according
to the correction factor F.

For optically anisotropic caustics the situation becomes more complicated due to the additional anisotropy term in the equations and the
resulting formation of double caustics. It can be shown, however, that
the outer caustic increases with crack velocity in a similar manner as
the single caustic for optically isotropic materials.[9] The increase
in size of the inner caustic with crack velocity is somewhat larger than
that for the single caustic. Thus, the isotropic correction factor F(v)
in equation (30) also represents a good approximation for double caustics
if applied to the outer caustic.

4.3 The Influence of Higher Order Effects on Caustics

So far only the near field stress distributions around crack tips
have been considered. These stress field solutions are strictly valid as
$r \rightarrow o$, i.e. the range of applicability of these solutions is restricted
to the direct vicinity of the crack tip. In the intermediate and far
field regions additional higher order terms of the stress field solution
will have to be included. Since the caustic is the image of the initial
curve with finite radius r_o around the crack tip, these higher order
terms may have an influence on the formation of the caustic. In general
no practical problems are encountered for long cracks in large specimens.
However, for short cracks and for cracks approaching the boundary of the
specimen, the possible influence of higher order terms must be considered.
The complete solution of the crack tip stress field is given by a series
of terms, with the first order term representing the near field stress
distribution. For tensile (mode I) loading the first six terms are[20]

$$\sigma_r = \frac{K_I}{\sqrt{2\pi r}}\left\{\frac{5}{4}\cos\frac{1}{2}\varphi - \frac{1}{4}\cos\frac{3}{2}\varphi\right\} + a_2\cos^2\varphi$$

$$+ r^{1/2}\, a_3\left\{\frac{3}{4}\cos\frac{1}{2}\varphi + \frac{1}{4}\cos\frac{5}{2}\varphi\right\} + a_4\left\{\cos\varphi + 3\cos 3\varphi\right\}$$

$$+ r^{3/2}\, a_5\left\{\frac{5}{4}\cos\frac{3}{2}\varphi + \frac{15}{4}\cos\frac{7}{2}\varphi\right\} + r^2 a_6\left\{\frac{7}{4}\cos\frac{3}{2}\varphi + \frac{33}{4}\cos\frac{7}{2}\varphi\right\}$$

$$\sigma_\varphi = \frac{K_I}{\sqrt{2\pi r}}\left\{\frac{3}{4}\cos\frac{1}{2}\varphi + \frac{1}{4}\cos\frac{3}{2}\varphi\right\} + a_2\sin^2\varphi \qquad (36)$$

$$+ r^{1/2}\, a_3\left\{\frac{5}{4}\cos\frac{1}{2}\varphi - \frac{1}{4}\cos\frac{5}{2}\varphi\right\} + r\, a_4\left\{\cos\varphi - \cos 3\varphi\right\}$$

$$+ r^{3/2}\, a_5\left\{\frac{35}{4}\cos\frac{3}{2}\varphi - \frac{15}{4}\cos\frac{7}{2}\varphi\right\} + r^2 a_6\left\{\frac{48}{4}\cos\frac{3}{2}\varphi - \frac{48}{4}\cos\frac{7}{2}\varphi\right\}$$

$$\tau_{r\varphi} = \frac{K_I}{\sqrt{2\pi r}}\left\{\frac{1}{4}\sin\frac{1}{2}\varphi - \frac{1}{4}\sin\frac{3}{2}\varphi\right\} - a_2\frac{1}{2}\sin 2\varphi$$

$$+ r^{1/2}\, a_3\left\{\frac{1}{4}\sin\frac{1}{2}\varphi - \frac{1}{4}\sin\frac{5}{2}\varphi\right\} + r\, a_4\left\{\sin\varphi - 3\sin\varphi\right\}$$

$$+ r^{3/2}\, a_5\left\{\frac{15}{4}\sin\frac{3}{2}\varphi - \frac{15}{4}\sin\frac{7}{2}\varphi\right\} + r^2 a_6\left\{\frac{18}{4}\sin\frac{3}{2}\varphi - \frac{42}{4}\sin\frac{7}{2}\varphi\right\}$$

In general, the most important term in expanding the range of applicability of a first order solution is the next following term, i.e. the second order term. This term represents a constant stress in x-direction, $\sigma_x = a_2$, which is superimposed to the singular near field stress distribution. However, the shadow optical effect is sensitive to stress gradients only. Constant stresses do not give rise to changes in the optical path length (see eqs. (1), (2) and (7)). Thus, the second order term does not have any influence on the formation of the caustic. In practical applications this can be a significant advantage of the shadow optical method of caustics over other stress optical techniques (e.g. photoelastic, moiré, or holographic techniques) which directly measure stresses or strains. Figure 21 illustrates the influence of the second order term on the shadow optical and the photoelastic pictures of a crack tip. For single-edge-notch (SEN) specimens the sign of the coefficient a_2 is negative causing the isochromatic fringes in the photoelastic picture to lean forward, whereas in double-cantilever-beam (DCB) specimens the sign of a_2

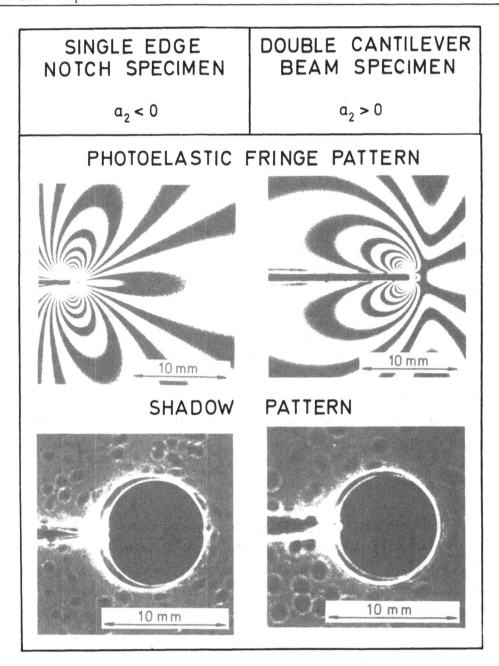

Fig. 21   Influence of the second order term of the crack tip stress field
           equations on photoelastic fringes and shadow optical caustics

is positive so that the isochromatic fringes lean backward. These changes
in the isochromatic patterns can cause severe difficulties in the evaluation procedure of photoelastic crack tip fringe patterns. The shadow
optical pictures, however, are identical, independent of the sign of the
second order coefficient.

Terms of orders higher than two have an influence on the formation
of the caustic. Figure 22 shows the resulting changes by including the
next four higher order terms. For arbitrary values of the individual
coefficients and both signs, i.e. positive and negative coefficients,
changes in the shape of the caustics are shown in comparison to the near
field caustic. Since in practice the coefficients of the higher order
terms are generally not known, a quantitative estimate of the influence
of higher order effects on the caustics is difficult to make. For practical applications it will be appropriate to compare the shape of the experimentally observed caustic with the theoretically calculated near
field caustic. Deviations of a significant amount would be an indication
of the nonnegligible influence of the higher order terms. In these cases
caution must be used in evaluating stress intensity factors from such
caustics.

4.5 Elastic-Plastic Caustics

For cracks in ductile materials, e.g. structural steels, the linear
theory of elasticity is not applicable any more. The crack tip stress
field is described by elastic-plastic relations. The stresses are bounded
by the yield stress, and consequently the strains near the crack tip become unbounded when the crack tip is approached. Thus, instead of an
elastic stress concentration problem a plastic strain concentration problem is encountered. But analogous to elastic stress concentrations, caustics also result for plastic strain concentrations.[6,7]

With regard to an application to steels the following consideration
applies to the reflection case only. Due to the deformation of the spe-

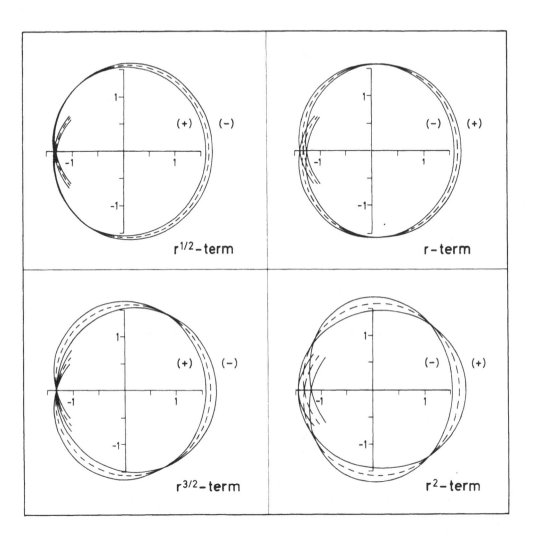

Fig. 22   Influence of higher order terms of the crack tip
stress field equations on the caustic curve

cimen surface the change in the optical path length of light rays according to equation (3) is given by

$$\Delta s = - \Delta d \tag{37}$$

where d is the total thickness of the specimen. The reduction in specimen
thickness is obtained from the plastic strain ε as

$$\Delta d = \varepsilon_{zz} \, d \tag{38}$$

Elastic strains ε^e are neglected relative to the plastic strains ε^p,
since $\varepsilon^e \ll \varepsilon^p$. Due to the incompressibility of the material under
plastic deformation (constancy of volume)

$$\varepsilon_{zz} = - (\varepsilon_{rr} + \varepsilon_{\varphi\varphi}) \tag{39}$$

Consequently the mapping equations (1), (2) take the form

$$\vec{r'} = \vec{r} - z_o \, d \, \overrightarrow{\text{grad}} \, (\varepsilon_{rr} + \varepsilon_{\varphi\varphi}) \tag{40}$$

The strains in the near field region around a crack tip have been
calculated by Hutchinson, Rice, Rosengren[21-23]. For an elastic-perfectly-
plastic material, i.e. a non hardening material with n $\rightarrow \infty$ (see Fig. 23),
the strains are unbounded in a fan region ($-79.7° < \varphi < +79.7°$) ahead
of the crack tip (see Fig. 24). Outside this fan region the strains are
bounded by finite values. Quantitatively the strains are given by the
following expressions

$$\left.\begin{array}{l} \varepsilon_{rr} = 0 \\[2mm] \varepsilon_{\varphi\varphi} = \dfrac{J}{2.57 \, \sigma_o \, r} \dfrac{3}{2} \cos \varphi \end{array}\right\} \quad -79.7° < \varphi < +79.7°$$

$$\left.\begin{array}{l} \varepsilon_{rr} = \text{bounded} \\[2mm] \varepsilon_{\varphi\varphi} = \text{bounded} \end{array}\right\} \quad \begin{array}{l} -180° < \varphi < -79.7° \\[1mm] +79.9° < \varphi < +180° \end{array} \tag{41}$$

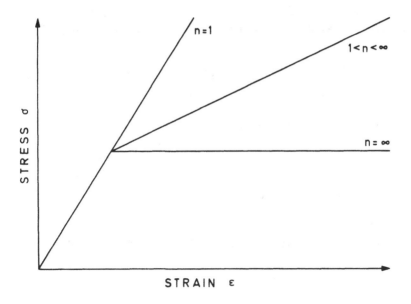

Fig. 23   Elastic-plastic stress strain behavior

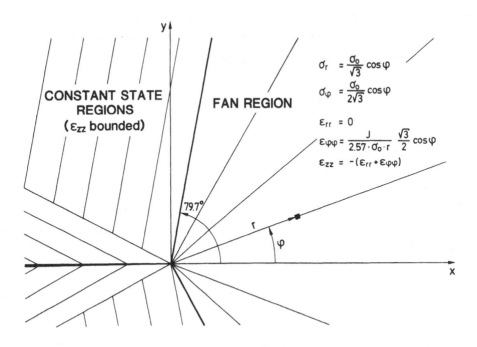

Fig. 24   Elastic-plastic stress strain field at the tip of a crack,
           HRR-field

where J is the so-called J-Integral and σ_o is the tensile yield stress.

With equations (41) the mapping equations (40) become

$$x' = r \cos \varphi + \frac{J z_o d \sqrt{3}}{2.57 \sigma_o 2} r^{-2} \cos \varphi$$

$$-79.7^\circ < \varphi < +79.7^\circ \quad (42)$$

$$y' = r \sin \varphi + \frac{J z_o d \sqrt{3}}{2.57 \sigma_o 2} r^{-2} \sin \varphi$$

The bounded strains outside the fan region $-79.7^\circ < \varphi < +79.7^\circ$ cannot contribute to the caustic. Application of equation (10) yields the initial curve

$$r = \left[\frac{J |z_o| d \sqrt{3}}{2.57 \sigma_o} \right]^{1/3} \equiv r_o \quad (43)$$

which again is a circle of fixed radius r_o around the crack tip. The mapping of the initial curve (43) onto the reference plane by equations (42) gives the equations of the caustic curve

$$x' = r_o (\cos \varphi + \text{sgn}(z_o) \frac{1}{2} \cos 2\varphi)$$

$$-79,7^\circ < \varphi < +79,7^\circ \quad (44)$$

$$y' = r_o (\sin \varphi + \text{sgn}(z_o) \frac{1}{2} \sin 2\varphi)$$

The caustic curve in a reference plane $z_o > 0$ is graphically shown in Figure 25. Analogous to the elastic case the characteristic length parameter is defined as the maximum diameter D of the caustic in y-direction which is related to the radius of the initial curve by

$$D = 2.38 \ r_o. \quad (45)$$

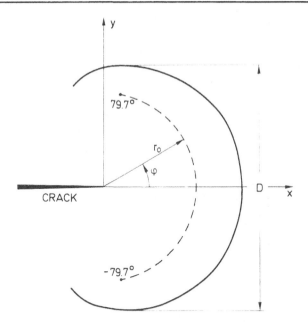

Fig. 25   Elastic-plastic crack tip caustics (n = ∞)

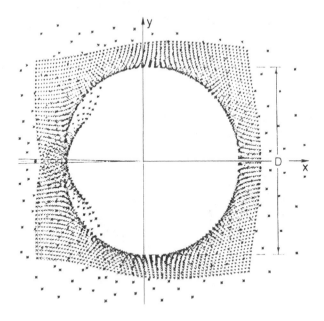

Fig. 26a   Shadow optical light distribution for a crack
in an elastic-plastic material, (n = 1)

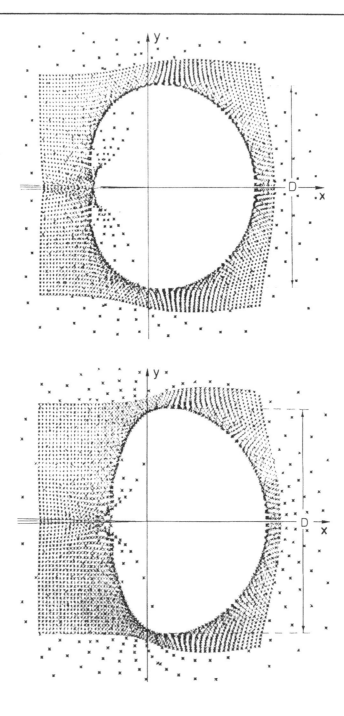

Fig. 26b,c   Shadow optical light distribution for a crack
              in an elastic-plastic material, (n = 3, 9)

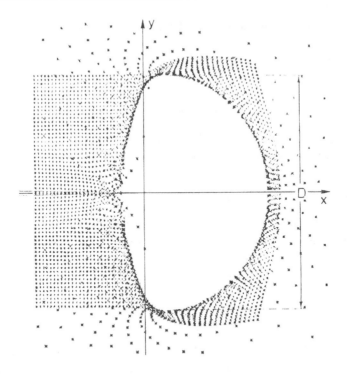

Fig. 26d   Shadow optical light distribution for a crack
            in an elastic-plastic material, (n = 25)

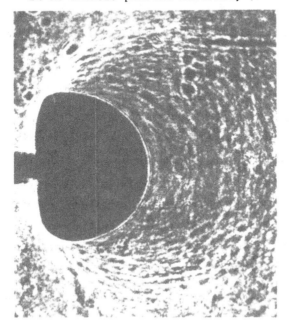

Fig. 27   Experimentally observed elastic-plastic shadow pattern

With equations (43) and (45) one obtains

$$J = \frac{\sigma_o}{13.5 \, z_o \, d} \, D^3 \tag{46}$$

This equation allows the determination of the J-integral value from elastic-plastic crack tip caustics generated by an initial curve located well within the elastic-plastic stress strain field around the crack tip.

For strain hardening materials, $1 < n < \infty$ (see Fig. 23), the strain fields become more complicated and the caustics cannot be calculated analytically any more. A numerical calculation of the deflections of an array of light rays around the crack tip, however, gives an illustrative picture of the light distribution in the reference plane. Results obtained for $n = 25, 9, 3$, and 1 are shown in Figure 26.[7] The caustic curve becomes visible as the boundary curve between areas of zero density (shadow area) and high density (light concentration).

It can be seen that the caustic curve for $n = 25$ resembles the analytically determined caustic curve for nonhardening materials, $n = \infty$. Furthermore, for the limiting case $n = 1$, i.e. an elastic material, the obtained caustic curve is identical to the elastic crack tip caustic curve. The caustic curves for $n = 3$ and $n = 9$ represent intermediate stages between these two limiting cases.

Figure 27 shows an experimentally observed elastic-plastic shadow pattern photographed with a tool steel showing a low rate of strain hardening.[7] In spite of the formation of slide lines which necessarily disturb the reflection conditions of the polished specimen surface the shadow pattern is of high quality. The resulting caustic curve is in good agreement with the theoretically derived caustic curve and demonstrates the applicability of the shadow optical method of caustics for investigating plastic strain concentration problems.

5. EXPERIMENTAL TECHNIQUES

Experimental arrangements used for shadow optical investigations are generally quite simple. Special sophisticated equipment is not needed. The only essentials are a suitable light beam for illumination of the specimen and a device for recording the shadow optical patterns.

5.1 Illumination of Specimens and Recording of Shadow Patterns

In the description of the physical principle of the shadow optical method a parallel light beam has been considered. Shadow optical pictures, however, can also be obtained with divergent or convergent light beams. The parallel light beam has only been considered for simplicity reasons in the theoretical considerations. In practice parallel light beams are seldom used. Shadow optical arrangements, with divergent, parallel, and convergent light beams are schematically shown in Figure 28. The illustrations show the transmission case but they apply in an analogous manner to the reflection case also. Regardless of what arrangement is utilized the light beam has to fulfill only one, but very stringent requirement: the light beam has to be exceptionally parallel, divergent, or convergent. In order to achieve this property the light source must have the essential features of a point source, i.e. a small aperture and a large distance from the object. If these conditions are not sufficiently fulfilled the shadow optical pattern gets a blurry "non-sharp" appearance. The boundary line between the regions of darkness and light concentration, i.e. the caustic curve, is not represented by a distinctly marked line any more. Thus, the quantitative evaluation of shadow patterns would become difficult or even erroneous.

Real caustics can be directly recorded on a photographic film positioned in the reference plane. This direct recording of shadow patterns is possible in transmission arrangements as well as in reflection arrangements. In the latter case the light beam has to be slightly tilted with regard to the normal of the specimen to get a separation of the reflected beam from the impinging beam. The direct recording of shadow patterns, however, is seldom used in practice. Mostly shadow patterns are recorded with a photographic camera (e.g. a conventional 35 mm camera). When shadow optical arrangements with divergent or parallel light beams are utilized, an additional field lens is needed to focus the light into the lens of the camera. In the case of convergent light beams, the camera is positioned directly in the focal point of the light beam. Recording arrangements for divergent, parallel, and convergent light beams are shown together with the respective illumination arrangements in Figure 28. For the registration of shadow optical pictures in a certain reference plane the photographic camera is focused on this special plane. The use of a screen positioned in the reference plane is not needed. The recording of shadow patterns with a photographic camera has the advantage in that not only real but also virtual shadow patterns can be registered. In the latter case it is only necessary to focus the camera on the respective virtual reference plane. (For simplicity reasons the virtual image planes are not marked in Fig. 28).

5.2 Scaling-Factor for Non-Parallel Light Beams

When shadow optical arrangements with divergent or convergent light rays are utilized the mapping equations and the caustic curves for parallel light beams do not apply any more in their simple form and have to be modified. A non-deflected ray of a non-parallel light beam traversing the object plane at the point $P(\vec{r})$ would not hit the image plane at the point P_{nd} with $\vec{r}_{nd} = \vec{r}$, as is the case for a parallel beam (see Section 3.1, Fig. 5), but it would hit the image plane at P_{nd} with $\vec{r}' = m\,\vec{r}$ (see Fig. 29)

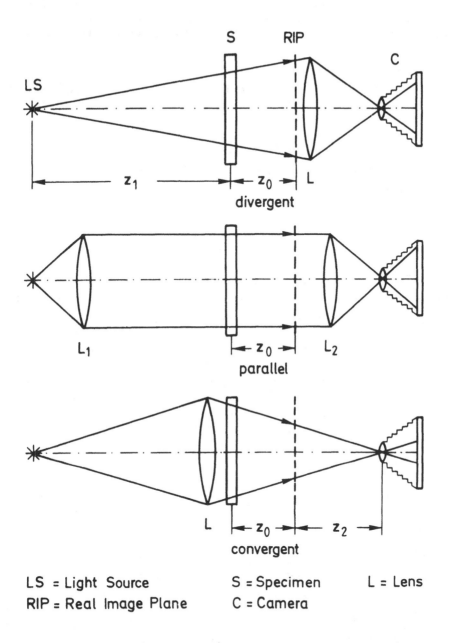

LS = Light Source   S = Specimen   L = Lens
RIP = Real Image Plane   C = Camera

Fig. 28   Shadow optical arrangements with divergent, parallel,
           or convergent light beams

where
$$m = \frac{z_o + z_1}{z_1} > 1 \qquad \text{for divergent light beams}$$

(47)

$$m = \frac{z_2}{z_o + z_2} < 1 \qquad \text{for convergent light beams}$$

with the distances z_0, z_1, and z_2 defined in Figure 28. Consequently the basic shadow optical mapping equation (1) for non-parallel light beams has to be modified to

$$\vec{r}' = m \, \vec{r} + \vec{w} \tag{48}$$

With this modified mapping equation the relations between the size of the shadow pattern and the respective generating load parameter for the three examples considered in Section 3.2 (see Fig. 7) are

a)
$$P = \frac{1}{m^2} \frac{\pi}{4 \, (2.6)^3 \, z_o \, c \, d_{eff}} D^3$$

b)
$$p-q = \frac{1}{m^3} \frac{1}{12 \, (2.66)^4 \, z_o \, c \, d_{eff} \, R^2} D^4 \tag{49}$$

c)
$$K_I = \frac{1}{m^{3/2}} \frac{2 \sqrt{2\pi}}{3 \, (3.17)^{5/2} \, z_o \, c \, d_{eff}} D^{5/2}$$

As can be easily seen, for divergent and parallel arrangements the characteristic length parameter D steadily increases in size with increasing distance z_o between specimen and reference plane. This is not true any more for convergent arrangements if large distances z_o are utilized, since the scaling factor m for this arrangement decreases with increasing z_o. For a fixed distance z_o+z_2 between the specimen and the photographic camera the largest caustic is obtained in a reference plane with a distance z_o given by

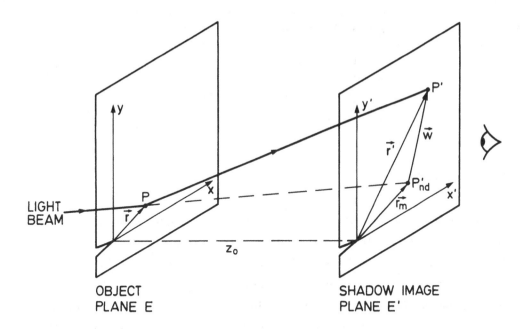

Fig. 29   Light ray deflection for a non-parallel light beam

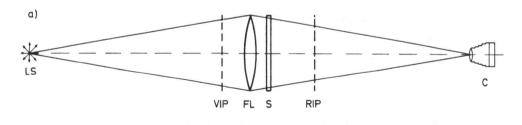

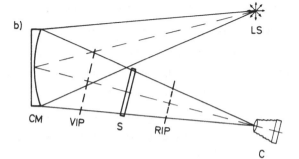

SYMBOLS

LS   LIGHT SOURCE

C   CAMERA

S   SPECIMEN

FL   FIELD LENS

CM   CONCAVE MIRROR

RIP   REAL IMAGE PLANE

VIP   VIRTUAL IMAGE PLANE

Fig. 30   Shadow optical arrangements utilizing a field lens (a)
           or a concave mirror (b)

a) $$z_o = \frac{1}{3} (z_o + z_2)$$

b) $$z_o = \frac{1}{4} (z_o + z_2)$$ (50)

c) $$z_o = \frac{2}{5} (z_o + z_2) \, .$$

5.3 Instrumentation

Shadow optical arrangements with convergent light beams are mostly utilized in practice since they require only one lens and since the view field obtained with this lens is largest. Instead of lenses some shadow optical arrangements use concave mirrors. Figure 30 shows a shadow optical arrangement with a concave mirror in comparison to an arrangement with a lens. Concave mirrors have the advantage that they are available with larger apertures and larger focal lengths then lenses. Thus, larger view fields can be obtained. Furthermore, large distances between the light source and the specimen can be realized. As a consequence, the quality of the illuminating light beam is improved and the distance z_o between the specimen and the reference plane can be varied over a larger range (see also Section 5.2).

All arrangements using mirrors as well as reflection arrangements with non-transparent specimens require light beams which are slightly tilted with regard to the optical axis. Consequently the influence of astigmatism due to the tilting of the light beams has to be considered. Errors in the evaluation procedure are minimized if the following pre-cautions are taken: Optical arrangements should be used for which the direction of the characteristic length parameter of the caustic is per-pendicular to one of the focal lines. This focal line and the correspon-ding focal length should be used in the evaluation procedure.

In order to experimentally determine the absolute size of the caustic curve which is recorded with a photographic camera it is appropriate to photograph together with the caustic a scale which is mounted on the specimen. The length parameter D of the caustic is then easily determined by a comparison with the scale. Since a shadow optical image is photographed the scale is not sharply reproduced, but the symmetry of the obtained line patterns of the scale nevertheless allows a sufficiently accurate evaluation. With convergent (divergent) arrangements the size of the caustic thus determined is too large (small) since the caustic exists in the reference plane whereas the scale is positioned in the object plane. Consequently the real size of the caustic is obtained by multiplication with the scaling factor m.

The application of the shadow optical method for investigating dynamic processes requires the recording of time dependent shadow patterns with high speed cameras. Figure 30 shows the experimental arrangements of a Cranz–Schardin 24 spark high speed camera for recording shadow patterns in transmission or in reflection. Basically the Cranz–Schardin system consists of a photographic recording unit which contains 24 individual cameras with optical axes that meet in the object. The light for illuminating the specimen is generated by 24 sparks which are assembled in a special light unit. The light emitted from one individual spark is focused into the lens of the corresponding individual camera via a lens or a mirror, the latter is used in Figure 30. The sparks are triggered at different pre-determined times (minimum picture interval time = 1µs) via an electronic control unit. For simplicity only two of the total 24 light beams are shown in Figure 30. A picture of the experimental set-up is given in Figure 31. For further details of the Cranz Schardin system see the Appendix (Chapter 9).

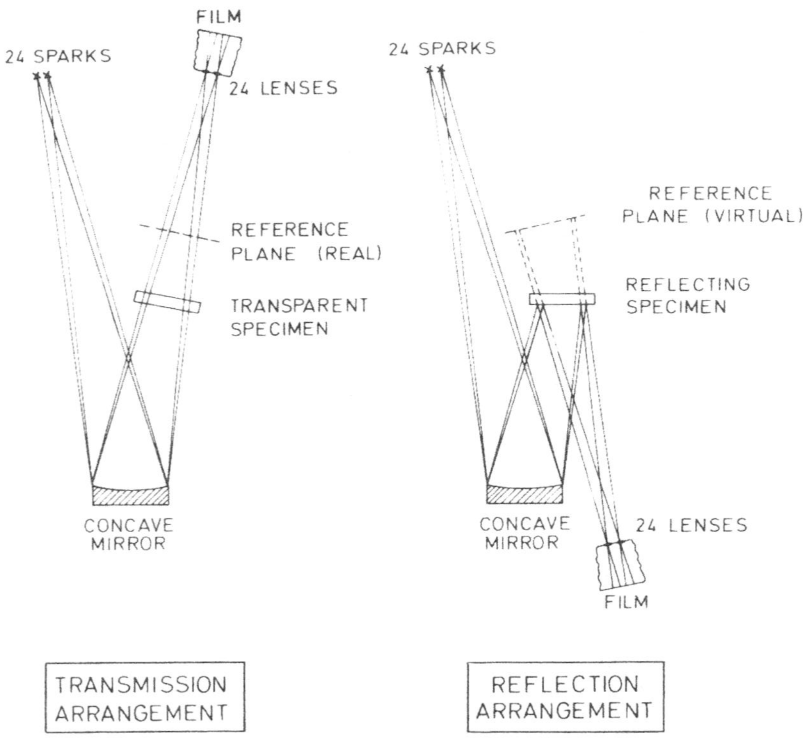

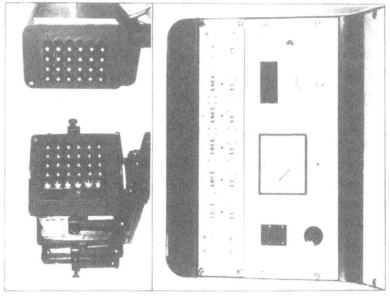

Fig. 32   IWM Cranz–Schardin 24 spark
           high speed camera

Fig. 31   Cranz–Schardin 24 spark high speed camera
           in transmission and reflection arrangement

5.4 Model Materials and Preparation of Specimens

Specimens for shadow optical investigations have to meet relatively high
standards concerning their optical quality. In particular care must be
taken that the specimens used in transmission arrangements are made from
materials that are free of local variations in density and in thickness.
Specimens used in reflection arrangements, on the other hand, must have
an optically planar surface at the illuminated side. If these conditions
are not sufficiently fulfilled then disturbing influences on the shadow
patterns may appear resulting in erroneous interpretations.

Various model materials are available for machining transparent specimens (see Table 1): Polymethylmethacrylat (PMMA) has a high shadow optical constant c. It is almost optically isotropic ($\lambda \approx 0$) and thus
yields single caustic curves. The material, however, exhibits strong
visco elastic effects, so that for time dependent loads its relaxation
behavior must be considered. For example, shadow patterns obtained with
PMMA do not immediately disappear when the applied load is removed. Depending on the time during which the load has been active the shadow patterns can decay over time ranges as long as several minutes. The materials
Homalite 100 and CR 39 also show a high shadow optical sensitivity and
relatively small effects of anisotropy, i.e. double caustics become only
visible in shadow optical arrangements of high resolution. Similar as
PMMA the material CR 39 has the disadvantage of being very visco elastic.
Homalite 100, however, shows considerably smaller visco elastic effects.
A material well suited for shadow optical investigations is the epoxy
resin Araldite B. The shadow optical sensitivity of the material is high,
but visco elastic effects are negligible. The material is highly anisotropic and thus exhibits clearly pronounced double caustics (see also
next Section).

In reflection arrangements the shadow optical sensitivity is controlled by the size of the surface deformation produced by the generating

load and thus is determined by the elastic properties of the material,
i.e. Youngs's modulus E and Poisson's ratio ν. The deformation at the
rear surface of the specimen and changes in the density of the material
(resulting in changes of the refractive index for transparent materials)
do not contribute to the shadow optical effect in reflection. Therefore,
reflection arrangements are generally less sensitive than transmission
arrangements which integrate over all of the effects discussed above. The
front surface of the specimen must be optically planar and mirrored, at
least in the localized area which generates the caustic. For steel specimens best results are obtained by subsequently preparing the front surface of the specimen by grinding, lapping, and polishing.

5.5 Influences of Local Plasticity and State of Stress

Stresses or strains obtained in stress concentration problems can be very
high, theoretically often unbounded. Thus in real materials plastically
deformed regions result. For the investigation of linear problems of
elasticity it is necessary, therefore, to assure that these plasticity
effects do not have a disturbing influence on the formation of the caustic. In practice this is usually achieved by utilizing shadow optical arrangements with an initial curve being larger and lying outside the plastically deformed region. The caustic then is generated from areas where
the linear elastic stress–strain–relations remain valid. This requirement
is experimentally realized by choosing a sufficiently large distance z_o
between the specimen and the reference plane so that

$$r_o > r_{pl} \tag{51}$$

where r_{pl} is the size of the plastically deformed region.

The values of the shadow optical constant c and of the anisotropy
coefficient λ are different for plane stress and plane strain (see

Table 1). Therefore, in equations (14), (15) and (22), (23), (35) for the evaluation of caustics those values must be utilized that apply for the state of stress which prevails along the initial curve. Usually this is a state of plane stress. With very high stress gradients, however, even in plates a state of plane stress cannot develop in the localized area around stress concentration points due to the very large differences in the constraints. Consequently instead of plane stress a mixed state of stress is obtained. In these cases the distance z_o between the specimen and the reference plane again must be chosen sufficiently large so that the initial curve lies outside this region of mixed state of stress

$$r_o > r_{ps} \tag{52}$$

where r_{ps} is the smallest radius around the center of stress concentration outside which a state of plane stress applies.

With optically anisotropic materials the correct choice of the distance z_o between the specimen and the reference plane can be easily verified experimentally. If the experimentally observed split-up of the double caustic, $(D_o - D_i)/D_i$, agrees with the theoretically predicted split-up for plane stress conditions then the initial curve lies outside the region of mixed state of stress, i.e. in the region of plane stress. However, if the experimentally observed split-up is smaller, then the initial curve lies within the region of mixed state of stress and an evaluation of the caustic would be erroneous.

6. APPLICATIONS

Some typical applications of the shadow optical method of caustics for investigating dynamic problems of practical interest are presented in this Chapter. All the pictures reported were photographed with a Cranz-Schardin 24 spark high speed camera. The shadow patterns were either re-corded in transmission arrangements with specimens made from the model material Araldite B or they were photographed in reflection arrangements with specimens made from the high strength maraging steel X2 NiCoMo 18 9 5*.

6.1. Dynamic Loading of Bend Specimens

The history of the load input into a specimen which is impacted by a falling knife edge is investigated. An experiment is performed with a bend specimen measuring $412 \times 72 \times 10 \text{ mm}^3$ made from the material Araldite B. The support span is 260 mm. The wedge of 5.3 kg mass impacts the specimen at a velocity of 1 m/s. The knife edge tip radius ($\rho_0 = 1$ mm) is small compared to the specimen dimensions and thus produces a point like load-ing condition. The virtual compression caustic formed around the point of impact is photographed with a transmission arrangement in a reference plane at $z_0 > 0$. A schematic illustration of the experimental set-up is given in Figure 33.

A series of shadow optical photographs is shown in Figure 34. The times given under each of the subsequent photographs are measured from the moment of first contact between the wedge and the specimen. Quanti-tative data on the impact load $P_c(t)$ which were obtained according to equation (14a) from the recorded caustics are given in Figure 35. In

---

* Produced by Stahlwerke Südwestfalen and designated HFX 760. Nominal composition: 18% Ni, 9% Co, 4.8% Mo and < 0.03% C. Heat treatment: 480°C for 4 hrs in air. This steel is similar to the American desig-nation 18 Ni maraging grade 300.

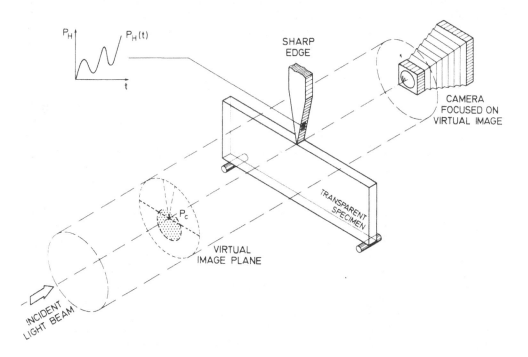

Fig. 33   Experimental arrangement for investigating the load input
into a specimen by a falling knife edge

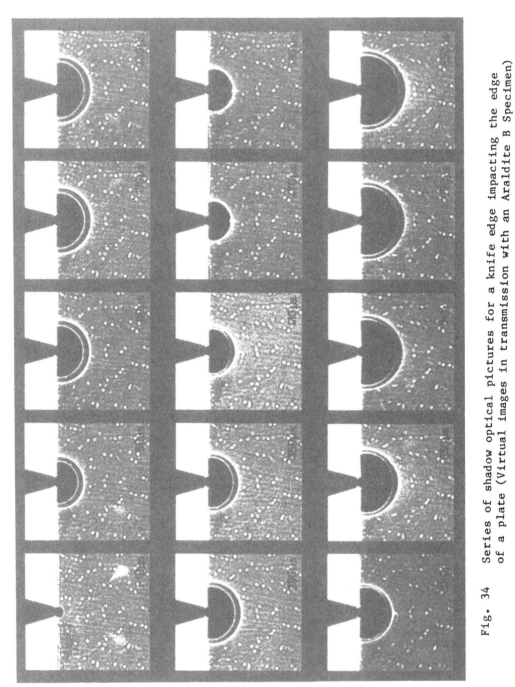

Fig. 34   Series of shadow optical pictures for a knife edge impacting the edge
of a plate (Virtual images in transmission with an Araldite B Specimen)

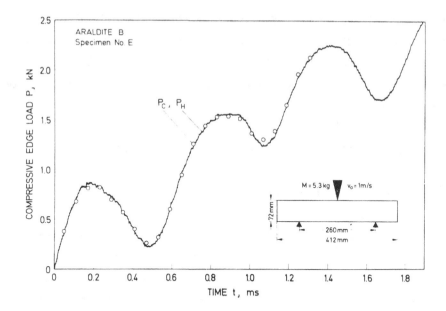

Fig. 35   Load input into a specimen by a falling knife edge

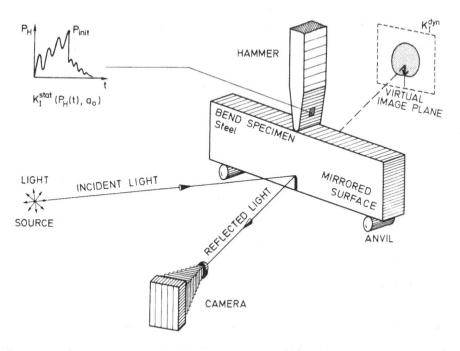

Fig. 36   Experimental arrangement for investigating the behavior of cracks under impact loading

addition to the shadow optically determined load values P_c also the
load curve $P_H(t)$ is given which is measured by a strain gage near the
tip of the striking knife edge. The two measurements are in good agreement. It is recognized that the load input into the specimen is not given
by a steadily increasing function of time. The load shows an oscillating
behavior with an overall increasing tendency. This is caused by eigenvi-
brations of the specimen that are excited by the impact process.

6.2. Cracks under Impact Loading

Impact tests with precracked bend specimens loaded in a pendulum or
a drop weight type test machine are widely used for measuring dynamic
material strength values. The response of the specimen at the crack tip
during the impact event is investigated (see also References 24-26). An
experiment is performed with a precracked bend specimen subjected to dynamic loading by a drop weight. The experimental arrangement is shown in
Figure 36. The specimen, made from the high strength steel X2 NiCoMo 18 9 5
measures $330 \times 60 \times 10$ mm^3. The specimen is supported by anvils with an
opening of 260 mm. The initial crack in the specimens has a length of
20 mm and is slightly blunted to increase the load carrying capacity of
the specimen and thus the observation time before failure. Dynamic loading is achieved by a drop weight of 90 kg mass impacting the specimen at
a velocity of 0.5 m/s.

A series of 12 virtual shadow optical pictures, photographed in reflection with $z_0 > 0$ is presented in Figure 37. Only the central part
of the impacted specimen is shown. Quantitative data obtained according
to equation (14c) from such shadow patterns are given in Figure 38. The
shadow optically determined dynamic stress intensity factors K_I^{dyn} are
plotted as a function of time. In addition, the stress intensity factors
$K_I^{stat}(P_H)$ are shown. These values were determined from the load signal
P_H registered at the instrumented tup of the striking hammer utilizing
the conventional stress intensity factor formula for a precracked specimen
under an equivalent quastistatic loading[27]

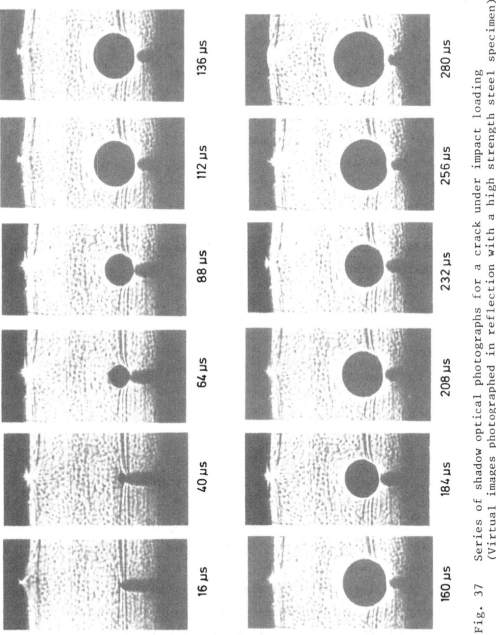

Fig. 37   Series of shadow optical photographs for a crack under impact loading (Virtual images photographed in reflection with a high strength steel specimen)

$$K_I^{stat} = \frac{P_H \, S}{d \, W^{3/2}} \left[\frac{1.99 - \frac{a}{W}(1 - \frac{a}{W})(2.25 - 3.93\frac{a}{W} + 2.7\frac{a}{W}2)}{2(1 + 2\frac{a}{W})(1 - \frac{a}{W})^{3/2}} \right] \tag{53}$$

where P_H = load registered at the tup of the striking hammer

   d   = specimen thickness

   S   = support span

   W   = width of the specimen

   a   = crack length.

The times in Figure 38 are given in absolute units and also in relative units by normalization with the time τ, where τ is the period of the oscillation of the impacted specimen. Values of τ are obtained from the approximation formula[28]

$$\tau = 1.68 \, (SWdCE)^{1/2}/c_o \tag{54}$$

where

   S   = support span

   W   = specimen width

   d   = specimen thickness

   C   = specimen compliance

   E   = Young's modulus

   c_o = sound wave speed (5000 m/s for steel).

The K_I^{stat}-values show a strongly oscillating behavior, whereas the actual dynamic stress intensity factors K_I^{dyn} show a more steadily increasing tendency. In the time range, $t < 3\tau$, these differences are very pronounced. The differences become smaller with increasing time, but even for times larger than 3τ the influences of dynamic effects obviously have not vanished and there are still marked differences between K_I^{stat} and K_I^{dyn}.

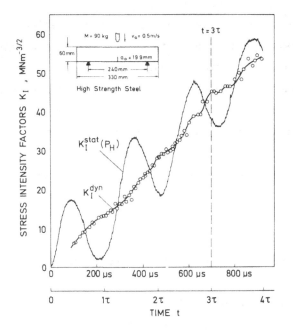

Fig. 38   Stress intensity factor for a crack under impact loading
by a drop weight

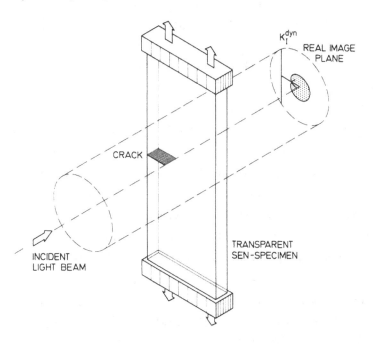

Fig. 39   Experimental arrangement for investigating the behavior
of propagating cracks

The data indicate that influences of dynamic effects on the crack
tip stress intensity factor must, in principle, be taken into account
for measuring dynamic material strength data in impact tests. In particular these dynamic influences can be very large for specimens failing in
a brittle manner in the early time regime. Only for larger crack initiation times resulting for specimens which fail in a more ductile manner
an evaluation procedure based on a quasi-static analysis of the dynamic
test will yield sufficiently reliable material strength data.

6.3. Propagating Cracks

The dependence of the crack tip stress intensity factor K_I from the
crack velocity is investigated. The experimental arrangement is shown in
Figure 39. In a single-edge-notch (SEN) specimen under uniform tensile
loading a propagating crack is initiated from a preexisting notch. With
increasing crack length and reduction of the remaining ligament the crack
tip stress intensification increases and the crack therefore accelerates.
The experiment is performed with a specimen made from Araldite B.

Figure 40 shows real shadow optical patterns photographed in transmission with $z_o < 0$. Only three of the total 24 pictures are reproduced
here showing the early crack propagation phase. The momentary crack velocity is given with each picture. Quantitative data obtained according to
equation (14c) from such photographs are given in Figure 41. The dynamic
stress intensity factor K_I is shown as a function of the crack velocity
v. For small and intermediate crack velocities (v < 250 m/s) the stress
intensity factor increases only slightly with crack velocity, but in the
higher crack velocity range (v > 250 m/s) the increase in K_I with v becomes very steep and K_I obviously approaches unbounded values. The maximum
possible crack velocity therefore is limited. For the material Araldite B
the resulting terminal crack velocity is about 400 m/s. (The theoretically
highest possible terminal crack velocity[18] is the Rayleigh wave speed c_R.
For Araldite B, c_R = 970 m/s.)

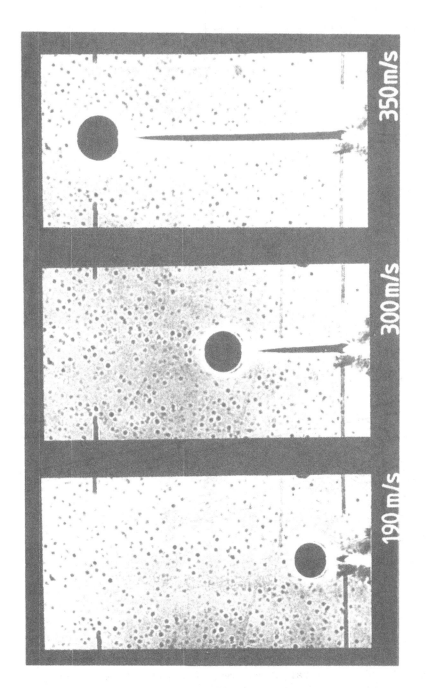

Fig. 40   Series of shadow optical photographs of a propagating crack
(Real images photographed in transmission with an Araldite B specimen)

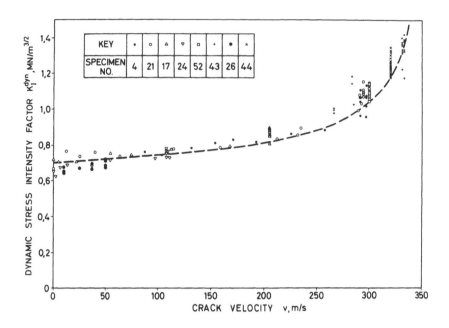

Fig. 41   Dynamic stress intensity factors for propagating cracks

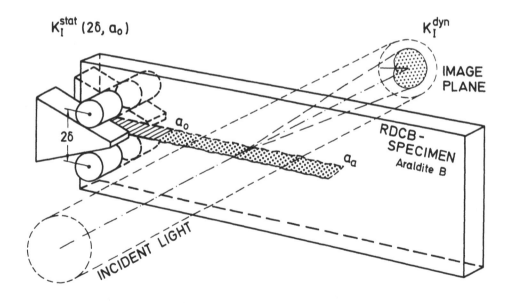

Fig. 42   Experimental arrangement for investigating the
behavior of an arresting crack

6.4. Arresting Cracks

Similar to the crack initiation toughness K_{Ic}, i.e. the critical stress intensity factor for failure of the material, the crack arrest toughness K_{Ia} is the critical stress intensity factor at the moment of arrest. K_{Ia} describes the ability of a material to arrest a propagating crack. Crack arrest toughness values are measured in special laboratory experiments. In order to get information on the stress condition at the tip of an arresting crack the general behavior of a propagating and subsequently arresting crack is investigated (see also References 29-32). One speculation is that the stress condition at arrest must be static since the crack velocity is zero at the moment of arrest.[33]

Arrest of a propagating crack is achieved in a specimen with a decreasing stress intensity field, e.g. a wedge loaded double-cantilever-beam (DCB) specimen (Figure 42). The propagating crack is initiated from a preexisting notch of length a_o at a critical displacement 2δ. The notch tip is blunted in order to store sufficient elastic energy in the specimen before crack initiation so that the crack can accelerate to high velocities. During crack propagation the displacement 2δ stays constant because of the stiffness of the wedge loading system. The specimen, however, becomes more compliant due to the increase of crack length. Consequently, the crack tip stress concentration decreases with increasing crack length and the crack finally comes to arrest at an arrest crack length a_a. The stress intensity factor history for propagating and subsequently arresting crack is investigated with a DCB-specimen measuring $321 \times 127 \times 10$ mm^3 made from Araldite B.

A series of 6 real shadow optical patterns photographed in transmission with $z_o < 0$ are reproduced in Figure 43. The momentary crack tip position is given with each photograph. It can be seen that with increasing crack length the shadow patterns decrease in size, thus indicating decreasing stress intensity factors for the advancing crack. Quanti-

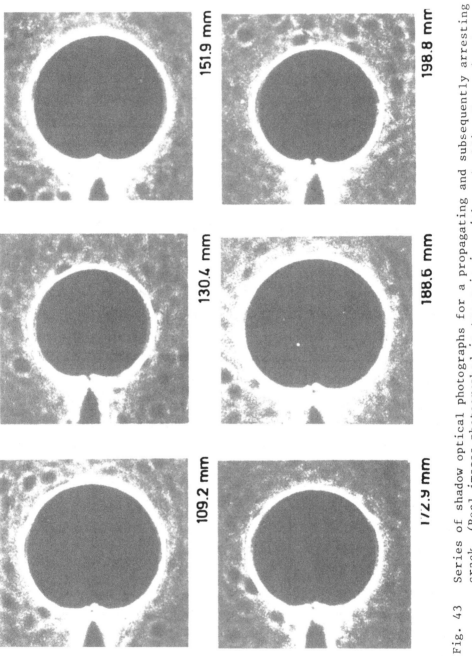

Fig. 43   Series of shadow optical photographs for a propagating and subsequently arresting crack. (Real images photographed in transmission with an Araldite B specimen)

tative data obtained from such photographs with the evaluation formula
(14c) are shown in Figure 44. Results for cracks initiated at different
values of the stress intensity factor at initiation (K_{Iq}) are summarized in the figure. Values of the dynamic stress intensity factor K_I^{dyn}
(experimental points) are shown as a function of crack length a, together
with the corresponding static stress intensity factor (K_I^{stat}) curve. The
static values were determined from the measured critical displacement 2δ
utilizing the conventional stress intensity factor formula[27] for a stationary crack

$$K_I^{stat} = \frac{\sqrt{3}}{4} \frac{E\ H^{3/2}}{a^2}\ 2\ \delta \left[\frac{1 + 0.64(\frac{H}{a})}{1 + 1.92(\frac{H}{a}) + 1.22(\frac{H}{a})^2 + 0.39(\frac{H}{a})^3} \right] \quad (55)$$

with

      E = Young's modulus

    2 H = height of the specimen

      a = crack length.

In addition to the stress intensity factors the measured crack velocities
are given in the lower part of the diagram. The following characteristics
of the crack arrest process can be deduced from the results:

At the beginning of the crack propagation phase the dynamic stress
intensity factor K_I^{dyn} is smaller than the corresponding static value
K_I^{stat}. At the end of the crack propagation phase the dynamic stress
intensity factor K_I^{dyn} is larger than the corresponding static value
K_I^{stat}. Only after arrest does the dynamic stress intensity factor K_I^{dyn}
approach the static stress intensity factor at arrest K_{Ia}^{stat}. Differences between the dynamic and the static stress intensity factor curves become smaller for cracks which are initiated at lower K_{Iq}-values and thus
propagate at lower velocities. The dynamic effects obviously decrease
with decreasing velocity, as one might expect.

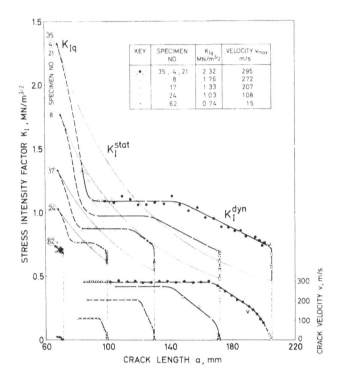

Fig. 44   Stress intensity factor for propagating and
            subsequently arresting cracks

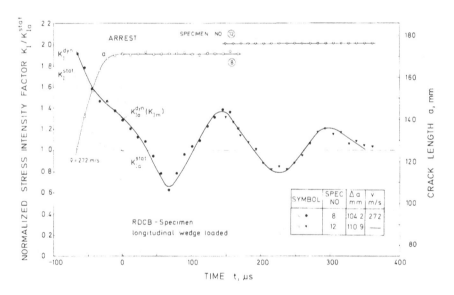

Fig. 45   Stress intensity factor behavior in the post arrest phase

For two experiments the behavior of the dynamic stress intensity factor in the post-arrest phase is shown in Figure 45. The dynamic stress intensity factors, K_I^{dyn}, are plotted as functions of time, t. It is recognized that K_I^{dyn} oscillates around the value of the static stress intensity factor at arrest, K_{Ia}^{stat}. Only for sufficiently large times after arrest the dynamic stress intensity factor approaches the static value.

The data indicate that a dynamic state of stress prevails at the moment of arrest, although the crack velocity is zero at this moment. A static state of stress is only obtained a long time after arrest. This behavior is in accordance with the concept of recovered kinetic energy.[34] Elastic waves are generated by the propagating crack, thus kinetic energy is radiated into the specimen. At the finite boundaries of the specimen these elastic waves are reflected and thus can later on interact with the crack tip again and contribute to the stress intensity factor. This process is illustrated in Figure 46 showing the shadow optical picture of a fast propagating crack (1000 m/s) in a high strength steel specimen. The picture was photographed in a reflection arrangement with $z_0 > 0$. In addition to the crack tip shadow pattern it shows the generation of elastic waves by the propagating crack, the reflection of these waves at the boundaries of the specimen and the subsequent interaction of the reflected waves with the crack tip.

The observed findings are summarized in Figure 47. The data indicate that dynamic effects do have an influence on the crack arrest process. These effects, therefore, in principle have to be taken into account for measuring the crack arrest toughness K_{Ia}. A static analysis can only yield correct arrest toughness data if the crack propagation velocity prior to arrest is sufficiently small.

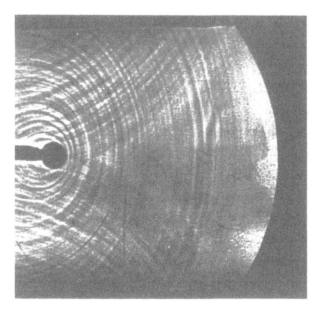

Fig. 46   Shadow optical photograph of a fast propagating
           crack in a high strength steel specimen.
           (Virtual image photographed in reflection)

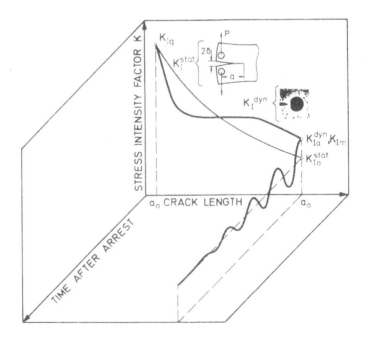

Fig. 47   Schematic representation of the crack arrest behavior

6.5. Stress Field in a Specimen under Projectile Loading

The stress condition in a specimen which is dynamically loaded by an impinging projectile is investigated (see also References 35, 36). The loading arrangement is shown in Figure 48. A projectile impinges on a specimen of twice the length of the projectile. Both, projectile and specimen are made from the same material, Araldite B. The impact process generates a compressive wave which propagates into the specimen thus creating a compressive state of stress. At the free end of the specimen the compressive wave then is reflected as a tensile wave. Similar processes take place in the projectile. A compressive wave is generated in the projectile which afterwards is reflected at its free end. These two tensile waves after having travelled the same distance meet in the middle of the specimen and create a tensile state of stress which then spreads over the entire specimen. Due to subsequent wave reflection processes, compressive stresses again result for later times, etc.

The stress history in the specimen has been made visible by the shadow optical hole field technique. A row or an array of small holes is drilled into the specimen. The stresses at the position of the hole generate a stress concentration around the holes which is registered by the shadow optical technique, here applied in transmission with $z_o < 0$. According to the results presented in Section 3.2, in particular see Figure 8, a tensile stress field will generate a shadow pattern with the intersection line pointing in the direction of the applied stress field, whereas for a compressive stress field the intersection line is oriented perpendicular to the stress field, as shown in Figure 49. Furthermore, the size of the caustic is a measure of the magnitude of the stress.

A series of 24 shadow optical pictures of a row of holes (1 mm diameter) positioned in length direction of the specimen (see Fig. 48, Row A) is reproduced in Figure 50. The recording times of the photographs are given with each picture. The time at which compression changes into tension has been set equal to zero in the figure. The first pictures show

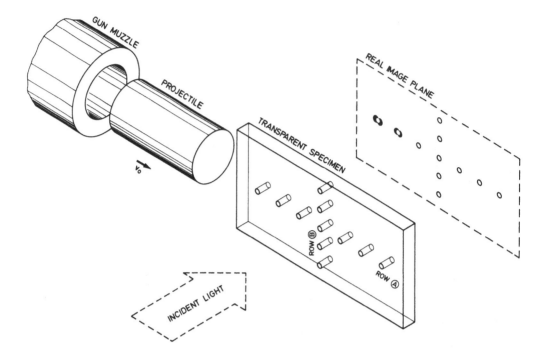

Fig. 48   Experimental arrangement for investigating the stress history
           in a target plate impacted by a projectile

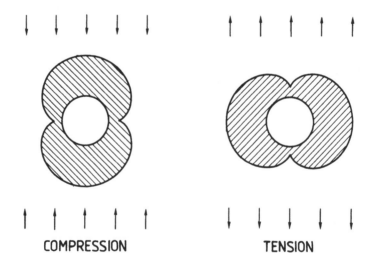

Fig. 49   Shadow patterns around circular holes under
           compression and tension

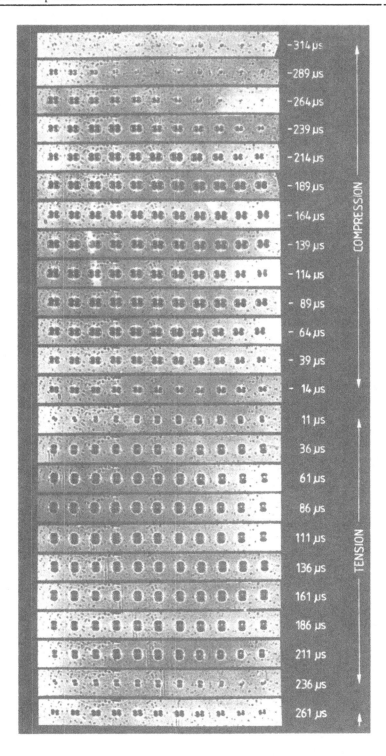

Fig. 50   Series of shadow optical photographs for a specimen with a row of holes impacted by a projectile. (Real images photographed in transmission with an Araldite B specimen)

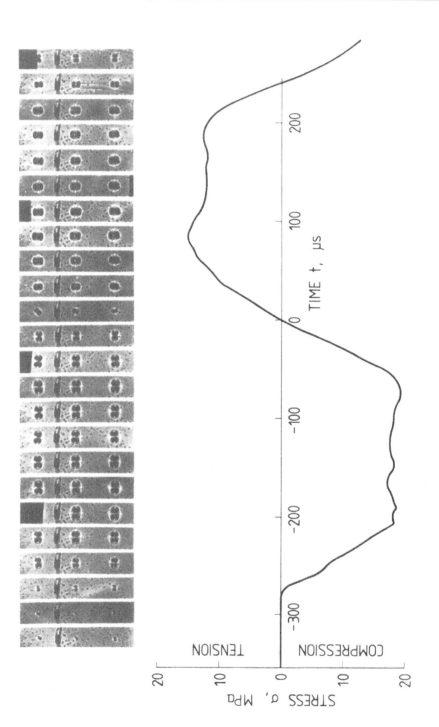

Fig. 51   Series of shadow optical photographs for a specimen with a row of holes impacted by a projectile. Comparison with strain gage signal. (Real images photographed in transmission with an Araldite B specimen)

the propagation of the compressive wave into the specimen (frames 1-5) creating a compressive loading field (frames 6-12). The compressive stresses then are reduced in magnitude and change into tensile stresses as is indicated by the change in the direction of the shadow patterns (frames 13-14). A tensile stress field is observed for the following time interval (frames 15-22), later on the stresses change into compression again (frames 23-24).

Figure 51 shows the results for a row of three holes positioned at half length and 1/4, 1/2, and 3/4 of the width of the specimen (see Fig. 48, Row B). In addition to the shadow optical recordings the stresses were directly measured by a strain gage located between two of the holes. The strain gage result is shown in comparison to the shadow optical pictures in Figure 51. It is recognized that the shadow optical patterns change their directions and their sizes analogous to the strain gage signal. In particular between pictures No. 13 and 15 the compression shadow patterns change into tension shadow patterns, in agreement with the strain gage measurements. The last shadow optical picture again indicates compressive stresses.

The data demonstrate the applicability of the shadow optical hole field technique for providing a simple overview of stress distributions in specimens.

6.6. Double Crack Configuration Loaded by a Tensile Stress Pulse

The stress intensity factor history of two mutually interacting parallel cracks under dynamic loading by an asymmetrically impinging tensile stress pulse is investigated (see also References 36, 37). A specimen made from Araldite B measuring $400 \times 100 \times 10$ mm^3 contains two single-edge-notches of length $a_o = 25$ mm and distance $d = 20$ mm. The holding fixture consists of two blocks, one is at a fixed position, the other is suddenly accelerated by an impinging projectile in a special loading device, i.e. $s = 0$ for $t < t_o$ but $s \neq 0$ for $t > t_o$. Thus a tensile

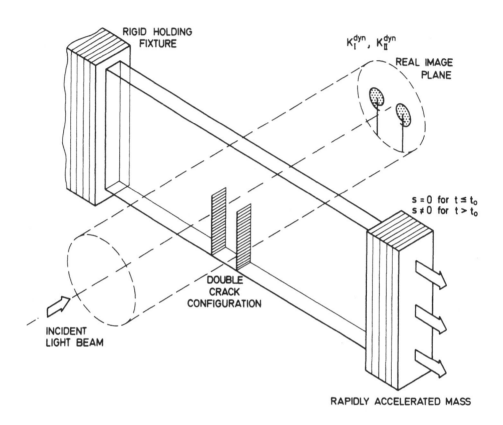

Fig. 52   Experimental arrangement for investigating the interaction of a
          double crack configuration loaded by an asymmetrically impinging
          stress pulse

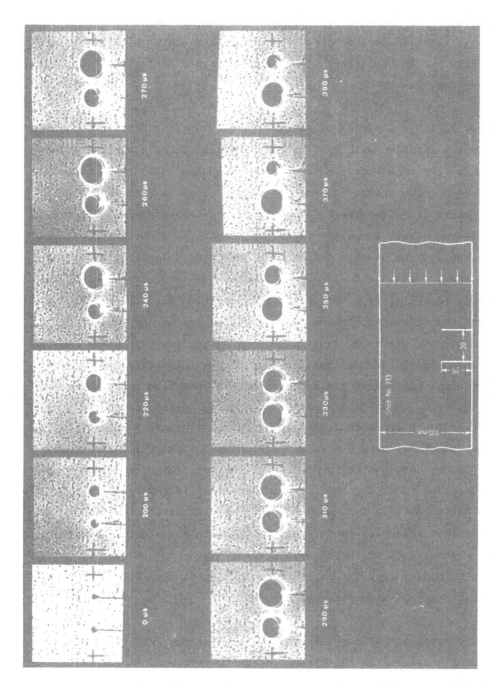

Fig. 53   Series of shadow optical photographs for a dynamically loaded double crack configuration. (Real images photographed in transmission with an Araldite B specimen)

stress pulse is initiated and propagates into the specimen. The crack
which is hit first is denoted crack "A", the other crack which is hit
somewhat later is denoted crack "B". The experimental arrangement is
shown in Figure 52.

Figure 53 shows a series of six real shadow optical pictures, photographed in transmission with $z_o < 0$. The recording times are given with
each picture. The time when the tensile stress pulse reaches the center
line between the two cracks "A" and "B" is set equal to zero in the figure. For early times the crack "A" exhibits the larger shadow pattern,
i.e. the larger stress intensity factor, since crack "A" is hit first by
the stress pulse. The caustic indicates a pure mode I loading, whereas
crack "B" shows a mixed mode I mode II loading due to disturbances of the
stress pulse resulting from the previous interaction with crack "A". For
later times the situation changes. Both caustics are of the same size,
and even later the crack "B" caustic becomes larger than the crack "A"
caustic, which indeed becomes smaller. Due to the interaction processes
of the two crack tip stress fields the crack "A" now also shows a mixed
mode loading type. Quantitative data on the mode I stress intensity factor histories for a crack configuration of similar geometry (crack length
a_o = 15 mm, crack distance d = 20 mm) are shown in Figure 54. The stress
intensity factor curves $K_I^A(t)$ and $K_I^B(t)$ of the two cracks oscillate
around each other. Only for early times the stress intensity factor of
crack "A" is larger than for crack "B", $K_I^A > K_I^B$. At later times the
crack "B" stress intensity factor can be larger than the crack "A" stress
intensity factor, e.g. at the time intervals 95 μs $< t <$ 110 μs,
200 μs $< t <$ 260 μs, etc. For large times the stress intensity factors
of both cracks approach the same values.

Thus, inspite of the more exposed position of the crack "A" towards
the impinging stress pulse the stress intensity factor of the hidden
crack "B" can for certain time intervals be larger than for crack "A".

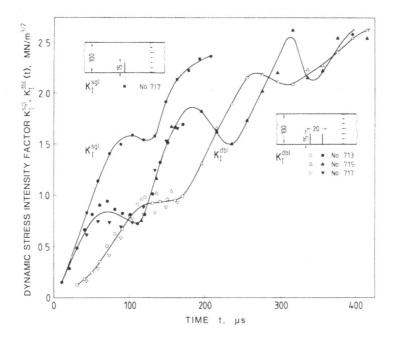

Fig. 54   Stress intensity factor history for dynamically loaded
           double crack configuration

SHADOWOPTICAL PICTURE   PHOTOELASTIC PICTURE

OF CRACK TIP STRESS DISTRIBUTION

Fig. 55   Comparison of shadow optical and photoelastic crack tip patterns

7. SUMMARY AND DISCUSSION

An overview has been given on the shadow optical method of caustics.
The basic physical principles of the method, the mathematical description
of shadow optical imaging processes, and the derivation of caustic evaluation formulas have been presented. The realization of the technique in
the laboratory has been discussed and the applicability of the method for
investigating practical problems has been demonstrated.

It has been shown that the shadow optical method of caustics is
based on the deflection of light rays due to stress gradients. The method,
therefore, is very well suited for the investigation of any kind of
stress-strain problem which shows large gradients, e.g. stress concentration problems around holes, notches, cracks, contact areas etc. For typical examples the shadow optical formulation is presented. It is shown that
the resulting shadow optical patterns are simple and easy to evaluate. In
general only one characteristic length parameter is taken from the measured caustic in order to determine the generating load parameter.

Most other experimental techniques of stress analysis are based on
effects which are directly proportional to stresses or strains. The specific advantages and disadvantages of the shadow optical method due to
these differences in the dependence from the stress become evident by the
comparison of a shadow optical pattern and a photoelastic pattern obtained for the same crack tip stress field shown in Figure 55. The shadow spot represents a direct quantitative measure of the stress intensification at the crack tip. The photoelastic pattern is more complicated
than the shadow optical pattern due to the large number of isochromatic
fringes. Thus the evaluation of photoelastic stress concentration patterns naturally becomes a laborious task. In particular in the near field
region around the crack tip the reduced resolution makes an evaluation
difficult and the derived results necessarily are of reduced accuracy.
Furthermore, data on the loading condition in the direct vicinity of the

crack tip can only be obtained by an extrapolation of data measured at
distances further away from the crack tip. The shadow optical picture
does not suffer from these disadvantages. The simplicity and clearness of
the shadow optical picture result from the fact that stress distributions with small variations of stress do not become visible. This, however, implies also certain disadvantages. Far field distributions around
stress concentrations, e.g. around the crack tip in Figure 55, do not result in shadow optical effects because of the small stress gradients and
therefore cannot be investigated. The photoelastic pattern, however,
yields accurate information in particular in this far field range. The
shadow optical method of caustics and the photoelastic method of isochromatic fringes, therefore, are not to be considered as compatitive experimental tools which can be used with the same success and efficiency
in obtaining the same kind of data. Depending on the specific property of
interest and problem to be investigated the one or the other measuring
technique is more appropriate. Both techniques have their specific ranges
of applicability and complement each other.

Because of its naturally inherent simplicity and the ease in evaluation of the recorded light patterns the shadow optical method of caustics
yields reliable data even if the problem to be investigated is of very
complicated nature. The technique, therefore, is very well suited for investigating complex phenomena, e.g. dynamic processes. Most applications
of the shadow optical method of caustics so far have been in the field of
fracture dynamics. Fast variations in crack tip stress intensifications
can be easily and accurately measured. Shadow optical analyses are available not only for the most important case of cracks under tensile loading, but also for shear types of loading and mixed mode loading conditions. The applicability of the derived evaluation formulas for the investigation of dynamic fracture problems has been discussed. In addition
to the analyses of linear elastic stress intensifications the shadow optical formulation has also been developed for elastic-plastic strain concentration fields around cracks in materials which show ductile behavior.

Caustic evaluation formulas are presented for determining the relevant fracture mechanics parameters, i.e. the linear elastic stress intensity factors K_I, K_{II}, and K_{III} or the elastic-plastic J-integral value.

The applicability of the shadow optical method of caustics for investigating problems of practical relevance in the field of fracture dynamics has been demonstrated by several examples. For many other stress concentrations problems the shadow optical method of caustics lends itself as an appropriate tool for investigation.

8. REFERENCES

1. Manogg, P., "Anwendung der Schattenoptik zur Untersuchung des Zer-
    reißvorgangs von Platten", Dissertation, Freiburg, Germany, 1964

2. Manogg, P., "Schattenoptische Messung der spezifischen Bruchenergie
    während des Bruchvorgangs bei Plexiglas", Proc. Int. Conf. on the
    Physics of Non-Crystalline Solids, Delft, The Netherlands, 1964,
    481-490

3. Theocaris, P.S., and Joakimides, N., "Some Properties of Generali-
    zed Epicycloids Applied to Fracture Mechanics", Journ. Appl. Mech.,
    22, 1971, 876-890

4. Theocaris, P.S., "The Reflected Caustic Method for the Evaluation
    of Mode III Stress Intensity Factor", Int. Journ. Mech. Sci., 23,
    1981, 105-117

5. Theocaris, P.S., "Stress Concentrations at Concentrated Loads",
    Experimental Mechanics, 13, 1973, 511-528

6. Rosakis, A.J., Freund, L.B., "Optical Measurement of the
    Plastic Strain Concentration at a Tip in a Ductile Steel Plate",
    Journ. Engr. Mat. Tech., 104, 1982, 115-125

7. Rosakis, A.J., Ma, C.C., Freund, L.B., "Analysis of the Optical
    Shadow Spot Method for a Tensile Crack in a Power-Law Hardening
    Material", Journ. Appl. Mech., 50, 1983, 777-782

8. Kalthoff, J.F., Beinert, J., Winkler, S., "Analysis of Fast Running
    and Arresting Cracks by the Shadow-Optical Method of Caustics",
    I.U.T.A.M. Symposium on Optical Methods in Mechanics of Solids, Ed.
    A. Lagarde, University of Poitiers, France, Sept. 10-14, 1979,
    Sijthoff and Nordhoff, 1980, 497-508

9.   Beinert, J., Kalthoff, J.F., "Experimental Determination of Dynamic
     Stress Intensity Factors by Shadow Patterns", in: "Mechanics of
     Fracture, Vol. 7, Experimental Fracture Mechanics", Ed. G.C. Sih,
     Martinus Nijhoff Publishers, The Hague, Boston, London, 1981,
     280-330

10. Kalthoff, J.F., "Stress Intensity Factor Determination by Caustics",
     Proc. Int. Conf. on Experimental Stress Analysis, organized by
     Japan Society of Mechanical Engineers (JSME) and American Society
     for Experimental Stress Analysis (SESA), Honolulu-Maui, Hawaii,
     May 23-29, 1982, 1119-1126

11. Kalthoff, J.F., Böhme, W., Winkler, S., "Analysis of Impact Frac-
     ture Phenomena by Means of the Shadow Optical Method of Caustics",
     Proc. VIIth Int. Conf. on Experimental Stress Analysis, organized
     by SESA, Haifa, Israel, Aug. 23-27, 1982, 148-160

12. Born, M., Wolf, E., "Principles of Optics", Pergamon Press,
     Oxford, London, Edinburgh, New York, Paris, Frankfurt, 1970

13. Broek, D., "Elementary Engineering Fracture Mechanics", Martinus
     Nijhoff Publishers, The Hague, Boston, London, 1982

14. Paris, P.C., Sih, G.C., "Stress Analysis of Cracks", Fracture
     Toughness Testing and its Application, ASTM STP 381, American
     Society for Testing and Materials, Philadelphia, 1965, 30-83

15. Theocaris, P.S., "Complex Stress Intensity Factors of Bifurated
     Cracks", Journ. Mech. Phys. Solids, 20, 1972, 265-279

16. Seidelmann, U., "Anwendung des schattenoptischen Kaustikenverfah-
     rens zur Bestimmung bruchmechanischer Kennwerte bei überlagerter
     Normal- und Scherbeanspruchung", Bericht 2/76 des Fraunhofer-Insti-
     tuts für Festkörpermechanik, Freiburg, 1976

17. Sih, G.C., "Handbook of Stress Intensity Factors", Institute of
     Fracture and Solid Mechanics, Lehigh University, Bethlehem, Pa.,
     1973

18. Freund, L.B., "Crack Propagation in an Elastic Solid Subjected to
     General Loading - I. Constant Rate of Extension", Journ. Mech.
     Phys. Solids, 20, 1972, 129-140

19. Kalthoff, J.F., "Zur Ausbreitung und Arretierung schnell laufender
     Risse", Fortschritt-Berichte der VDI-Zeitschriften, Reihe 18, Nr. 4,
     VDI-Verlag, Düsseldorf, 1987, 1-95

20. Williams, M.L., "On the Stress Distribution at the Base of a
     Stationary Crack", Journ. Appl. Mech., 24, 1957, 109-114

21. Hutchinson, J.W., "Plastic Stress and Strain Fields of a Tensile
     Crack", Journ. Mech. Phys. Soldids, 16, 1968, 13-31

22. Rice, J.R., Rosengreen, G.F., "Plane Strain Deformation Near a
     Crack Tip in a Power Law Hardening Material", Journ. Mech. Phys.
     Solids, 16, 1968, 1-12

23. Hutchinson, J.W., "Plastic Stress and Strain Fields of a Crack
     Tip", Journ. Mech. Phys. Solids, 16, 1968, 337-347

24. Kalthoff, J.F., Winkler, S., Beinert, J., "The Influence of Dynamic
     Effects in Impact Testing", Int. Journ. of Fracture, 13, 1977,
     528-531

25. Kalthoff, J.F., Böhme, W., Winkler, S., Klemm, W., "Measurements of
     Dynamic Stress Intensity Factors in Impacted Bend Specimens", CSNI-
     Specialist Meeting on Instrumented Precracked Charpy-Testing,
     Electric Power Research Institute, Palo Alto, Calif., Dec. 1-3,
     1980, 1-17

26. Kalthoff, J.F., Winkler, S., Böhme, W., Klemm, W., "Determination
     of the Dynamic Fracture Toughness K_{Id} in Impact Tests by Means of
     Response Curves", Proc. 5th Int. Conf. on Fracture, Cannes, March
     29 - April 3, 1981, in "Advances in Fracture Research", Pergamon
     Press, 1981, 363-373

27. ASTM E 399, "Standard Test Method for Plane-Strain Fracture Tough-
     ness of Metallic Materials", Annual Book of ASTM Standards, Part 10,
     American Society for Testing and Materials, Philadelphia, 1983

28. Ireland, D.R., "Critical Review of Instrumented Precracked Charpy
     Testing", Proc. Int. Conf. Dynamic Fracture Toughness, London, July
     5-7, 1976, 47-62

29. Kalthoff, J.F., Winkler, S., Beinert, J., "Dynamic Stress Intensity
     Factors for Arresting Cracks in DCB Specimens", Int. Journ. of
     Fracture, 12, 1976, 317-319

30. Kalthoff, J.F., Beinert, J., Winkler, S., "Measurements of Dynamic
     Stress Intensity Factors for Fast Running and Arresting Cracks in
     Double-Cantilever-Beam Specimens", Fast Fracture and Crack Arrest,
     ASTM STP 627, Eds. G.T. Hahn and M.F. Kanninen, American Society
     for Testing and Materials, Philadelphia, 1977, 161-176

31. Kalthoff, J.F., Beinert, J., Winkler, S., Klemm, W., "Experimental
     Analysis of Dynamic Effects in Different Crack Arrest Test Speci-
     mens", Crack Arrest Methodology and Applications, ASTM STP 711,
     Eds. G.T. Hahn and M.F. Kanninen, American Society for Testing and
     Materials, Philadelphia, 1980, 109-127

32. Kalthoff, J.F., "Bruchdynamik laufender und arretierender Risse",
     Int. Seminar über Bruchmechanik, Schadensanalyse für die Praxis,
     Bruchsicherheit, Ed. H.P. Rossmanith, Wien, 12./13. Juni 1980, K 1-22;

published in "Grundlagen der Bruchmechanik", Ed. H.P. Rossmanith, Springer Verlag, Wien, New York, 1982, 191-219

33. Crosley, P.B., Ripling, E.J., "Characteristics of a Run-Arrest Segment of Crack Extension", Fast Fracture and Crack Arrest, ASTM STP 627, Eds. G.T. Hahn and M.F. Kanninen, American Society for Testing and Materials, Philadelphia, 1977, 203-227

34. Hahn, G.T., et al., "Critical Experiments, Measurements and Analyses to Establish a Crack Arrest Methodology for Nuclear Pressure Vessel Steels", Reports BMI-1937, 1959, 1985 prepared for U.S. Nuclear Regulatory Commission, Battelle Columbus Laboratories, Ohio, 1975-1978

35. Kalthoff, J.F., Winkler, S., "Fracture Behavior under Impact", First Annual Report prepared for United States Army, European Research Office, IWM Report W 8/82, Fraunhofer-Institut für Werkstoffmechanik, Freiburg, 1983

36. Kalthoff, J.F., Winkler, S., "Fracture Behavior under Impact", Second Annual Report prepared for United States Army, European Research Office, IWM Report W 10/82, Fraunhofer-Institut für Werkstoffmechanik, Freiburg, 1983

37. Kalthoff, J.F., "On Some Current Problems in Experimental Fracture Dynamics", Workshop on Dynamic Fracture, Ed. W.G. Knauss, California Institute of Technology, Pasadena, Calif., Feb. 17-18, 1983, 11-35

9. APPENDIX

CRANZ-SCHARDIN HIGH SPEED CAMERA
RECENT DEVELOPMENTS

Siegfried Winkler

A photographic high speed recording system was invented in 1929[1] by
Carl Cranz und Hubert Schardin (late director of the author's institute).
This camera assembly has one great advantage among the many different
types of photographic registration systems for fast events: There are no
moving mechanical parts. This is the main reason for the high quality of
the pictures obtained with this camera in comparison with those from
other systems. The separation principle for consecutive pictures is shown
in Figure 1. A large-format camera with a certain assembly of n lenses

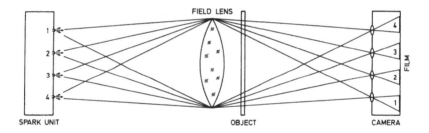

Fig. 1   Principle of optical picture separation
(after Cranz and Schardin[1])

for n pictures is accompanied by the same assembly of n light sources.
The light of any particular source will be focused exactly on just one
lens. Every camera lens, however, is focused onto the object. The light
sources are realized by an array of spark gaps which are specially designed for short (≈ 0.3 µs half width) light pulses. Cinematographical
picture sequences are achieved by firing the individual spark gaps in a
predetermined time sequence. The sequence frequency is bounded only by
electronic limitations. It can, therefore, be significantly high, which
makes the camera very fast. Even in 1929 the inventors realized the possibility of recording several million pictures per second[1]. The original electrical firing circuit is given in Figure 2 for three spark gaps.
The elements of this circuit originate from 1885 and are named "Mach's
circuit"[2]. They are combined by induction coils to form a delay circuit. This is the reason for a sequential firing of all n gaps. The in-

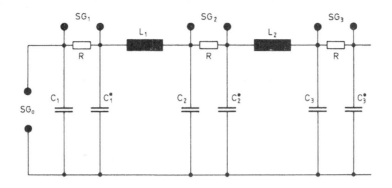

list of symbols·

SG_0	=	trigger spark gap
SG_1....	=	spark gap light sources
$C_1, C_1^\bullet$..	=	discharge capacitors
R	=	resistors
L_1....	=	timing induction coils

Fig. 2 Discharge circuit for multiple spark gaps
(after Cranz and Schardin[1])

ductance L_i and the capacitance C_i determine the delay from one gap
to the next governed by Thomson's formula. The calculation, however, is
approximative since properties of the gap (e.g. the ignition threshold)
must also be taken into account. For accurate measurements, therefore,
time records must be provided.

The number of pictures which can be taken from one event with the
Cranz–Schardin camera is small. Assemblies with 8 to 35 sparks are reported or available. Increasing this number also increases the physical size
and decreases the handling capability. The 24 spark type is therefore
considered to be the most effective camera.

This limited number of pictures of the Cranz–Schardin system is sometimes considered to be disadvantageous and the term "cinematography" not
accurately descriptive. However, in high speed measurement applications,
a small number of high quality pictures can deliver a satisfactory amount
of experimental data. Since the Cranz–Schardin camera can perform such
measurements, it is therefore a valuable scientific tool.

One important feature of the Cranz–Schardin assembly is its limitation
to a certain class of objects. These can be either transparent or non-
transparent, but in any case they have to fit into the illumination ray
tracing as is shown in Figures 3 and 4. This restriction in application
is due to the special picture separation principle demonstrated in
Figure 1. A luminous object or an object which scatters light diffusely
would illuminate all 24 pictures at one time. Therefore, care must be

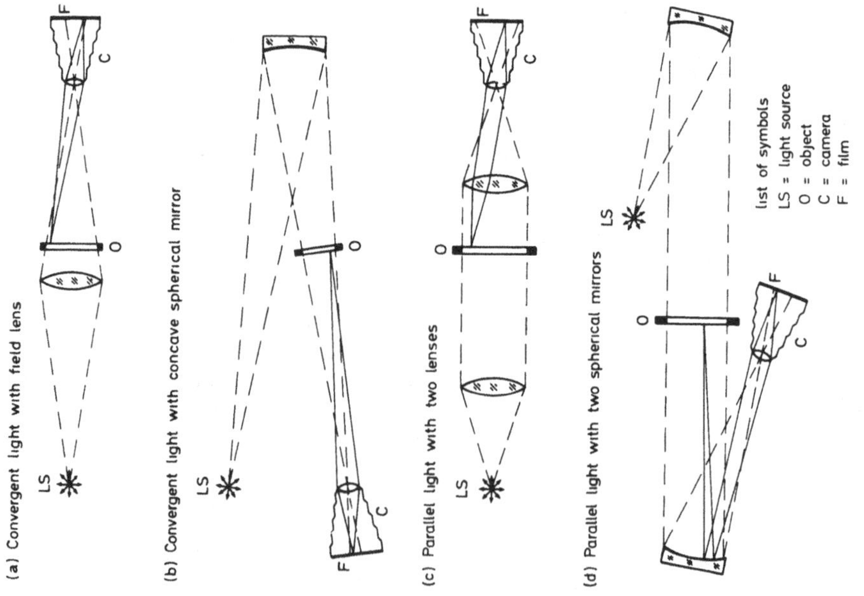

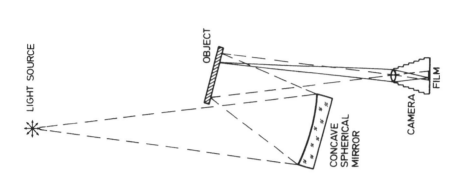

Fig. 4   Example for an illumination of
         a reflecting object

Fig. 3   Examples of illumination modes for
         transparent objects

taken in order to achieve an appropriate illumination ray tracing. This
can be done in various ways, as is shown by the examples of Figure 3 for
transparent objects. As an imaging device, either a lens or a concave
spherical mirror can be used. A mirror is the preferred instrument for
this purpose because it exhibits less optical imperfections. The illumination ray tracing can be either convergent (Fig. 3a,b), parallel (Fig.
3c, d), or divergent. The latter configuration is of less importance and
not shown in the figures. The arrangement which is utilized depends on
several circumstances, i.e., object size, view field, etc.

In principle, the same illumination modes can be applied in the case
of non-transparent objects (for one example see Fig. 4). However, the experimental preparation for this reflection case can be more troublesome
than that necessary for the transmission arrangements. This complication
arises because the object must become an optical component (usually a
mirror) for this special opaque-object test configuration. The required
optical properties are very stringent and this leads to extended machine
shop work (grinding, lapping, and polishing with optical quality).

The rapid development of electronics during the last decades also gave
rise to some improvements of the electronic part of the Cranz-Schardin
camera system. The original discharge chain (Fig. 2), though simple and
straightforward, is limited in application and not easy to handle. This
is caused not only by heavy induction coils. The requirement for accurate
measurements also demands extended procedures to adjust and to control or
record the framing rate.

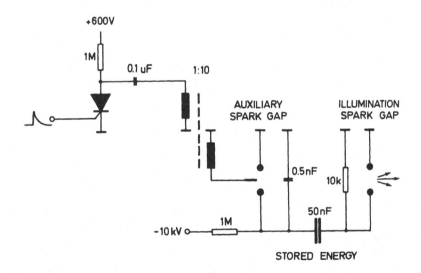

Fig. 5 Spark gap unit (schematically)
        (after Stenzel[3,4])

Newer developments add more versatility to the Cranz-Schardin system. These devices, although more sophisticated, are much lighter and less bulky due to semiconducter technology. A necessary first step for further developments was the design of a triggerable spark gap by Stenzel in 1959[3,4]. Its principle (one spark of 24) is shown in Figure 5. This can be fired independently from others, which gives the advantage of using pulse generators for timing purposes.

A total experiment demonstrating the use of a modern Cranz-Schardin camera assembly is sketched in Figure 6. The object is represented by a fracture mechanics bend specimen, the dynamic event by the fast propagating crack. The photographic arrangement is that of Figure 3b. Convergent

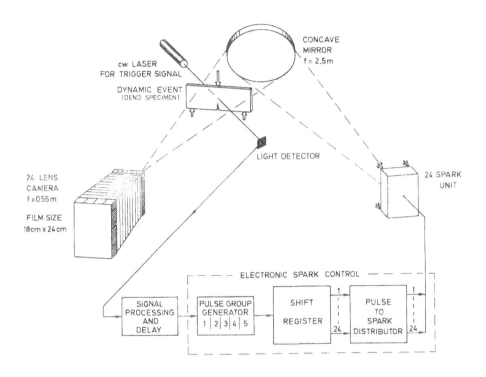

Fig. 6   Complete experimental configuration

light is formed by a concave mirror of focal length f = 2,5 m. The 24 sparks and the 24 camera lenses are therefore located at 2f = 5 m apart from the mirror in order to get a 1:1 image of each spark on its corresponding camera lens. The diameter of that part of the film which is illuminated through one lens is determined by the size of the mirror. An appropriate mirror diameter is 0.25 m. To avoid overlapping of neighbouring pictures, the mirror size can be limited by a rectangular mask. The position of the object (distance from the camera) determines the size of the view field.

A detection device provides a signal to trigger the camera. This signal must be matched to the subsequent electronics and sometimes also delayed, depending on experimental requirements. As an example, an optical detection system has been drawn in Figure 6. A light beam (e.g., from a laser) is interrupted or scattered by the event. In the case of a propagating crack in a transparent material, total reflection at the crack surface is utilized. The signal processing is demonstrated with more detail in Figure 7. This method first delivers a very fast analog signal.

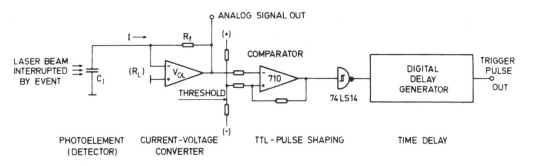

THEORY: (1) $R_L = \dfrac{R_f}{V_{OL}}$ (V_{OL} = open loop gain)

(2) If R_L sufficiently small, $I = I_{short\ circuit} = const. \ I_{phot}$

it follows Analog signal is proportional to light intensity

(3) Time constant of input circuit $\tau = R_L \cdot C_j$ responsible for fast signal rise

(C_j = junction capacity)

(4) Example: For $V_{OL} \sim 10^3$ and $R_f = 10^4$ Ohm follows $R_L \sim 10$ Ohm

If $C_j \sim 10^{-8}$ Farad, the time constant $\tau \sim 10^{-7}$ seconds

Fig. 7   Optical detector configuration

A rise time of about 100 ns can be easily achieved. A comparator with adjustable threshold and a Schmitt-trigger form a TTL-signal suitable to start a time delay. This is represented by a quartz-controlled digital generator with 100 ns increments. From this generator either a delayed or an undelayed output pulse is available to trigger the sparks of the high speed camera.

Figure 8 is a photograph of the camera assembly. The 24 lens camera and the spark unit are shown in the upper part. The lower part shows electronic spark control and high voltage supply. The spark control will now

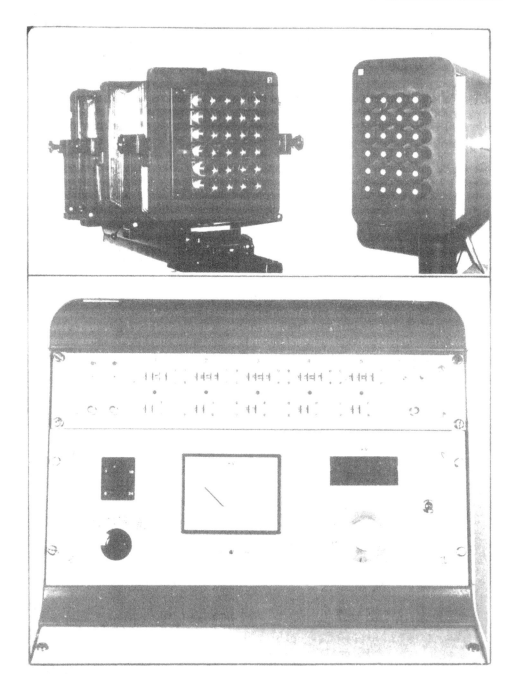

Fig. 8   Modern Cranz-Schardin assembly; 24 lens camera (upper left),
         24 spark unit (upper right), and supply and control unit
         (lower part)

be described in somewhat more detail. It consists mainly of three parts
(see Fig. 6). A generator (pulse group generator) produces a predetermined series of pulses, counted in time. These have to be transferred
into a time-space distribution which is accomplished by a 24 bit shift
register. Each of the 24 output connectors of this device carries one and
only one pulse of the input series, i.e., the connector of number j
carries the pulse of number j. A third device distributes these 24 pulses
to the 24 sparks.

The development of the pulse group generator enables the experimenter
to optimally match the recording to the event. This is of importance when
experiments are expensive or reproducibility is limited. Certain parts of
the event can be emphasized, others of minor interest ignored. Even
though the number of pictures is small, the amount of information can be
large when this generator is in use.

The principle of this generator is simply to combine several individual pulse generators into a series, each triggering the next one. The
number of pulses of each individual generator can be digitally chosen by
two digits and their spacing in the form

$$xy \cdot 10^{-z} \text{ seconds,}$$

where the tens x and the units y can be set from 0 to 9 and the power z
from 1 to 6. It is possible, therefore, to set framing rates from

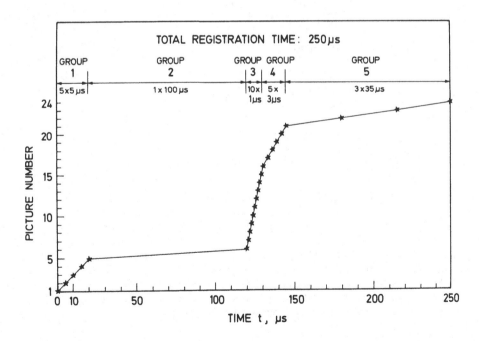

Fig. 9   Arbitrary sequence of pictures

1 microsecond to 9.9 seconds, totally independent from what is set in the
neighbouring generators (groups).

An example for a certain experiment is shown in the diagram of
Figure 9. Some parts (groups 1,3,4) of the total registration time
(250 μs) are photographed with a higher time resolution. A delay time of
exactly 100 μs is inserted by group 2. Finally, the post-event is recorded by group 5 with a very low resolution just to make certain that no
important part of the event has been missed.

An electronic schematic of the pulse group generator is given in
Figure 10. All groups (the individual generators) are of the same type,

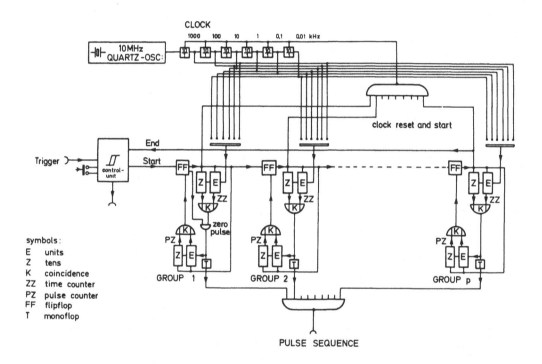

Fig. 10 Pulse group generator (schematically)

except for the first one. There is an additional device providing a "zero
pulse" which must appear at time 0, i.e., the arrival of the trigger signal. All groups are operated by the same clock generator which is reset
to zero after the generating period of each group. The clock is controlled by a 10 MHz quartz oscillator. This reduces the jitter to 100 ns
maximum. The width of the output pulses of each group is about 300 ns.
All output pulses are collected by a single gate.

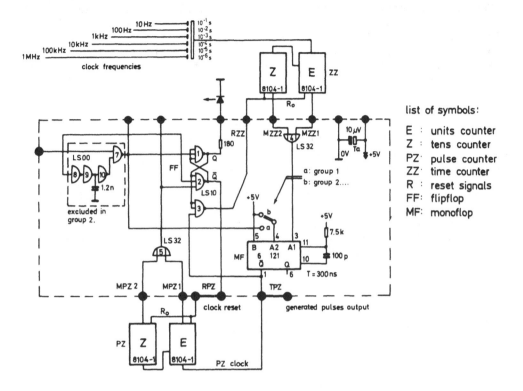

Fig. 11   Group generator element

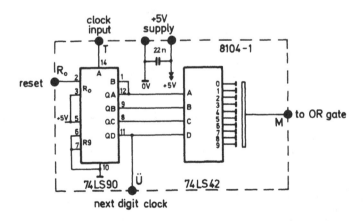

Fig. 12   Tens counter

More details of these circuits are shown in Figures 11 and 12. A time counter ZZ counts the clock pulses. The output pulses are counted by the pulse counter PZ. A flipflop FF controls the generating period, a monoflop MF the width of the output pulses. Because of the relatively low maximum frequency of 1 MHz, the counters itself (Fig. 12) are of a quite simple design.

The spatial arrangement of the sparks in a 4 x 6 matrix (see Fig. 8, upper part) could allow several sequences of the sparks. In some cases it is advantageous when the moving event is "followed" by the sparks, i.e., the sparks move in the same direction as the event. This happens if overlapping of pictures cannot be avoided in a large view field. In order to illustrate the pulse to spark distribution, the sparks have been arbitrarily numbered as shown in Figure 14, left side. A very universal, but nevertheless simple, method to correlate the sparks to the pulses is shown in Figure 13. Any desired distribution can be realized by setting the 24 pins into their places in a matrix connector. However, this requires a considerable amount of work and can cause some problems.

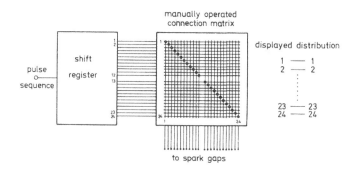

Fig. 13 Universal pulse to spark distributor

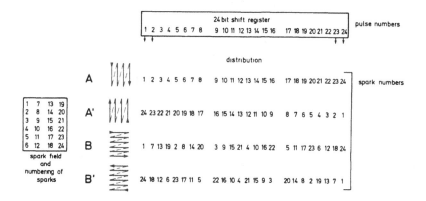

Fig. 14 Scheme of predetermined spark sequences

On the other hand, experience shows that the number of possible distributions can be drastically reduced. Only the four distributions of
Figure 14 should be selectable. These four distributions have, therefore,
been printed on boards together with octal line drivers as shown in
Figure 15. The line drivers can be enabled, one at a time, by a single
contact. Handling of this distributor is, therefore, most simple by rotating a switch, as can be seen in Figure 8, left side.

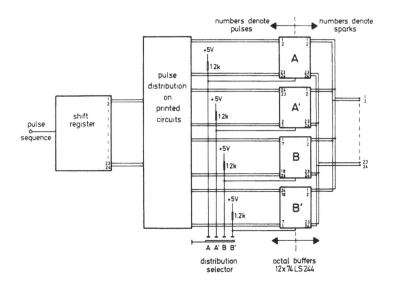

Fig. 15   Distributor for predetermined spark sequences

The Cranz-Schardin camera assembly is - like other high speed cameras -
inherently a laboratory measuring system. This is most obvious by the
fact that the laboratory must be completely darkened during the test. The
components, camera, mirror, and spark unit are still heavy and stable
stands must be used. It is still time consuming to fix the light ray
tracing and to match this recording system to the experiment.

However, the time necessary to run an experiment is drastically reduced due to the described developments of the electronic part. Quartz-
controlled picture spacing delivers high accurate motion versus time measurements and renders time records superfluous.

This camera, though still an assembly of several parts, has become
more versatile, more precise, and easier to handle.

References

1.   Cranz, C. und Schardin, H., "Kinematographie auf ruhendem Film und mit extrem hoher Bildfrequenz", Z. Phys. 56, 1929, 147-183

2.   Mach, E., Berichte der Wiener Akademie. Abs. 2a, 92, 625, 1885

3.   Stenzel, A., "Elektronisch gesteuerte Cranz-Schardin-Funkenzeitlupe", Proc. IV. Int. Congr. High-Speed Phot. Köln, Hellwich, Darmstadt, 1959, 136-138

4.   Vollrath, K., "Funkenlichtquellen und HF-Funkenkinematographie", Kurzzeitphysik, Ed. K. Vollrath und G. Thomer, Springer, 1967, 76-157

Printed in the United States
By Bookmasters